METHODS OF BIOCHEMICAL ANALYSIS

Volume 35

METHODS OF
BIOCHEMICAL ANALYSIS

Series Editor

Clarence H. Suelter

Volume 35

PROTEIN STRUCTURE DETERMINATION

Edited by
CLARENCE H. SUELTER
Department of Biochemistry
Michigan State University
East Lansing, Michigan

An Interscience® Publication
JOHN WILEY & SONS, INC.

New York • **Chichester** • **Brisbane** • **Toronto** • **Singapore**

An Interscience® Publication

Copyright © 1991 by John Wiley & Sons, Inc.

All rights reserved. Published simultaneously in Canada.

Reproduction or translation of any part of this work
beyond that permitted by Section 107 or 108 of the
1976 United States Copyright Act without the permission
of the copyright owner is unlawful. Requests for
permission or further information should be addressed to
the Permissions Department, John Wiley & Sons, Inc.

Library of Congress Catalog Card Number: 54-7232

ISBN 0-471-51326-1

Printed in the United States of America

10 9 8 7 6 5 4 3 2

SERIES PREFACE

Methods of Biochemical Analysis was established in 1954 with the publication of Volume 1 and has continued to the present with volumes appearing on a more or less yearly basis. Each volume deals with biochemical methods and techniques used in different areas of science. Professor David Glick, the series' originator and editor for the first 33 volumes, sensed the need for a book series that focused on methods and instrumentation. Already in 1954, he noted that it was becoming increasingly difficult to keep abreast of the development of new techniques and the improvement of well-established methods. This difficulty often constituted the limiting factor for the growth of experimental sciences. Professor Glick's foresight marked the creation of a unique set of methods volumes which have set the standard for many other reviews.

With Professor Glick's retirement from the series and beginning with Volume 34, I have assumed editorship. Because the rationale used in 1954 for the establishment of the series is even more cogent today, I hope to maintain the excellent traditions developed earlier. The format of Volume 34 and later volumes, however, is changed. Rather than cover a variety of topics as previous volumes did, each volume will now focus on a specific method or the application of a variety of methods to solve a specific biological or biomedical problem.

CLARENCE H. SUELTER

East Lansing, Michigan

v

PREFACE

Volume 35 of Methods of Biochemical Analysis presents an in-depth discussion of several issues involved in determining secondary and tertiary structure of proteins.

The first chapter presents both a theoretical and an empirical approach to predicting protein structure. The unique distribution of the amino acids between the interior and the surface of a folded protein molecule affects its stability, dynamics, and function such as receptor binding, antigen-antibody recognition and catalysis. Chapter two provides a review of the use of protein ligand interactions to study surface properties of proteins. An additional method involving the use of fluorescence techniques for studying the structure and dynamics of proteins in solution is presented in Chapter three. Finally, Chapter four describes the use of biological tools to understand how specific structural features confer biological function. This chapter focuses on two methods which are frequently complementary, namely limited proteolysis and the use of monoclonal antibodies. Other methods such as x-ray crystallography, NMR spectrometry, hydrogen exchange and calorimetry are planned for future volumes.

The Editor planned a compilation of chapters which can be used as a reference text for a course in protein structure and function.

CLARENCE H. SUELTER

East Lansing, Michigan
September, 1990

CONTENTS

Theoretical and Empirical Approaches to Protein-Structure Prediction and Analysis

G.M. Maggiora, B. Mao, K.C. Chou, and S.L. Narasimhan, *Computational Chemistry, Upjohn Laboratories Kalamazoo, Michigan*

Methods of Biochemical Analysis, Volume 35: Protein Structure Determination, Edited by Clarence H. Suelter.
ISBN 0−471−51326−1 © 1991 John Wiley & Sons, Inc.

1. INTRODUCTION

The relationship of a protein's function to its underlying structure has been an important and persistent theme in biochemical research for many years. The early crystallographic work of Pauling and Corey (1) provided our first glimpse of this structure. It was here that the existence of "regular" structures, namely α helices and β sheets, were first described, setting the stage for the elegant work of Perutz and his colleagues (2) on the complete three-dimensional (3D) structure of myoglobin. Subsequent to the publication of this work Lindestrom-Lang and Schellman (3) presented a scheme for characterizing the hierarchical nature of protein structure in terms of primary, secondary, tertiary, and quaternary levels of structure. While the basic scheme remains intact today, it has been embellished considerably with additional levels of structural detail (Section 2.1). This is due in large measure to the success of crystallographic methods in determining the structures of a significant number and variety of proteins. Today x-ray crystallography remains the most important source of 3D structural data on proteins, essentially all of which is contained in the Brookhaven Protein Data Bank (PDB) (4). However, multidimension NMR (nuclear magnetic resonance) methods, most notably 2D-NMR, have begun to challenge crystallography's preeminent position, at least for small (<100 amino acid residues) proteins (5–7). An NMR structure database equivalent to that of the PDB is also being set up (8), and should provide another useful source of protein structure information.

Within the last decade theoretical and empirical procedures, driven by the greater availability of faster and more powerful computers, have begun to play a role in protein-structure determination. For example, heuristic methods, such as homology model building (Section 5.1), are becoming widespread, although their reliability, in general, remains somewhat questionable. Improved a priori procedures for predicting 3D structure have also been developed recently, but robust methods that can reliably fold proteins do not at present exist. Considerable research is taking place that is aimed at improving all of these methods. Thus, in the near future it should be possible to take greater advantage of the vast amount of primary structural data becoming available from the widespread application of modern protein and DNA sequencing techniques: More than 15 000 primary structures are currently available compared to just over 400 3D structures (9), and the gap is widening each year.

The present work reviews theoretical and empirical approaches to protein-structure prediction and analysis, especially secondary and tertiary structure, in globular protein. Neither fibrous nor membrane proteins are considered in the current review to keep the scope to manageable proportions. Analysis of primary structural features is not addressed, even though a considerable amount of interesting work has continued to emerge in the analysis of protein sequences (see, e.g., 10) especially in the area of deducing sequence motifs and relating these motifs to higher-order structural features (see, e.g., 11, 12). Ligand—protein interactions, dynamical features such as low-frequency motions and electrostatics and solvent effects, though important subjects, are not treated in a substantive way. Ligand—protein interactions are well covered in a recent book by Dean (13), while dynamical effects are discussed in great detail in the books by McCammon and Harvey (14) and Brooks et al. (15), and the review of Chou (16). Where appropriate, material on electrostatics and solvation is included, but a detailed review of these subjects is not given here because a number of excellent reviews are available that may be consulted for further details (17—21).

Section 2 provides an overview of the "modern" hierarchy of protein structure, including a brief discussion of the chiral features that pervade all levels of structure. Section 3 describes many of the methodologies used in current theoretical studies, particularly potential energy functions and their applications in molecular mechanics, Monte Carlo, and molecular dynamics calculations. A brief discussion of free-energy methods is also included. Section 4 covers a variety of empirical secondary-structure prediction schemes, emphasizing the popular approaches of Chou and Fasman (22) and of Garnier et al. (23), but also covering a number of other approaches including those designed to improve the predictions of the more traditional ones. The packing of secondary-structural elements leading to the formation of supersecondary structures is also treated in Section 4, in terms of both the structural determinants and energetics of packing. Section 5 addresses tertiary-structure prediction. A lengthy discussion of heuristic methods is presented, emphasizing homology/comparative model building procedures. A priori approaches are also discussed and, while they are not yet capable of truly predicting tertiary structure, they have, nevertheless, undergone significant improvements, and are beginning to show some promise. Section 6 provides a brief discussion of quaternary structure, highlighting allosteric features and protein—protein recognition. Section 7 concludes with an overview of the material covered, and a short discussion of future prospects for applying computers to the prediction and analysis of protein structure.

2. ELEMENTS OF PROTEIN STRUCTURE

Protein structure possesses two distinct characteristics: It is hierarchical and it exhibits a number of distinct chiral features. Its early classification into primary, secondary, tertiary, and quaternary levels of structure (3) provides a framework from which to understand the architectural principles that govern protein structure. A striking feature of all proteins is the way in which chirality (i.e., handedness) arises at each level of the protein−structure hierarchy. Interestingly, a number of proteins exhibit not only the more familiar types of *geometric chirality* but also a number of less well-known forms of *topologic chirality* (24−26), as is discussed in Section 2.2.

2.1. Hierarchical Character of Protein Structure

Primary structure characterizes the order or sequence of amino acids in a polypeptide chain beginning at the N-terminal amino acid, which contains the free amine, and proceeding toward the C-terminal amino acid, which contains the free carboxyl group at the other end of the chain, as depicted in Fig. 1*a*. While primary structure identifies in what order the amino acids are linked together in a protein, it does little to identify the 3D structural features that give a protein its unique functional properties. *Secondary structure* characterizes the quasi-regular features of polypeptide chains, including α helices (Fig. 1*b*) and extended chains that form the strands of parallel and antiparallel β sheets (Figs. 1*c* and *d*, respectively). As is apparent from Figs. 1*b*, *c* and *d*, hydrogen bonding plays a major role in stabilizing these structures. *Tertiary structure* characterizes the complete 3D structure of a single polypeptide chain, including the overall geometry of its chain fold, a feature that endows a protein with its distinctive 3D architecture (Fig. 2*a*, see color insert), and the geometry of its side chains, which endows a protein with its specific catalytic, binding, or other functional qualities (Fig. 2*b*, see color insert). *Quaternary structure* characterizes the manner in which individual subunits of a multisubunit protein are organized.

As more 3D structure data became available, it was clear that certain patterns of association of secondary-structural elements (typically 2−6 elements) occur frequently. This has given rise to an additional level of resolution in the structural hierarchy, the *supersecondary structure*. Supersecondary-structure motifs are characterized by specific orientations of their secondary-structural elements that are dictated by packing and energetic considerations. Examples of supersecondary structures include stacked α helices (αα), antiparallel β sheets (ββ), parallel β sheets (βαβ), and stacked β sheets (see, e.g., 27, 28). A discussion of the topographic and energetic factors that determine supersecondary structure is presented in Section 4.5.

The next level in the hierarchy of protein structure is the *structural domain*. An analysis of the 3D structures of globular proteins revealed the presence of distinct globular units; each unit consists of a single, continuous polypeptide chain and is only loosely connected to other globular units. In early work recognizing this as a general feature of protein structure, Wetlaufer (29) hypothesized that these compact, globular units are independent sites of nucleation of protein folding. These globular units, now called structural domains, are a recognized level of the protein structure hierarchy, typically 50—150 residues in size (28), made up of secondary and supersecondary structural elements which themselves make up proteins/subunits. Their existence is important in simplifying the folding process into separable, smaller steps, especially for large proteins, and they can serve as either moving parts for a protein that requires hinge motion to function or as modular bricks to aid in efficient assembly of proteins with complex enzymatic activities (30, 31).

However, there is no clear consensus on precisely what defines a domain, and consequently how to identify the domains of a given protein in an objective manner. There are three concepts of domains, the first of these being that they are globular, semi-independent units, with contiguous intrachain connectivity, as identified by Wetlaufer (29). Such domains may be called *Wetlaufer domains*. While there are examples where these are independently stable fragments as in immunoglobulins (32), or segments capable of autonomous folding as in chymotrypsin (33), or hinge units as in hexokinase (34), there are also cases where domains identified under this concept do not fall into such functional classifications (30).

In an attempt to identify domain-like structures objectively, Crippen (35), Rose (36), and Wodak and Janin (37) developed automated methods for analysis of proteins. While the parameters they used to identify substructures were different, the conclusions they obtained were similar: Wetlaufer domains represent only the upper levels in a hierarchy of simpler structures that descends all the way to simple secondary-structural elements. This is the second concept of domains, namely, that they form a hierarchy, in which the highest levels correspond to qualitative definitions of folding domains, and the lower levels suggest possible folding pathways.

Studies of immunoglobulins (38, 39) and nucleotide binding domains in various dehydrogenases (40) have led to a third concept of domains, as units of evolution, that are building blocks of enzymes. In the immunoglobulins the variable domains of the light and heavy chains are homologous to each other. In addition, the three constant domains of the heavy chain are homologous to each other and to the constant domain of the light chain. It has been shown that the three constant domains and the hinge region between domains 1 and 2 of the heavy chain are encoded in separate DNA segments, supporting the hypothesis that the domains of immunoglobulins arose through duplication of a common "domain gene" (41). Among the dehydrogenases, the common NAD binding domain is structurally conserved, while other portions of the chains have other functions, such as substrate binding and catalysis (40). Similar commonality is observed in the substrate binding domains of pyruvate kinase and triose-phosphate isomerase (42). The aspartic acid proteases are made up of two domains, each contributing identical residues to the active

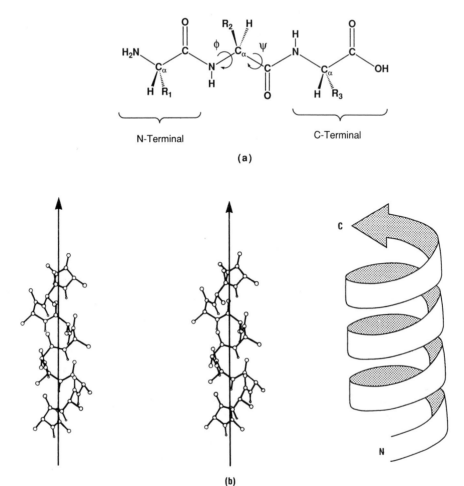

Fig. 1 a—b

site, suggesting that the domains originated from a common precursor, through gene duplication and fusion (43). The recent characterization of the aspartic acid protease from HIV-1 where the active site is made up of not two domains, but a pair of identical subunits, bolsters this hypothesis (44). Much of the work on the concept of domains has been reviewed (31, 38, 45) and an excellent graphic coverage of several observed domains is available (30).

The concept of a *tertiary fold* has also been introduced into the protein structure hierarchy (9); examples include the four-α-helical bundle (αααα), the Rossmann fold (βαβαβ), the β meander (βββ), and the Greek key (ββββ). Unfortunately, the distinction between tertiary fold and structural domain in some instances becomes blurred. For example, classifying a four-α-helical bundle as a tertiary fold is reasonable, but since four-α-helical bundles are

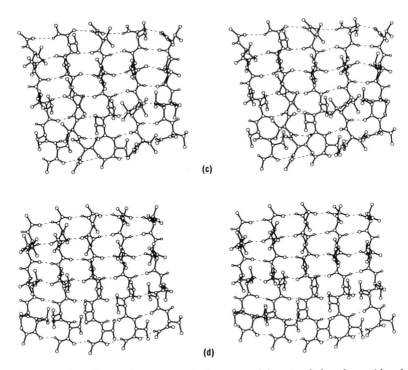

Fig. 1. Examples of protein structural elements. (*a*) extended polypeptide chain showing the N- (*amino*) and C- (*carboxyl*) terminal residues. By convention the direction of the chain runs from its N-terminal residue (R_1) to its C-terminal residue (R_3). The (Φ, Ψ) rotation angles are also indicated on the figure. (*b*) Stereoscopic representation of an α helix. The arrow points in the direction of the chain, and the coil ribbon diagram on the right side of the figure depicts the right-handed sense of the helix. N and C indicate the N- and C-termini, respectively. (*c*) Stereoscopic representation of an antiparallel β sheet formed from five polyvaline chains. (*d*) Stereoscopic representation of a parallel β sheet formed from five polyvaline chains. [(*b*) is reproduced from (297), and (*c*) and (*d*) are reproduced from (285).]

self-folding units it appears more appropriate to classify them as structural domains. On the other hand, a β meander is not a self-folding unit and is clearly best classified as a tertiary fold. Currently, a universally accepted scheme does not exist for describing protein-structure hierarchy, although Schulz and Schirmer (27) (Fig. 5−1), (cf. 28) have presented one that captures the essential features of the structure hierarchy.

Globular proteins and their domains may be divided into four main classes:

1. α proteins (domains), which possess only α-helical secondary structure

2. β proteins (domains), which possess only β-sheet secondary structure

3. α/β proteins (domains), which possess mixed or approximately alternating α-helical and β-strand segments

4. α + β proteins (domains), which possess α-helical and β-strand segments that do not mix but rather are grouped within particular regions of the polypeptide chain (46, 47)

Richardson (48) has presented an excellent graphical description of the structural motifs found in these classes. Schulz and Schirmer (Table 5−2 in (27)) and Robson and Garnier (Table 3.4 in (28)) have also provided extensive compilations of proteins (domains) that fall into the different classes.

In addition to the hierarchical view expressed above, supersecondary structures and domains can be viewed as possessing well-defined chain topologies. An examination of the various chain topologies by a number of different authors (27, 29, 46, 49−57) has shown that regularities exist and, according to Chothia (47), can be codified into three rules:

1. Pieces of secondary structure that are adjacent in sequence are also in contact in three dimensions.

2. The connections in β−X−β units are right-handed (β's are parallel, though not necessarily adjacent strands in the same β sheet and X is an α helix, a strand in a different β sheet , or an extended segment of polypeptide chain) (see also discussion in Sections 2.2 and 4.5.1).

3. The connections between secondary structures neither cross each other nor form knots in a chain.

Other proposed rules are, according to Chothia (47), either the same as or arise from a combination of the three rules given above. Thus, considerations of chain topology have been quite useful in elucidating underlying regularities of protein structure. Interestingly, different but related topological attributes of particular types of chain folds and disulfide bonding patterns have also been described by a number of authors (24, 58−60) as giving rise to novel chiral features in proteins (see Section 2.2 for further details).

2.2. Chiral Features of Protein Structure

In chemical systems, chirality is usually related to the nonsuperimposability of a molecule with its mirror image, an essentially geometric concept. This distinguishes it from the less well-known chirality based on the topological properties of molecules. (The topological properties of an object relate its invariant features under very general types of transformations, namely transformations that can bring about deformations of the object without tearing and rejoining it.) Both types of chirality are encountered in proteins.

Chiral features of proteins, as is the case with structural features where lower-level structures can be said to induce higher-level ones, also can induce higher-level chiral features (26). For example, the L-stereochemistry of the α-carbon of each amino acid (except glycine) tends to induce a right-handed twist in a fully extended polypeptide chain. When extended chains containing such right-handed twists come together to form β sheets, the sheets are

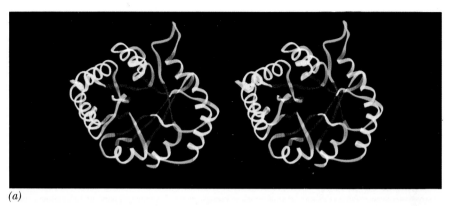

(a)

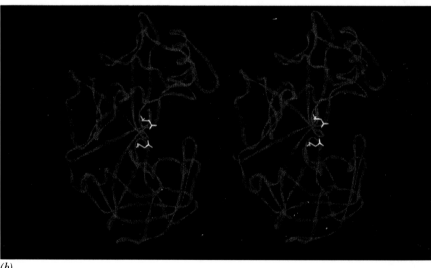

(b)

Fig. 2 Examples of the 3D architecture of proteins. *(a)* Stereoscopic ribbon diagram of the TIM barrel structure of triose phosphate isomerase. The red, yellow, and blue segments of the ribbon represent β strands, α helix, and loops, respectively. *(b)* Stereoscopic ribbon diagram of the aspartic acid proteinase *Rhizopuspepsin*. The two catalytically important aspartic acid residues are shown in yellow. [Coordinate data for triose phosphate isomerase and *Rhizopuspepsin* were obtained from the PDB. The former was based on the work of Banner, et al., *Biochem. Biophys. Res. Commun.* **72,** 146 (1976), and the latter on the work of Suguna et al., *J. Mol. Biol.* **196,** 877 (1987).]

observed to possess an overall right-handed twist when viewed along the direction of the strands. Interestingly, if the N- and C-terminal residues of α helices, which are all observed to possess a right-handed helical sense, are slowly pulled apart to yield an extended chain, the resulting chain has the expected right-handed twist (26; cf. 51). In addition to the right-handed character of α helices and β sheets, a number of other chiral features have been observed, including the right-handed arrangement of βαβ crossovers, the right-handed tilt in β barrels, and the left-handed twist of the set helices of four-α-helical bundle proteins. Section 4.6 presents detailed discussion of the geometric and energetic basis of the chirality observed in many of the secondary, supersecondary-, and tertiary-fold structures found in proteins. Interestingly, although protein structures are highly complex and can undergo substantial changes during evolution, the underlying chiral features remain highly conserved — they are a common feature in all proteins.

Recently, the discovery of several novel chiral features related to helix packing in α-helical domains of proteins was reported (26, 61). The work is based on the elegant solution to the helix packing problem presented by Murzin and Finkelstein (62; cf. 63). These authors showed that the packing of helices in helical domains could be represented to a very good approximation by the placement of *nonoriented* helices, represented as cylinders, along specific edges of regular, quasi-spherical polyhedra, as depicted in Fig. 3a for three, four, and five helices.

A close examination of the three helix case shows that its mirror image is not superimposable with the original image (Fig. 3b). Thus, the packing arrangement of three helices exhibits chirality, even when the natural orientation of a helix due to the directionality of its polypeptide chain is disregarded. Moreover, it was shown (26, 61) that the case of three-helix packing can be generalized to include greater numbers of helices by "decomposing" the problem into sets of helix triples which can be analyzed in the same fashion as shown above for three helices.

Although the work of Murzin and Finkelstein provided critical insights into helix packing that were instrumental in the development of the theory, the theory does not require that the helices be placed on idealized polyhedra, and thus is quite general. If the orientation of the helices is also considered, a number of additional chiral features emerge (26) that may provide, along with the nonoriented case, new modes for classifying protein structure.

The chiral properties of topological stereoisomers have fascinated chemists for a number of years (64). In biochemistry work on topological stereoisomers has been confined primarily to nucleic acids where the study of topoisomerases has elucidated a number of interesting topological stereochemical phenomena (64, 65). Although the folding of polypeptide chains in proteins, with or without disulfide linkages, has been studied by several authors (58−60, 66, 67), the topological stereochemistry of polypeptide chains has been analyzed in detail only recently. Many of the topological stereoisomers of multiply disulfide-linked polypeptide chains are also chiral (24).

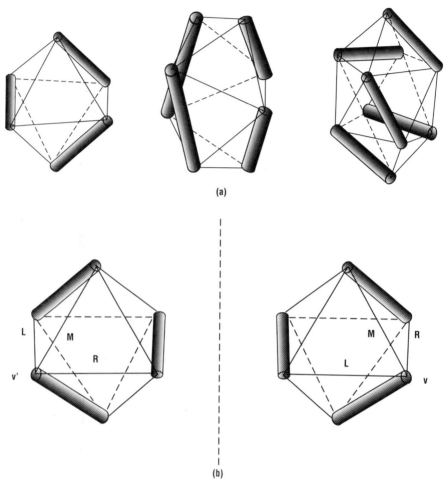

Fig. 3. Packing of nonoriented helices, depicted as cylinders, on regular, quasi-spherical polyhedra based on the method of Murzin and Finkelstein [*J. Mol. Biol.* **204**, 749–769 (1988)] for representing helix packing in α-helical domains of proteins. (*a*) Examples of three-, four-, and five-helix packings on (moving from left to right) an octahedron, a dodecahedron, and a hexadecahedron, respectively. (*b*) Assignment of the handedness of mirror image three-helix packings. The arrangement on the left of the figure corre-spondes to a left-handed packing, while the nonsuperimposable mirror image on the right corresponds to right-handed packing. Details concerning the method of assigning handedness are given in (26, 61). [Figures are reproduced from (26).]

An example of topological stereospecificity in the folding of a polypeptide chain is the D form of the neurotoxins from *Centruroides sculpturatus* Ewing and *Andructonus australis* Hector. In this case, the topological stereospecificity is due to the particular connections of the four disulfide bonds in the proteins. Figure

4 depicts the covalent graph (24) of the D form of these proteins and its topologically enantiomeric L form. The recent work of Mao (24) should be consulted for details. A similar approach has been used to analyze the patterns of hydrogen bonding in α helices and β sheets (25); this analysis reveals the topological chirality of these types of secondary structures. Whether such topologically based analyses of protein structure will lead to additional insights is unclear at this time. Nevertheless, it appears that such analyses may be useful in elucidating subtle aspects of protein structure that are refractory to more traditional approaches.

3. THEORETICAL METHODS

Ab initio and semiempirical quantum-mechanical methods have been instrumental in providing theoretical descriptions of the physical and chemical properties of simple molecular systems. Such methods are, however, precluded in studies of protein molecules because they contain hundreds to thousands of atoms. The development of empirical potential-energy functions (PEFs) in conjunction with molecular mechanics, molecular dynamics, and Monte Carlo methods (vide infra) has made the study of such large molecules possible by computer simulation, although their size and complexity pushes the limits of current computational approaches. Nevertheless, significant advances in understanding the structure, properties, and dynamics of proteins have occurred over the last two decades (see, e.g., 14, 15). And with the rapid growth in computer power, the next decade holds even greater promise.

Material included in the present section covers PEFs, molecular mechanics, molecular dynamics, Monte Carlo simulations, and the exciting new area of free-energy methods. The section is not meant to be inclusive, but rather to give the flavor of the currently available theoretical methods and some of their areas of application.

3.1. Potential-Energy Functions

Potential-energy functions play a central, critical role in essentially all theoretical studies of proteins, and are generally of the form

$$V = V_{\text{bond}} + V_{\text{ang}} + V_{\text{tor}} + V_{\text{vdw}} + V_{\text{hb}} + V_{\text{el}} \qquad (1)$$

where V represents the total potential energy of a molecular system as a function of an appropriate set of atomic coordinates. The first three terms describe potential energy changes due to bond stretching (V_{bond}), bond-angle bending (V_{ang}), and torsional motions about bonds (V_{tor}). The last three terms, usually called nonbonding interactions, describe potential energy contributions due to short-range repulsive and London attractive forces, the so-called van der Waals interactions (V_{vdw}), hydrogen bonding (V_{nb}), and

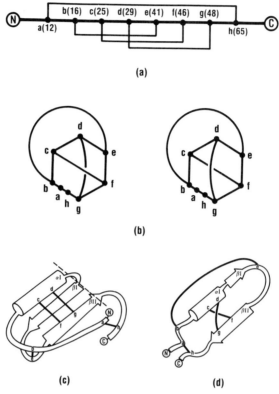

Fig. 4. Topological features of proteins. (*a*) Representation of the disulfide linkages in a scorpion neurotoxin from *Centruroides sculpturatus* Ewing (variant 3 toxin). The numbers in parentheses denote cysteines forming disulfide bonds in the protein. (*b*) Two topological isomeric embedded graphs of the disulfide-linked polypeptide chain shown in (*a*). The graph on the left corresponds to the L-topological stereoisomer and that on the right to the D-topological stereoisomer. The two graphs are enantiomorphic isomers. (*c*) Schematic drawing of the 3D structure of the neurotoxin described in (*a*) taken from the x-ray crystallographic work of Almassey et al. [*J. Mol. Biol.* **170**, 497 (1983)]. (*d*) Rearrangement of the polypeptide chain in (*c*) brings the C_α atoms of the cysteine residues into positions approximately corresponding to the vertices of the graphs in (*b*). (Note that this may be accomplished without changing the mathematical nature of the graph.) The relative positioning of the edges '*cf*' and '*dg*' corresponds to that of the D-topological stereoisomer shown in (*b*), and hence the scorpion neurotoxin is designated as a D-topoisomer. [Reproduced from (24, 26).]

Coulombic electrostatic interactions (V_{el}). Table I presents a concise summary of the typical functional forms used to represent the various terms in Eq. 1.

Sometimes an out-of-plane distortion term of the form $V_{op} = K_d (d - d^0)^2$ is added to a PEF to insure that the planarity of trigonal carbon atoms and the

TABLE I.

Terms Commonly Found in the Potential-Energy Functions of Proteins[a]

Bond stretch	$V_{\text{bond}} = \sum\limits_{\text{bonds}} K_r(r - r^0)^2$	K_r and r^0 are related to the harmonic force constant and equilibrium bond distance, respectively.
Angle bend	$V_{\text{ang}} = \sum\limits_{\text{ang}} K_\theta(\theta - \theta^0)^2$	K_θ and θ^0 are related to the harmonic force constant and equilibrium bond angle, respectively.
Torsion	$V_{\text{tor}} = \sum\limits_{\text{dihedrals}} \dfrac{V_n}{2} [1 + \cos(n\phi - \delta)]$	V_n is related to the energy barrier for rotation where n is the periodicity of the rotation. Multiple terms are possible for each dihedral.
van der Waals	$V_{\text{vdw}} = \sum\limits_{i} \sum\limits_{j>i} \left(\dfrac{A_{ij}}{R_{ij}^{12}} - \dfrac{B_{ij}}{R_{ij}^{6}} \right)$	A_{ij} and B_{ij} are van der Waals parameters and R_{ij} is the distance between interacting centers.
H bond	$V_{\text{hb}} = \sum\limits_{i} \sum\limits_{j>i} \left(\dfrac{C_{ij}}{R_{ij}^{12}} - \dfrac{D_{ij}}{R_{ij}^{10}} \right)$	C_{ij} and D_{ij} are H-bonding parameters and R_{ij} is the distance between interacting centers.
Electrostatic	$V_{\text{el}} = \kappa_Q \sum\limits_{i} \sum\limits_{j>i} \dfrac{Q_i Q_j}{\varepsilon R_{ij}}$	Q_i and Q_j are point charges, R_{ij} is the distance between them and ε is the dielectric constant. κ_Q is the unit conversion term.

[a] See text for further discussion.

stereochemistry about C_α-carbons is maintained: K_d is the force constant and, d is the out-of-plane distortion, generally referred to as an "improper torsion" angle (68), although other representations of d have been considered (69, 70).

The assumption that the PEF parameters given in Table I are *transferable* among similar atom types in similar molecular environments is crucial to the development of essentially all PEFs designed for use in proteins. However, while transferability of parameters can be demonstrated for many small molecular systems (see, e.g., 71, 72), its applicability to proteins needs further investigation (15).

An important feature of the PEFs in Table I is the existence of both harmonic (V_{bond} and V_{ang}) and anharmonic (V_{tor}, V_{vdw}, V_{hb}, and V_{el}) terms. Anharmonic terms are necessary due to the inherent anharmonicity of many regions of the potential-energy surface normally encountered by a protein. The need for anharmonic bond stretching and bending terms as well as cross terms that couple these motions with each other and with torsional motions was pointed out by Hagler and coworkers (68, 70, 73), particularly with regard to the accurate description of dynamical properties (e.g., normal-mode vibration frequencies).

Scheraga and coworkers (74) have espoused the desirability of excluding bond-stretching and angle-bending terms, primarily in the interest of reducing the already large number of degrees of freedom that must be dealt with in proteins. While such an approach does indeed reduce the magnitude of the problem, it has potential drawbacks that may or may not be of importance to a particular application. For example, dynamical properties associated with high-frequency motions related to bond stretching and angle bending are not taken into account (75), and, as discussed by Gelin and Karplus (76), the lack of flexibility brought about by constraining bond angles can have adverse effects on the results of conformational analysis, especially in sterically crowded systems. The effect of such bond length and angle constraints on free-energy calculations has been discussed by van Gunsteren (77).

A desire for simpler PEFs has given rise to the use of united-atom representations for certain aliphatic and aromatic carbon atoms (78—81). Even simpler PEFs, in which each amino acid is treated as one or at most two "interaction centers," have also been developed in an effort to facilitate energy-based protein-folding studies, details of which are presented in Section 5.3 along with an extensive description of a priori protein-folding methods.

As an alternative simplification, Hagler et al. (82) argued that specific terms to describe hydrogen bonding interactions were not required. In contrast, Weiner et al. (80) presented evidence that the inclusion of explicit lone pairs was desirable to describe hydrogen bonding at sulfur. However, the need for more detailed PEFs in certain instances remains an important question at this time. For example, should explicit three-body (see, e.g., 83) or polarizability (see, e.g., 84) terms be included?

In proteins, the large number of intramolecular nonbonded interactions suggests that solvent may have a greater effect on molecular structure and dynamics than in small molecules. Solvation effects have been accounted for *implicitly* by including a solvation term in the potential energy function (85, 86), by using a distance-dependent dielectric constant to modify electrostatic terms V_{el}, (79), or by modifying atomic partial charges and Lennard-Jones potentials (87, 88). Alternatively, the solvent can be included *explicitly* in a calculation. A number of water models have been developed (see, e.g., 89—93), and some have been used in molecular dynamics simulations of solvated proteins (e.g., 94, 95). While explicit treatment of water provides a greater level of detail than implicit approaches, the computational cost is high. In many cases, the level of detail provided by explicit consideration of solvent is not needed, and in those cases implicit approaches may provide an effective means of treating solvent effects without the high computational costs associated with more detailed treatments.

Determination of PEF parameters can be accomplished in a number of different ways. The usual approach involves iterative adjustment of the parameters until the calculated values of experimental observables such as molecular structure, crystal packing parameters, dipole moments, heats of sublimation, and normal-mode vibrational frequencies, to name a few, agree as well as possible with the corresponding experimental values (see, e.g., 72). Ideally, as in the consistent-force-field method (96, 97), the set of parameters is determined by a least-squares fit of the appropriate experimental quantities (vide supra).

Alternatively, since PEFs are designed to provide an optimal representation of the potential-energy surface of a given molecular system, one may generate an approximation to the surface at a number of selected points by evaluating the system's energy quantum mechanically at geometries that correspond to the points. This approach has the advantage of not requiring experimental data, and thus is especially useful when good quality data are difficult if not impossible to obtain. The accuracy of the quantum-mechanical methods used is, of course, an important consideration, but rapid improvements in the speed and accuracy of these methods hold considerable promise for their continued usefulness in PEF development.

Early work carried out by a number of authors using this approach (98—100) dealt only with intermolecular interactions. Recently, however, Hagler and coworkers (68—70) utilized information from first and second energy derivatives to develop highly accurate intramolecular PEFs. The use of derivatives provides data on the slope and the curvature of a potential-energy surface and produces a more efficient sampling of the surface features. Hagler and coworkers have also developed a hybrid approach in which initial estimates of the parameters from quantum-mechanical data are refined using experimental data. In addition, Dinur and Hagler (101, 102) have also addressed the problem of intermolecular interactions from a quantum-mechanical perspective.

Some of the PEFs discussed above have been used in comparative studies of polypeptide conformations (103, 104). Computer programs have been written that employ PEFs for studying the structure and properties of proteins (79, 105, 106). The development of improved potential-energy functions continues to be an area of active research (68, 101, 102, 107).

3.2. Molecular Mechanics Methods

Given a PEF that describes the intra- and inter-atomic interactions of a molecular system, a number of methods are available for studying the system's static, dynamic, and thermodynamic properties. In this regard, molecular-mechanics methods, many of which were developed for studying organic molecules (see, e.g., 72), have also shown usefulness in studying the *static* structural features of proteins. (Note that the term "mechanics," which has a much more general meaning in physics, technically is misused here, although its meaning in the present context is well understood by most chemists.) The most common use of molecular-mechanics methods is the determination of molecular structure, although other mechanical properties of proteins such as hinge bending between domains (88, 108) and rotation barriers of surface residues (76, 109) have also been investigated.

Structure determination by molecular-mechanics methods basically involves minimizing the PEF with respect to a suitable set of structural variables (vide infra). Numerous techniques exist (see, e.g., 110) for minimizing multivariable functions. A serious problem in the case of proteins and other large molecules is the large number of local minima that can "trap" the system in a structure away from the desired global minimum. This is known as the *multiple-minimum problem* (see, e.g., 111), and its relationship to the prediction of protein structure is considered further in Section 5. Recently, a new technique called *simulated annealing* (112–114) has been developed and shows promise as a means of overcoming, in some cases, the multiple-minimum problem. A number of obstacles must be overcome before a practical implementation suitable for proteins is available (cf. 115). Other approaches for circumventing the multiple-minimum problem are based on "buildup" procedures, various types of Monte Carlo sampling, or a novel self-consistent electric field method and were developed by Scheraga and coworkers (111). It is not clear, however, whether these methods are capable of treating proteins of more than about 50 residues. Lastly, Robson and Osguthorpe (116) employed a modified version of the classic simplex method using a simplified PEF that shows some promise, but based on their current work it does not appear that the method has fully overcome the multiple-minimum problem. Further discussion of these methods can be found in Section 5. Here the focus will remain on the more traditional gradient-based methods.

A PEF can be expanded in the region about point $\mathbf{x}$,

$$V(\mathbf{x} + \Delta\mathbf{x}) = V(\mathbf{x}) + \sum_i \left(\frac{\partial V}{\partial x_i}\right)\Delta x_i$$
$$+ \tfrac{1}{2}\sum_i \sum_j \left(\frac{\partial^2 V}{\partial x_i \partial x_j}\right)\Delta x_i \Delta x_j + O(\Delta x^3) \qquad (2)$$

where $\mathbf{x}$ represents a set of Cartesian or appropriate internal coordinates, and $O(\Delta x^3)$ represents anharmonic terms. The use of Cartesian coordinates is usually favored due to the relative simplicity of the required derivatives (117) and the absence of redundant coordinates that plague internal coordinate systems. Equation 2 may be represented in the more compact matrix notation as

$$V(\mathbf{x} + \Delta\mathbf{x}) = V(\mathbf{x}) + \mathbf{g}^T\Delta\mathbf{x} + \tfrac{1}{2}\Delta\mathbf{x}^T\mathbf{H}\,\Delta\mathbf{x} + O(\Delta x^3) \qquad (3)$$

where $\mathbf{x}$ and the gradient $\mathbf{g}$ are column vectors, the Hessian $\mathbf{H}$ is a matrix, and the superscript T indicates the transpose; the elements of $\mathbf{g}$ are given by $g_i = \partial V/\partial x_i$, and those of $\mathbf{H}$ by $H_{ij} = \partial^2 V/\partial x_i \partial x_j$. An early application of minimization techniques in the determination of molecular structure (118) employed the steepest-descent algorithm (119). In this procedure the gradient of the PEF, with respect to the Cartesian coordinates of each atom, is computed in every iterative cycle, and the atomic coordinates are adjusted by small amounts proportional to the negative of the gradient:

$$\Delta \mathbf{x}^{(n+1)} = \mathbf{x}^{(n+1)} - \mathbf{x}^{(n)} = \lambda^{(n)} \, \mathbf{u}^{(n)} \tag{4}$$

The superscripts represent the iteration number, $\mathbf{u}^{(n)}$ is a unit vector directed along the negative of the gradient at $\mathbf{x}^{(n)}$, i.e., $\mathbf{u}^{(n)} = -\mathbf{g}^{(n)}/|\mathbf{g}^{(n)}|$, and $\lambda^{(n)}$ is usually obtained by locating the minimum of $V(\mathbf{x}^{(n)} + \lambda^{(n)} \mathbf{u}^{(n)})$ through some type of line search. This method is a *first-order method* and is very efficient in lowering a molecule's potential energy in regions where the energy gradient is large, but relatively inefficient near minima where these derivatives are small.

In contrast, the Newton–Raphson, or simply Newton method, utilizes second derivatives to determine the new search direction, and thus exhibits better convergence in the relatively flat regions around the minima of a potential-energy surface. In Newton procedures the next step is determined by

$$\Delta \mathbf{x} = - \mathbf{H}^{-1} \mathbf{g} \tag{5}$$

which can be derived by substituting $\mathbf{x}_{\min} \approx \mathbf{x} + \Delta \mathbf{x}$ into Eq. 2 or 3 and noting that $\nabla V(\mathbf{x}_{\min}) = 0$.

The large storage and computation requirements of using the inverse Hessian matrix $\mathbf{H}^{-1}$ have prompted several variations. Allinger (120) modified the procedure so that only one atom is displaced in each cycle. In the block–diagonal Newton–Raphson method (72), the full matrix is divided into 3×3 blocks related to the Cartesian coordinates of each atom. The off-diagonal blocks are ignored for certain minimization calculations resulting in reduced storage requirements and reduced matrix-inversion computation. Another method for reducing the matrix computation is the adapted-basis Newton–Raphson procedure described by Brooks et al. (79), which employs a basis that spans the subspace where past moves have made significant progress. Recently, Ponder and Richards (121) have described another variant, the truncated-Newton method, which shows particular promise for large systems such as proteins. In their work, second derivatives below an adjustable threshold were discarded, and particular care was taken in utilizing the sparsity of the resulting Hessian to implement efficient storage and compu-tational procedures. These authors note that the method shows quadratic convergence near minima, a feature that rigorously holds for the full Newton–Raphson method.

In addition to Newton–Raphson procedures, the quadratically convergent conjugate-gradient method of Fletcher and Reeves (122), in which only gradients are computed, has been used considerably in structure optimization. In each iterative cycle, a line search is performed along the direction given by $\mathbf{p}^{(n+1)}$, which is the sum of the current negative gradient vector, $\mathbf{g}^{(n+1)}$, and the vector of the previous search direction $\mathbf{p}^{(n)}$ weighted by the ratio $\beta^{(n)}$,

$$\mathbf{p}^{(n+1)} = -\mathbf{g}^{(n+1)} + \beta^{(n)} \, \mathbf{p}^{(n)} \tag{6}$$

where

$$\beta^{(n)} = \frac{|\mathbf{g}^{(n+1)}|^2}{|\mathbf{g}^{(n)}|^2}$$

For a quadratic function of N variables, the first N search directions are shown to be conjugate vectors with respect to the Hessian matrix, and the gradient at the Nth step is guaranteed to be zero. A slight variation of this method has been applied to conformational studies of normal hydrocarbon molecules (123) and bovine pancreatic trypsin inhibitor (124).

Originally designed for conformational studies of organic molecules in which covalent interactions dominate the interatomic interactions, these methods have variable convergence characteristics when applied to larger molecules in general (125, 126), and to biological macromolecules in particular, where nonbonded interactions are as important as covalent terms. The overall efficiency of the methods in determining *local* minimum-energy structures depends on the molecular systems studied. Thus, hybrid methods have been developed to improve computational efficiency, combining, for example, steepest descent with Newton−Raphson (125, 126) or conjugate-gradient procedures. Using the latter approach, van Gunsteren and Karplus (124) carried out studies on proteins applying bond length and angle constraints.

Given that these methods are gradient based, and that there are many independent variables in a protein molecule, local structural changes can be studied efficiently. Large-scale motions, on the other hand, are best studied by considering the physicochemical and structural characteristics of the system. For example, hinge bending in lysozyme was modeled by incrementally "bending" and energy minimizing the region linking the two domains (127). A difference-refinement application of the procedure was used to calculate the adiabatic bending energy of the bacterial L-arabinose binding protein and to study the flexibility of the hinge region and energetic and structural features of hinge bending (88, 108). Similar interdomain motions have been studied in other proteins (128, 129). Other, more refined, techniques for calculating energy changes, in general, have been proposed (130−132).

3.3. Molecular Dynamics and Monte Carlo Simulations

Systems with the complexity of biopolymers such as proteins cannot be fully understood in terms of their static structural properties alone. Information on their dynamic and thermodynamic properties is also needed. Molecular dynamics (MD) and Monte Carlo (MC) simulation methods furnish suitable means for obtaining such information.

A. MOLECULAR DYNAMICS SIMULATIONS

The molecular-dynamics simulation methods currently used for proteins were based on procedures originally developed for statistical mechanical studies of simple liquids (89, 133). Their application to proteins, pioneered by Karplus and McCammon (134, 135), has led to the elucidation of a considerable number of interesting dynamical and thermodynamical properties. A brief description of some of these studies is presented in the following sections; more comprehensive descriptions, along with extensive references to the original literature, can be found in two recent books (14, 15).

Molecular dynamics is based on Newton's classical equations of motion:

$$m_i \frac{d^2 \mathbf{r}_i(t)}{dt^2} = \mathbf{F}_i(t) \qquad i = 1, 2, \ldots, N_{\text{atoms}} \tag{7}$$

where m_i is the mass of the ith atom, $\mathbf{r}_i$ is its position in Cartesian coordinates, and $\mathbf{F}_i$ is the force, which is equal to the gradient of V,

$$\mathbf{F}_i = -\nabla_i V(\mathbf{r}_1, \mathbf{r}_2, \ldots, \mathbf{r}_{N_{\text{atoms}}}) \tag{8}$$

The solutions of Eq. 7, $\mathbf{r}_i$, can be obtained by numerical integration (136, 137) given an appropriate set of initial atomic positions and velocities, and represent the trajectory followed by each atom in the system.

Using this method, McCammon et al. (127, 138) first studied the atomic motions of bovine pancreatic trypsin inhibitor (BPTI), a small protein containing 51 amino acid residues. Molecular dynamics simulations have since been extended to other proteins and peptides (14, 15), including proteins in solution (139, 140) and in crystalline environments (141, 142).

Results of MD simulations of atomic motions in globular proteins have provided insights into the internal flexibility of proteins (143—147). For example, results from early simulations of BPTI were used to characterize the magnitude, correlations, and decay fluctuations of atomic motions, and to suggest a diffusional and liquid-like character for the interior of a protein (127, 138). In other protein systems that have been studied since, protein structures were found to lie very close to observed x-ray structures, and the fluctuations of atoms about their average structure positions were close to those obtained from thermal factors in x-ray diffraction data (148, 149). Atom fluctuations were also found to be anisotropic (148) and anharmonic (150, 151). In addition, the motions of protein secondary-structural elements during MD simulations have been characterized (152, 153). Structures from MD trajectories recently have been incorporated into structure refinement methods in protein crystallography (154, 155) and in NMR structure determination (156—158).

A number of modified MD methods have also been investigated, DiNola et al. (159) conducted simulations at elevated temperatures as a means of

sampling molecular conformations. Protein systems under constant temperature and constant pressure conditions have also been studied by coupling the
systems to a proper external heat bath and scaling periodic boundary conditions
(160—162). Reactions in enzyme active sites, and the role of solvent in such
reactions, have also been investigated (163—165). Even for rare events such as
the "ring flipping" of tyrosine residues, methods have been developed for
calculating transmission coefficients (166, 167). In quasi-harmonic dynamics,
atomic displacements from MD simulations are used to construct a force-
constant matrix which can be diagonalized to yield normal-mode frequencies
(168).

Stochastic dynamics (SD), often called Brownian dynamics, provides a
means for including microenvironmental effects *implicitly* into the dynamics of
large systems such as proteins (169). In SD simulations Langevin's equations
of motion, Eq. (9), are solved to again yield the atomic trajectories, $\mathbf{r}_i$:

$$m_i \frac{d^2 \mathbf{r}_i(t)}{dt^2} = \mathbf{F}_i(t) + \mathbf{R}_i(t) - \gamma_i \frac{d\mathbf{r}_i(t)}{dt} \qquad i = 1, 2, \ldots, N_{\text{atoms}} \qquad (9)$$

In this case the trajectories are influenced not only by deterministic forces, $\mathbf{F}_i$,
but also by stochastic forces, $\mathbf{R}_i$, which along with the dissipative frictional
forces $-\gamma_i(d\mathbf{r}_i/dt)$, simulate the effect of the surrounding microenvironment
(162).

While not as widely employed as more traditional MD simulations, SD
simulations have nevertheless been used to address a number of important
questions that are difficult to handle with current MD procedures. These
include free diffusion of individual proteins, diffusional encounters between
proteins and between an enzyme and its substrate, certain activated-rate
processes, and long-term dynamics of polypeptide chains. An interesting
variant of the SD approach is the use of stochastic boundary conditions.
Localized regions of proteins are treated by traditional MD methods while the
effects of the remainder of the protein and the surrounding solvent are incorporated implicitly. Excellent overviews of SD methodology and its application
to proteins are given by McCammon and Harvey (14) and Brooks et al. (15),
and should be consulted for further details and for references to the growing
literature in this area.

A number of new MD approaches are being developed that promise to
extend the capabilities of current MD procedures for studying proteins.
Recently, Thacher et al. (170) developed an interesting hybrid MD procedure
that, similar to the stochastic-boundary method, divides a protein into two
regions, one that is treated as a deformable continuum and another treated as
a collection of interacting atoms. Thacher and Rabitz (171) successfully applied
this method to BPTI. The general applicability of the method, however, needs
to be established by further tests on a number of diverse protein systems.

The need to treat certain degrees of freedom quantum mechanically, as in
the study of enzyme-catalyzed reactions, has posed a significant problem for

classical MD simulations. However, new methods that combine quantum-mechanical procedures with classical MD methods are being developed and applied to a number of problems primarily in enzyme catalysis (172−177). Recently, Zheng et al. (178) have investigated the application of path integral techniques to problems requiring the treatment of both classical and quantal degrees of freedom.

B.   MONTE CARLO SIMULATIONS

In contrast to MD simulations, where time evolution of the system configuration is followed, MC simulations do not furnish time-dependent information. Nevertheless, both methods provide the means for generating the statistical-mechanical ensembles required in the calculation of thermodynamic properties. Monte Carlo methods generate configurations that are weighted by the Boltzmann factor $\exp(-\beta V)$, where V is the potential energy and β is equal to $(k_B T)^{-1}$, k_B being Boltzmann's constant and T the absolute temperature. Metropolis et al. (179) have modified the method to accept randomly generated configurations on the basis of Boltzmann factors for canonical ensembles. As a further modification of the MC approach, a Brownian dynamics simulation (see Section 3.3A) has also been developed in which random displacements during configuration generation are biased by atomic forces (180). For proteins, however, these methods are inefficient for ensemble generation because trial structures are accepted with a relatively low probability due to the highly anisotropic nature of the PEF for systems containing covalently linked atoms (181). Wakana et al. (182) have nevertheless successfully applied the MC approach in their study of the local, small-amplitude conformational fluctuations in hen egg-white lysozyme. The key to their success was the use of a method, due to Gō and Scheraga (183), that allows a small number of nearby angles to be varied without significantly disturbing the overall structure of the protein.

Alternatively, Noguti and Gō (184) addressed the anisotropy problem through the use of *collective coordinates* — eigenvectors of a Hessian matrix whose elements are given by $H_{ij} = \partial^2 V/\partial\eta_i\partial\eta_j$, where η_i and η_j represent only torsional degrees of freedom. (Note that although the collective coordinates resemble normal coordinates, they do not possess the proper dynamical characteristics of normal coordinates.) The authors found that the efficiency of configuration sampling using such collective coordinates was greatly improved, when compared to the use of uncoupled torsional coordinates.

Metropolis MC sampling, assisted by energy minimization, has also been applied to searching the conformation space of short peptides (185), but it is doubtful whether such an approach can be extended to polypeptides the size of most proteins. The most extensive use of MC procedures in studies of protein structure has been in lattice models of protein folding, as discussed in Section 5.3B. Unfortunately, lattice-based approaches have yet to provide convincing evidence that they can deal realistically with proteins.

3.4. Free-Energy Estimations

Estimates of the free energies of molecular systems obtained by computer
simulation provide an important link between computer simulations and
experimentally measurable quantities such as binding constants and solubilities.
The development and implementation of methods for reliably predicting free
energies is currently a very active area of protein research. Two recent reviews by Beveridge and DiCapua (186, 187) provide an excellent, somewhat
pedagogical overview of research in this field. In addition, a volume edited by
van Gunsteren and Weiner (188) summarizes the work of many laboratories.
Books by McCammon and Harvey (14) and Brooks et al. (15) should also be
consulted for additional details and references to the original literature.

The statistical-mechanical ensembles of molecular configurations needed
for free-energy estimates may be generated probabilistically by MC or deter-
ministically by MD methods. Appropriate partition functions, as in Eq. 10,
can be constructed using either method.

$$Z_N = \int \ldots \int \exp[-\beta V(\mathbf{x}^N)] d\mathbf{x}^N \qquad (10)$$

$V(\mathbf{x}^N)$ is the potential energy of the system corresponding to configuration $\mathbf{x}^N$
of the N-particle system and $d\mathbf{x}^N$ indicates that integration is carried out over
the entire N-particle configuration space. Thermodynamic and statistical-
mechanical quantities such as the Helmholtz free energy, A, can then be
calculated apart from a constant which depends on the standard state chosen,

$$A = -k_B T \ln Z_N \qquad (11)$$

In actual calculations free-energy differences, and not absolute free energies,
are generally computed by one of two basic procedures, *thermodynamic integration* or
thermodynamic perturbation, although other approaches have been used (186,
187).

In thermodynamic integration the change in Helmholtz free energy may be
written as

$$\Delta A = \int_0^1 \frac{\partial A(\lambda)}{\partial \lambda} \, d\lambda \qquad (12)$$

where the *coupling parameter*, λ, is used to describe the system as it moves from
its initial state, $\lambda = 0$, to its final state, $\lambda = 1$. In this regard, it is important to
note that since free energy is a thermodynamic state function, the path taken
between the two states is immaterial and any computationally convenient path
may be used. In certain instances physically meaningful paths, such as a
reaction path that passes through the transition state of a reaction, may be
used. In these cases, information on the free energy of activation can be
obtained in a free-energy simulation. Mezei and Beveridge (189) have provided a

detailed account of what they call topographical-transition coordinates including a variety of physically meaningful λ coordinates (vide supra). It can be shown from Eq. 12 that

$$\Delta A = \int_0^1 \left\langle \frac{\partial V(\mathbf{x}^N, \lambda)}{\partial \lambda} \right\rangle_\lambda d\lambda \tag{13}$$

where $\langle \ldots \rangle_\lambda$ indicates an ensemble average at a specific value of λ. Numerical integration of the ensemble average over λ yields the Helmholtz free-energy differences. The details of this process are described in earlier references.

In the thermodynamic-perturbation approach, one state is considered a reference state, and the other is obtained by a perturbation of the reference state. Thus, the perturbed state is assumed to be similar to the reference state. Based on the reference state, the free-energy difference is computed as an ensemble average of the potential-energy difference, $\Delta V(\mathbf{x}^N)$, between the perturbed and reference states,

$$\Delta A = -k_B T \ln \left\langle \exp\left[-\frac{\Delta V(\mathbf{x}^N)}{k_B T} \right] \right\rangle_0 \tag{14}$$

In cases where the difference in the free-energy changes of two paralled processes is sufficient, a *thermodynamic cycle* (Scheme 1) can be constructed and the $\Delta\Delta A$ computed as the difference in free energy of the two processes, $\Delta\Delta A = \Delta A_2 - \Delta A_1 = \Delta A_4 - \Delta A_3 (190)$.

Scheme 1*

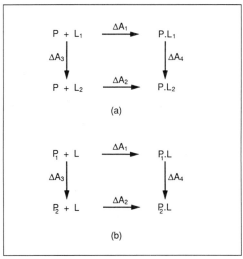

(a)

(b)

* P = Protein, L = Ligand

In Part (a) of Scheme 1, ΔA_3 can be determined by transforming the ligand L_1 ($\lambda = 0$) to L_2 ($\lambda = 1$). ΔA_4 can be determined by a similar change of the ligand bound to the protein, i.e., $P \cdot L_1 \rightarrow P \cdot L_2$. In contrast, the protein is changed in part (b), a situation that occurs in site-specific mutagenesis experiments. Thermodynamic cycle-perturbation calculations have been carried out on a number of systems from the solvation of ions (191) and amino acids (192) to acid dissociation (193) and ligand binding in proteins (194, 195). As has been pointed out recently, a large ensemble is necessary for computing free-energy differences between systems that differ significantly (196, 197).

The Gibbs free energy, G, calculated in a constant temperature−pressure ensemble, is more appropriate for the study of most biochemical phenomena, such as ligand−protein, enzyme−substrate (inhibitor), and protein−protein interactions, than Helmholtz free energy, A, which is calculated in a constant temperature−volume ensemble. Calculation of Gibbs free energy can be accomplished quite easily with MD methods by carrying out the simulations under constant temperature and pressure (see, e.g., Appendix 3 in (14) and references cited therein). Allen and Tildesley (198) provide a relevant discussion of the application of MC methods to the calculation of constant temperature−pressure ensembles (see also, e.g., (199)).

4. SECONDARY AND SUPERSECONDARY STRUCTURE

Even before the present explosive growth in the amount of protein sequence information, a number of efforts to develop secondary-structure prediction methods were undertaken. Robson and Garnier (28; Chapter 9) provide a very readable description of the history of these methods, whose development began in earnest in the early 1970s when sufficient 3D protein structure data became available. Since essentially all secondary-structure prediction methods rely heavily on structural information contained in the PDB, they should be classed as *knowledge-based* rather than as a priori methods (cf. discussions in Section 5).

When using secondary-structure predictions, one should consider the role prediction will play in the planned studies. For example, will the predictions be used as a basis for 3D folding calculations (see, e.g., Sections 5.2 and 5.3), or as a basis for determining the *sequence* of secondary-structure elements along a polypeptide chain? Each case places different demands on the prediction method. In the former case, reasonably accurate predictions of the location and extent of each secondary-structure element are required to insure that the folding processes behave properly and lead to a correctly folded structure (see, e.g., the work of Cohen, Sternberg, and coworkers discussed in Section 5.2 and that of Saito et al., discussed in Section 5.3). Moreover, if the boundaries of some secondary structures are accurately predicted, it may be possible to fill in incorrectly predicted gaps within those boundaries. In the latter case, such accuracy is not required, but all secondary-structure elements present should be predicted in their correct order. If this can be accomplished, the information can be used to search for proteins with identical or similar patterns of secondary-structural elements, and to identify supersecondary structure motifs such as the $\beta\alpha\beta\alpha\beta$ pattern found in Rossmann folds (49). For example, Crawford et al. (200) has predicted that the α subunit of tryptophan synthase possesses eight repeated β-loop motifs, and thus have classified it as a possible TIM barrel (see also Section 5.2).

Basically, three different approaches have been developed for secondary-structure prediction: (1) statistical, (2) pattern-based, and (3) statistical—mechanical, although this classification is somewhat arbitrary. Essentially all methods employ either a three- or four-state model to describe the secondary-structure states (i.e., conformational states) of a protein chain. These states are then used to partition a given amino acid sequence into its appropriate secondary-structural elements. Three-state models resolve sequences into α helices and β strands, plus coil elements, which include everything else. A four-state model is identical to a three-state one except that the coil elements are further resolved to distinguish β turns explicitly. Coils are sometimes referred to as random coils, but this is inappropriate because these moieties, while generally nonregular, are not random chain structures.

While both three- and four-state models provide a sensible basis for de-scribing, and hence predicting, the secondary-structural features of proteins, nevertheless some complications inevitably arise. One complication is due to difficulties encountered in reliably characterizing the secondary structural regions obtained from x-ray crystallographic, and now even NMR data. This arises from the fact that all structures in the PDB are not given to equal resolution (28, 201, 202). An important consequence of this is that what may, at low resolution, appear to be the end of an α helix may in fact be a 3_{10} helix; thus, the lengths and/or boundaries of secondary-structural elements may change significantly at higher resolution. In addition, methods for assigning the location and extent of even high-resolution structures are imperfect (cf. 203). Because all secondary-structure prediction methods rely on infor-mation contained in the PDB, changes in the nature and location of secondary-structural elements, albeit small ones, could adversely affect analyses based upon them. However, the likelihood of this having a major effect on the reliability of any prediction method is not very high. Rather, Schulz (202), among others (e.g., 28), has discussed the fact that parameters or other secondary structure-based information derived from the PDB have converged to the extent that additional structural data is not likely to improve the parameter set significantly. Nevertheless, work by several authors (204—209) has indicated that knowledge of supersecondary-structural motifs may aid secondary-structure predictions (see discussion in Section 4.4).

The book by Robson and Garnier (28) provides an excellent description of many of the most heavily used methods. Schulz and Schirmer (27) also provide a readable, but slightly more abbreviated, account of these methods. And a recent review (202) provides an excellent pedagogical overview of the current state of the field. Due to space limitations, the following discussion will address only the general features of the three classes of methods, namely, statistical, pattern-based, and statistical—mechanical, and the original references or those cited above should be consulted for further details.

4.1. Statistical Methods of Secondary-Structure Prediction

Statistical methods are based primarily on analyses of the frequency of occurrence of particular amino acids in specific secondary-structural regions of proteins found in the PDB. Information from these analyses is then used to estimate the propensity of an amino acid within a particular sequence to adopt a specific conformational state. Two statistical methods stand out in terms of their level of use: that developed by Chou and Fasman (22, 210, 211) and that developed initially by Robson (212) and Robson and Pain (213), and later refined by Garnier, Robson, and their coworkers (23, 214).

The Chou—Fasman (CHF) method utilizes the observed frequency of occurrence of individual amino acids, that is, singlet frequencies, in particular types of secondary structures, based upon structural information available in the PDB. Both three- and four-state models have been used with this method. Each amino acid is assigned a helix and sheet propensity based on its frequency of occurrence as part of the respective secondary-structural element as found in the PDB. In predicting the secondary structure for a given sequence, the occurrence of residues with certain limiting and cumulative helix/sheet probabilities over short ranges are interpreted by a set of rules that are used to determine secondary-structure prediction. These rules are intended to account for the effect that neighboring amino acids in the sequence have on the conformational state of a given residue. However, application of the rules, which are clearly stated, is nevertheless subject to individual interpretation, and thus it is difficult to apply them unambiguously. Different individuals may very well predict different secondary structures for the same sequence by the CHF method. This ambiguity also occurs when predictions are obtained from the many programs (203, 215—217) that apply these rules.

Argos and Palau (218) further analyzed the PDB in terms of amino acid composition at unique positions such as those at the N- and C-termini of α helices. In a following paper (219) this information was used in a variant of the CHF method developed by the authors. While the results obtained appear to lead to a modest improvement in prediction, it comes at the expense of additional parameterization. The authors point out that their results "suggest that the limit of prediction accuracy from singlet amino acid methodologies has nearly been reached," echoing the sentiments of other workers in this field (e.g., 28). Even earlier, Kotelchuck and Scheraga (220) used the results of energy calculations to assign probabilities that given amino acid residues would exist in either helix or nonhelix conformational state. These authors then applied a series of rules, reminiscent of those applied later by Chou and Fasman, to predict the locations of helical regions in polypeptides.

Doublet frequencies generally defined as the frequency, that a particular residue type occurs at position i in a particular type of Secondary Structure, given that a specific residue type is found at position $i \pm j$, have been employed by a number of workers (221–226) to improve secondary-structure predictions. Although the current size and diversity of the PDB is sufficient to support such an approach, the results obtained lead to only modest improvements over those obtained using singlet frequencies (see, e.g., discussion in Chapter 6 of Schulz and Schirmer's book (27)). Triplet frequencies have also been considered (55, 227), but the size of the PDB is not sufficient to produce statistically valid frequencies of occurrence. It has been suggested (202) that the use of triplet frequencies will not likely play a major role in secondary-structure prediction, even when sufficient information in the PDB is available (cf. 214).

The Garnier, Osguthorpe, Robson (GOR) method, while still probabilistic in nature, relies on the use of an information theoretic measure, $I(x, R_i + m)$, which represents the information that the type of residue R at position $i + m$ possesses about the conformational state x of the residue at position i (note that the information measure I is related to the logarithm of a ratio of probabilities, and thus is inherently a statistical method (28)). This type of information, called directional information by Robson and Pain (213), covers the neighboring eight residues on either side of the ith residue, i.e., $-8 \leq m \leq 8$. Thus, the GOR method, in contrast to the CHF method, takes account of neighborhood effects in an unambiguous manner. Basically, the GOR method determines the likelihood $L(x, i)$ that a residue at position i will adopt a particular secondary-structure state x, by the formula

$$L(x, i) = \sum_{m=-8}^{8} I(x, R_{i+m})$$

An advantage of the GOR method is that data not derived from sequences can be easily incorporated to improve secondary-structure prediction. For example, the amount of each secondary-structural type present in a protein can be estimated from CD measurements (228), and the appropriate likelihood function $L(x, i)$ can be modified to reflect this (229). In addition, while the GOR method is considerably more difficult to understand than the CHF method, it is easy to program unambiguously. Recently, Gibrat et al. (214) reevaluated the parameters of the GOR method based on the increased size of the PDB and the improved description of secondary structures available today. The results they obtained show an improvement of about 7% over the original implementation, bringing the overall accuracy of prediction to nearly 63%.

Both the CHF and GOR methods yield correct predictions in the range of ~55% (230) by the usual criterion of comparing the number of residues predicted correctly to the total number of residues predicted. As noted by many workers, while this measure is straightforward to calculate it can be very misleading. In fact, Schulz and Schirmer in their book (27) list no fewer than 10 quality indices that have been used to characterize the correctness of a secondary-structure prediction (cf. 216). Perhaps the best criterion, which has not been used much to date, is that described by Robson and Garnier (28, 214) based on an information theoretic measure (cf. 231). One final consideration when evaluating the success of any secondary-structure prediction method, is a comparison of the results to those that would be produced by a purely random prediction (202). For example, in the work of Kabsch and Sander cited above, the approximately 55% correct prediction based on the CHF, GOR, and Lim (see Section 4.2) methods should be contrasted with the considerably lower value of about 38% one would obtain from a purely random prediction.

4.2. Pattern-Based Methods of Secondary-Structure Prediction

The pattern-based approach is exemplified by the procedure developed by Lim (232) more than 15 years ago. Lim's procedure utilizes rules based upon sequence patterns derived from stereochemical features such as the hydrophilicity/hydrophobicity or steric bulk of each amino acid residue. For example, the clustering of triplets of hydrophobic residues in positions i, $i + 1$, and $i + 4$ and i, $i + 3$, and $i + 4$ are deemed important unless their side chains are too bulky. Large hydrophilic amino acids are viewed as residues that can potentially stabilize such hydrophobic clusters. Prediction of helical regions is done first, and the remaining nonhelical regions are then examined for the existence of potential β strands, which are divided into three classes, namely, surface, semisurface, and buried strands. Buried strands generally possess only hydrophobic residues, in contrast to the other two classes where both hydrophilic and hydrophobic residues are present depending upon the amount of exposed surface possessed by the strand. It is important to realize that while Lim's method does pay particular attention to the stereochemical (and physicochemical) characteristics of individual amino acids and their effect on secondary structure, his view was strongly influenced by the limited amount of structural information available in the PDB at that time.

As pointed out by Schulz and Schirmer (27), the number of detailed rules needed to fit the test dataset in Lim's method is quite large (22 for α-helix and 17 for β-sheet predictions), and questions the ability of the method to predict reliably the secondary structures of proteins not in the initial test set. To illustrate, early predictions using Lim's method produced reasonably satisfactory results, generally somewhat better than those of either the CHF or GOR methods. However, this method has recently been called into question by some workers in light of seemingly poorer performance (216, 230). Such behavior is not entirely unexpected, based on the fact that Lim's rules, which were developed in 1974, have not been updated and/or reevaluated to account for the additional structural information now available in the PDB. In spite of this, recent work by Sweet (233) indicates that results obtained by Lim's method are clearly competitive to those obtained by either the CHF or GOR methods.

A number of other pattern-based methods have also been developed recently (cf. review by Taylor (12); see Section 4.4 for additional discussion). In one approach, which is exemplified by the work of Levin et al. (234), Sweet (233), and Nishikawa and Ooi (235), sequences of 7—12 amino acids of a protein whose secondary structure is to be predicted are compared against similar sequences in the PDB. Here, as in all homologue methods, the notion of similarity plays an essential role, and each group employs a somewhat different measure of similarity (cf. 236). As identical short sequences of amino acid residues present in the PDB exhibit striking conformational diversity (237), a scoring system is used to determine the best conformation for the target sequence. In this regard, the approach of Nishikawa and Ooi (235), with a prediction of up to ~70% accuracy, appears to be superior to the other two. The average predictions of all three, however, are in the same range and compare favorably with the methods discussed in Section 4.4A.

Recently, Rooman and Wodak (238) have carried out a very thorough analysis of the ability of pattern-based methods to predict secondary structure reliably. These authors searched for sequence patterns consistently associated with particular secondary-structure types, and found that short sequence patterns are capable of predicting secondary structure, but with a reliability of only ~60% — no better than any of the other methods described in Section 4. They did find some short sequences to be strongly predictive, although they estimate that a database of 1500 proteins possessing on average 180 residues each will be needed to significantly raise the level of predictability of short sequences.

A number of pattern-based approaches have been developed to handle the prediction of turns and loops. A discussion of this work is presented in Section 4.4B.

4.3. Statistical-Mechanical Methods of Secondary-Structure Prediction

A number of methods based on classical statistical mechanics have been developed to deal with the problem of helix—coil transitions in biopolymers (for general discussions see (27, 28, 239)). These methods arose from theoretical studies of idealized systems (e.g., homopolymers), and a number of models based on the Ising model used to treat magnetic systems were proposed, the most famous being Zimm and Bragg's (240). Mattice (241) recently reviewed similar work on β-sheet—coil transitions. Perhaps the earliest application of the statistical-mechanical approach to secondary-structure prediction is the work of Lewis and coworkers (242, 243) on the prediction of the α-helical segments of globular proteins. These workers employed a two-state (helix and coil) model and considered only nearest-neighbor interactions. Each of the 20 amino acids was classified into one of three types with its own set of Zimm—Bragg parameters, s and σ. Based on helix probability profiles calculated with these parameters for several proteins, the predicted location of helices was found to be in reasonable agreement with their location in the native structures. Tanaka and Scheraga (244, 245) employed a more elaborate model involving more states (α-helix, extended structure, β-turn, and coil), but also considered only nearest-neighbor interactions. In their work each of the conformational states of the 20 amino acids was assigned a statistical weight based on x-ray crystallographic data.

Not surprisingly, the neglect of medium- and long-range interactions affects the stability of helices, sheets, and turns, which can lead to errors in secondary-structure predictions (see, e.g., the books by Schulz and Schirmer (27) and Robson and Garnier (28)). Wako et al. (246) adapted the statistical-mechanical method of Wako and Saito (247), which included medium-range interactions, to the calculation of α-helical and extended structures of proteins. As discussed by Wako et al. (246), the size of the existing PDB is not sufficient to provide statistically reliable data to treat medium-range interactions; some simplifications were introduced to deal with this lack of data. Extension of this approach to include long-range interactions is not likely in the near future.

Ptitsyn and Finkelstein (248) also have summarized the results produced by the statistical-mechanical method they developed during the late 1970s. Their theory takes into account both local and long-range interactions, incorporating them into an Ising-like model, which, however, differs from that used by Zimm and Bragg. An interesting feature of the Ptitsyn—Finkelstein model is that it uses features of the stereochemical approach to treat short-range interactions, while long-range interactions are approximated by the interaction of specific chain regions with an average hydrophobic template (the "floating-logs" model).

While these studies have provided a number of insights into the secondary-structural features of proteins, they have not significantly improved the prediction of secondary structure, nor have they been utilized in any substantive way. And thus, due to limitations of space, this topic will not be further pursued here. The books by Schulz and Schirmer (27) and Robson and Garnier (28) should be consulted for highly readable discussions of the major works in this area.

4.4. Miscellaneous Secondary-Structure Prediction Methods

A. IMPROVED METHODS

Numerous attempts have been made to improve secondary-structure predictions. In general, most attempts have been based on the incorporation of additional information into the prediction scheme (cf. the discussion on the GOR method in Section 4.1). In this regard, three types of information have been considered: (1) that obtained by using a combination of prediction methods, (2) that obtained from supersecondary and tertiary structure data, and (3) that obtained from data relating to conserved residues and sequences. As will become apparent, this last and newest approach appears to have yielded generally better results.

As early as 1974, a number of the principal workers developing secondary-structure prediction methods joined forces in an attempt to predict the structure of adenylate kinase using a combination of 11 methods (249). The prediction, which was based on an unweighted average of the methods, yielded results in better agreement with observed data than results obtained by any method alone. Several years following this work, Argos et al. (250) obtained improved predictions on all protein structures available at that time with a similar approach involving five methods. Somewhat later, Nishikawa (216) combined three prediction methods, but with somewhat mixed results. Recently, Biou et al. (251) also combined three methods including an updated version of the GOR approach along with a homologue-based approach (see below for further details), and a new method based on a bit-pattern representation of hydrophilic/hydrophobic sequence motifs. While the overall accuracy of the predictions is improved to about 65%, the most interesting aspect of their work is its seeming ability to predict the location of runs of regular secondary structure (i.e., helices and sheets) to about 75% accuracy. Results within this range may be sufficient for studies aimed at computer-assisted folding of proteins.

It has long been thought that some account of the interactions responsible for the packing of secondary-structural elements into supersecondary and tertiary structures must be incorporated into any method for the reliable prediction of secondary structure. Kabsch and Sander (237) presented evidence to support this view, in which they showed that specific pentapeptide sequences did not exhibit strong correlations with particular secondary structures. In fact, specific pentapeptide sequences exhibited a surprising diversity of conformational preferences, a characteristic that clearly suggests that improvements in the quality of secondary-structure prediction methods will have to incorporate information on the 3D structural context in which such short sequences reside — which is just another way of saying long-range interactions must be considered in some fashion (cf., 238). Recently, Garratt et al. (208) have addressed this question explicitly for the case of β sheets. These authors found that the quality of the prediction declined markedly with increasing solvent accessibility, which is related to the long-range characteristics of chain folding as well as to solvent effects (note that these two features are not entirely independent of each other).

As noted in Section 2.1, globular proteins may generally be put into four classes: α, β, α/β, and α + β (30, 46). Classifications of this type can be predicted with much greater accuracy (ca. 75−85%) (Nishikawa et al., 252−256) than is typical for secondary structures. Thus, one may first determine the class of a protein of interest, and then use this information with an appropriately modified set of secondary-structure prediction parameters and/or algorithms to achieve improved secondary-structure prediction.

An example of such an approach is examplified by the double-prediction method of Deleage and Roux (209). In this method, the class of a target protein is first determined from its amino acid composition by the method of

Nakashima et al. (257). Then, using modified CHF-like parameters, a tentative secondary structure prediction is made using a new algorithm which provides a means for implicitly incorporating features, such as the lengths of helices and sheet strands, pertinent to particular protein classes. The information gained from these two independent predictions is then used to optimize the prediction parameters with respect to (1) maximal accuracy of secondary-structure prediction, (2) maximal agreement between predicted and observed secondary-structure composition, and (3) improvement in secondary-structure prediction for the majority of proteins in a given class. Predictions obtained by this method do show a modest improvement over those that do not take into account the class to which a protein belongs (single-prediction methods). Furthermore, it has been suggested by the authors that division of proteins into their domains may lead to improvements in the quality of the predictions. This has not, however, been tested as yet, due mainly to the difficulty of predicting the location of protein domains with reasonable accuracy (see, however, Busetta and Barrans (255)).

An alternative but related approach was given earlier by Taylor and Thornton (206, 207), who developed a novel template-based method that provides a means for incorporating supersecondary-structure information, specifically $\beta\alpha\beta$ units, into secondary-structure predictions. These authors determined a prototypical sequence or template for a $\beta\alpha\beta$ unit based on an analysis of all such units existing in the PDB at the time of their work. Information obtained with standard secondary-structure prediction methods, such as CHF or GOR, along with information on specific hydrophobic residue patterns was then used in conjunction with prototypical $\beta\alpha\beta$ sequence information to enhance the probability of locating unknown sequences. Their results were encouraging: About 70% of the units on 16 $\beta\alpha\beta$-type proteins were correctly located, an improvement of 7–8% over the initial secondary-structure prediction. This particular approach, while promising, has not been extended to other types of proteins at this time, although applications of related pattern-based recognition procedures have been implemented by Cohen et al. (205).

The last class of methods to be considered here fall loosely under the rubric of homologue methods. In this approach, which is exemplified by the work of Zvelebil et al. (258) and Crawford et al. (200), information from multiple-sequence alignments of a set of proteins within the same family (i.e., similar functionality) is used to ascertain the existence of short runs of conserved residues. To insure that realistic alignments are obtained, equivalent, similar, or conserved amino acids (e.g., 236) must be considered in the alignment process.

Since structural homology is known to be conserved far better than sequence homology within a protein family (259) it is assumed that the conformations of conserved residues (or groups of residues) will remain relatively unchanged within the family. Thus, a variety of prediction methods (e.g., CHF and GOR) are then applied to the whole set of proteins, and the predictions are averaged. In addition, other features such as hydrophobicity and flexibility may be considered to help resolve cases where prediction is unclear. Both methods discussed here do lead to improvements over standard methods, and have an advantage over the first type of homology methods in that they do not require knowledge of the class to which a protein belongs for their predictions. They do, however, require that sufficient sequence data on related proteins exist, but this potential problem is rapidly diminishing due to the explosive growth in the number of protein sequences available.

B. TURN AND LOOP PREDICTION

The recognition of turns as distinct structural elements was first described by Venkatachalam (260) in 1968. Recently, Rose et al. (261) presented an extensive review of turns, and Milner-White and Poet (262) have also discussed turns and other related structural elements such as loops, bulges, and hairpins. Because turns on average constitute about 25% of the residues of globular proteins (202, 261), they undoubtedly play an important role in determining their overall 3D structure. Moreover, the average frequency with which turns occur along the polypeptide chain provides a means for reducing the difficulty of secondary-structure prediction through a divide and conquer strategy, which can be used to break the original problem up into a set of smaller subproblems (205).

The earliest turn-prediction procedure is probably that of Lewis et al. (243), which several years later was modified by Chou and Fasman (210, 211). Both the CHF and GOR methods are capable of predicting the existence and location of turns with modest accuracy (see papers cited in Section 4.1 for further details). Recently, Wilmot and Thornton (263) have improved the parameters used in the CHF procedure and have included information on turn type as well. These changes led to a corresponding improvement in turn prediction, with 60−80% of the turn residues being predicted correctly.

An alternative approach to turn prediction is based upon the fact that turns are generally found at the surface of proteins. This strongly implies the presence of hydrophilic amino acid residues in these secondary-structure elements. Information of this type has been used by a number of workers for turn prediction (264−266). A novel visual approach to the prediction of turns as well as α helices and β sheets was developed by Cid and coworkers (266). In their approach, information contained in hydropathy plots is analyzed in terms of prototypical shapes that correspond to α helices, β sheets, and reverse turns. The nature of the prototypical shapes is deduced from the expected properties of the protein chain possessing particular secondary-structural characteristics. For example, in the case of reverse turns one would most likely encounter a pattern of residues of the following form: hydrophobic−hydrophilic−hydrophilic−hydrophobic. Not expectedly, unambiguous application of this method is difficult in practice.

Recently, Cohen et al. (267) have developed a pattern-matching approach that shows considerable promise for turn (and loop) prediction of proteins within specific classes, although its general applicability remains to be fully characterized. These workers note success rates of 95% for proteins in their test set and 90% for proteins in homologous families. No other turn-prediction methods are even near the predictability claimed by these workers. Interestingly, Rooman et al. (268) investigated the amino acid sequence templates obtained from recurrent turn patterns. They concluded that although these patterns may provide a suitable means for improving predictions made by other methods, their predictive power is minimal when they are used alone.

C. NEURAL NETWORK APPROACHES TO SECONDARY-STRUCTURE PREDICTION

Recently, neural-net computers (or neural networks) have attracted considerable attention due to their reported pattern-recognition capabilities (269). Neural networks consist of numerous interconnected nodes that are analogous but not equivalent to the neuronal connections in the human brain. The strengths of the interconnections, which determine the nature of the neural network, are obtained by training the network on a number of appropriate examples. Thus, neural networks, in contrast to expert systems, do not require that the rules determining the system's behavior be known in advance. Rather, neural networks can, in principle, extract the relevant rules from information provided in the training sessions. Thus, it was hoped that neural networks might provide a novel solution to the secondary-structure prediction problem, without the need to define the rule set explicitly.

To date only four papers have been published. Qian and Sejnowski (270) and Holley and Karplus (271) applied neural nets in the context of a three-state (helix, sheet, coil) model and obtained results of similar accuracy (~63–64%) to those achieved by the best methods (see Section 4.4). Interestingly, both groups concluded that a simple Perceptron model (no hidden layers in the network) was all that was required, implying that higher-order information derived from proteins in their training sets was not incorporated into their networks. Bohr et al. (272) also showed that the Perceptron model may be used in a three-state model to predict the secondary structure of rhodopsin, a membrane-bound protein. The results of their study were in agreement with other data that suggest that rhodopsin is a seven-helix-bundle protein. Moreover, the neural-net approach appears to have better discriminatory power with respect to helix prediction than is obtained using the CHF method (273). McGregor et al. (274) confined their efforts to turn prediction, and were able to predict 71% of the turn residues correctly. This number is similar to that obtained by Wilmot and Thornton (263) and is considerably better than the value of about 63% attributed to many of the 'improved' methods described in Section 4.4, although it is not as good as that claimed by Cohen et al. (267) for their pattern-matching procedure.

4.5. Packing of Secondary-Structural Elements

In addition to primary and secondary structure, the packing arrangements of α helices and β sheets with each other is a major determinant of the 3D structure of proteins. Moreover, the packing rules, found either from an analysis of known protein structures or from the results of energy minimization studies, derive, at least in part, from features related to the handedness that occurs in secondary structures, as discussed in Section 2.2.

A. CHIRAL DETERMINANTS OF PROTEIN STRUCTURE

a. Alpha Helices. All α helices found in proteins are right-handed with the exception of a very short left-handed α helix in thermolysin. In addition, polymers formed from the L-amino acids exhibit identical chirality to those naturally found in proteins (275). The observed preference for right-handed screw sense (Fig. 1*b*) of the α helix has been accounted for in terms of nonbonded interactions involving the C_β and H_β atoms (276). Energy calculations have also predicted that left-handed α helices can occur in certain synthetic homopolyamino acid esters. This follows from the fact that in such synthetic molecules, electrostatic interactions between side-chain dipoles and the back-bone dominate over nonbonded interactions, and thus stabilize the left-handed α-helical form over the right-handed one (276). Results of the theoretical calculations have subsequently been verified experimentally (277, 278).

b. Beta Sheets. In proteins all β sheets, when viewed along the strand direction, have a right-handed twist (279), rather than the planar form proposed initially by Pauling and Corey (1). It has been suggested that this preference may be due to energetic and entropic factors operating at the single-residue level (279, 280) or to the geometric constraints of hydrogen bonds (281–283). Extensive energy calculations on β sheets have been carried out (284–287), and a detailed analysis of the results (286, 287) indicates that the preference for right-handedness arises from intrastrand side-chain—backbone interactions (284, 285), and from interstrand side-chain—side-chain interactions (286). For example, favorable intrastrand interactions in an isolated, extended strand of poly-Ile would cause the left-handed twist of the chain to be favored energetically, but the interactions are overcome by interstrand interactions that lead to the proper right-handed twist of the β sheet (286). Interestingly, backbone—backbone hydrogen bonding between strands does not give rise to a preference for either twist sense, and actually favors planar sheet structures (283, 285). Increasing the bulk of the side chains also increases the magnitude of the twist, as seen from a comparison of β sheets containing Gly, Ala, Abu, and Val residues (287). The two residues yielding the largest computed twist, Val and Ile, together with Leu, are the three most frequently occurring residues in the β sheets of proteins (288), which explains why the observed twists in β sheets are generally quite large. The extent of twisting can be enhanced by interchain interactions that operate within the β-sheet. This is demonstrated by the strongly twisted two-stranded β-sheet consisting of residues 14–38 in bovine pancreatic trypsin inhibitor (289). Prototypical right-handed twisted antiparallel and parallel β sheets derived from energy minimizations are shown in Fig. 1c and d, respectively.

c. βαβ **Crossovers**. The crossover connecting two parallel strands in a β sheet is nearly always right-handed (50, 54, 291). In addition, the chain never, in known protein structures, lies in the plane of the strands, a structural feature that if present would remove the handedness of the structure. Moreover, only one clear-cut case of a left-handed βαβ crossover has been found (viz., residues 45–95 of subtilisin). Figure 5 shows the two possible types of crossover connections schematically. An α helix occurs frequently in the crossover, which may occur in a simple βαβ structure, as part of a more complex βαβ structure such as is found in Rossmann folds (49) (βαβαβ), or as a component of parallel α/β-barrels.

The preference for a right-handed crossover has been interpreted in terms of the preference of β sheets for a right-handed twist, which tends to lessen the strain on the segment of the chain connecting the β strands (54), in terms of favorable complementarity of the surface of the sheet and, where applicable, that of the crossover helix (55), or in terms of possible folding mechanisms that could couple the handedness to the right-handedness of the central α helix. (50). Obviously, all of these hypotheses, which are based on geometrical factors, apply only to relatively short crossover connections. Therefore, it has been assumed that very long and convoluted crossover connections are right-handed probably because they evolved from shorter βαβ loops.

Recently, a comparison of right and left-handed crossovers has been made based on conformational energy calculations. The results of the calculations show that right-handed crossovers are much more energetically favorable than left-handed ones. For example, a right-handed crossover structure consisting of two (L-Val)$_6$ β strands and an (L-Ala)$_{12}$ α helix, connected by two flexible

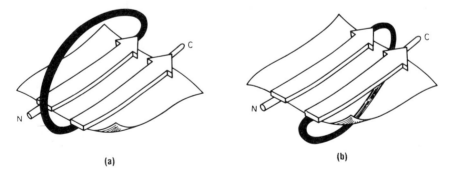

(a)   (b)

Fig. 5. Schematic representation of a βαβ-crossover structure: (a) a right-handed crossover, and (b) a left-handed crossover. N and C indicate the N- and C-termini, respectively. [Adapted, with permission, from (30).]

(L-Ala)$_4$ links (292) was shown to be ~15 kcal/mol more stable than the corresponding left-handed one. The right-handed crossover is strain-free, as indicated by the large right-handed twist of the β sheet shown in Fig. 6a and by the absence of high-energy conformations in the connecting links. In contrast, the left-handed crossover depicted in Fig. 6b is significantly strained.

    d. Beta Barrels. Beta sheets may also be rolled up into approximately cylindrical objects that are "closed" by hydrogen bonding between the first and last strands of the sheet: such structures have been termed β barrels (30). As a result of the inherent twisting of β sheets, the formation of β barrels results in a tilting of the strands with respect to the major axis of the barrel. The strands of the resultant β barrel possess a right tilt, that is, the strands follow a right-handed helical path about the major axis of the barrel shown in Fig. 7a. In a right-tilted β barrel, the hydrogen bonds between two neighboring chains must possess a staggered pattern, as illustrated in Fig. 7b and c for parallel and antiparallel β barrels, respectively (293).

    Conformational energy calculations on antiparallel β barrels composed of eight H-(L-Val-L-Gly)$_3$-OH strands have shown that a barrel with a right-handed tilt is much more stable than one with either a left-handed tilt or no tilt at all (294). The relative energies for the three cases are 0.0, 8.6, and 46.1 kcal/mol, respectively. A stereodrawing of the lowest-energy right-tilted anti-parrel β barrel thus obtained is given in Fig. 8. Tilted β barrels are preferred over nontilted ones because there is relatively more room for a tilted β barrel to accomodate the steric bulk of side chains located in the interior of the barrel (293). This suggests that a conformational change from a nontilted β barrel to a tilted one would ease the repulsion among the crowded internal side chains and improve side-chain packing.

    From the above discussion, it is clear that α helices, β sheets, βαβ crossovers, and β barrels all possess significant right-handed structural features as summarized below:

Right-handed α helices ≫ Left-handed α helices
Right-handed twist β sheets ≫ Left-handed twist β-sheets
Right-handed βαβ crossovers ≫ Left-handed βαβ crossovers
Right-tilted β barrels ≫ Left-tilted β barrels

The symbol ≫ is adopted here to indicate that essentially all of the structures follow the stated relationship, although in some very special and rare cases the structures in the right column might exist. The fact that both geometric and energetic analyses provide a consistent explanation of the observed chiral features of these structures is gratifying. More importantly, it suggests that rules based on geometric properties or conformational energy calculations may provide a reasonable account of the chiral features of these structures, and that such calculations can be used as a basis for predicating the 3D structures of proteins.

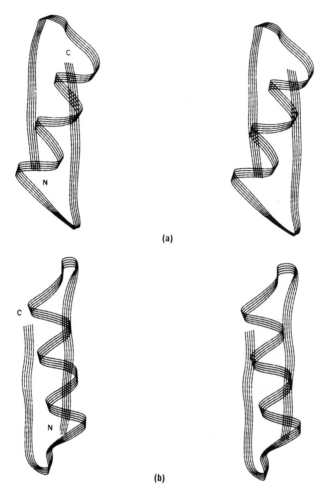

Fig. 6. Stereoscopic ribbon diagram of energy-minimized right-handed (*a*) and left-handed (*b*) βαβ-crossover structures. [Reproduced from (292).]

In the following subsections, an examination of the packing of secondary-structural elements in the supersecondary structures of proteins will be described. In addition, the results of energy calculations will be examined in terms of their ability to predict the observed packing arrangements and thus to provide a means for explaining the underlying physics of protein folding.

B. PACKING PARAMETERS

It is convenient to express the packing arrangements of structural units in terms of parameters that are independent of the coordinate system used. A useful set of such parameters consists of the distance D_M between the centers

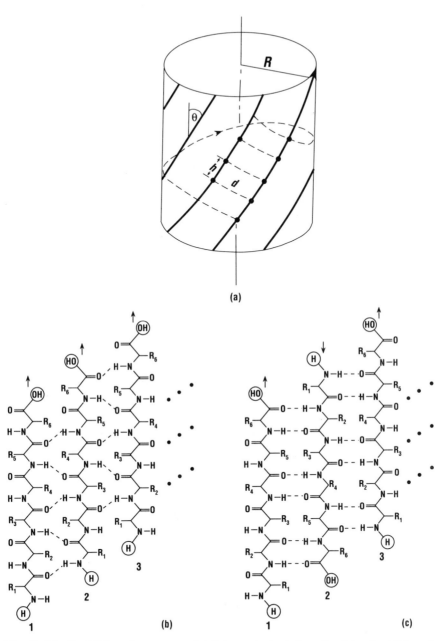

Fig. 7. Idealized β-barrel structures. (*a*) Schematic depiction of a right-tilted β barrel, where *h* is the advance per residue along a β strand, *d* is the distance between two neighboring strands, *R* is the radius of the barrel, and θ is the tilt angle of a strand. (*b*) The staggered hydrogen-bonding pattern of a right-tilted parallel β barrel. (*c*) The staggered hydrogen-bonding pattern of a right-tilted antiparallel β barrel. In (*b*) and (*c*) each strand consists of six residues. Dashed lines represent the hydrogen bonds between strands. [Reproduced from (293).]

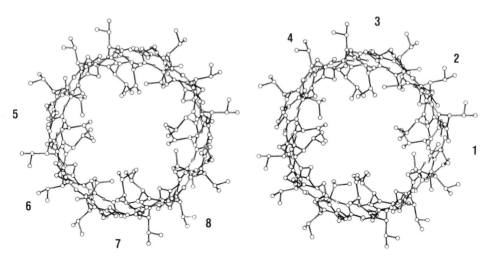

Fig. 8. Stereoscopic diagram of the minimum-energy right-tilted β barrel formed by eight H-(L-Val-L-Gly)₃ strands, viewed along the axis of the barrel. Only heavy atoms and amide hydrogens are shown. [Reproduced from (294).]

of the two structures and one or more orientational angles, as illustrated in Fig. 9. For two helices (or for two individual strands in a β sheet), the orientation angle Ω_o is a measure of the tilting of the axes of the two helices, with $\Omega_o = 0°$ for parallel and $180°$ for antiparallel orientation (Fig. 9a). An alternative but equivalent parameter, the projected torsion angle Ω_p, is the angle between the projections of the two helix axes on their contact plane (286, 287, 295−298). An analogous parameter, the horizontal projected angle, $\Omega_{\alpha\beta}$ or $\Omega_{\beta\beta}$, respectively, is used in helix−sheet or sheet−sheet packing to describe the orientation of the axis of the helix or of the second β sheet, relative to the axis of the β sheet or the first β sheet, which serves as the reference structure (Fig. 9b and c). A second orientation angle, which measures the tilting of the helix (or sheet) relative to the reference plane defined by the β sheet (or the first β sheet), viz., the vertical projected angle for a helix−sheet packing (290) or the inclination angle for sheet−sheet packing (299), is also quite useful in packing studies.

C. GEOMETRIC AND ENERGETIC APPROACHES

Studies of packing in proteins can be classified into two categories: (1) geometric approaches, which are based on the notion of optimal complementary fits of the surface topographies of secondary-structural elements (46, 281, 282, 295, 300−303), and (2) energetic approaches, which are based on the optimal interaction energy amongst the secondary-structural elements (286, 287, 289, 290, 297−299, 304). Considerable information about the packing preferences of secondary structures can be obtained from an analysis of the geometry of

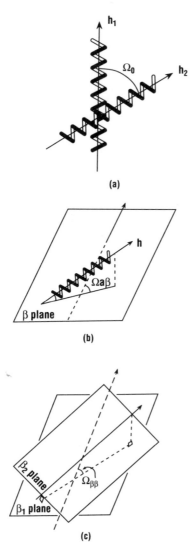

Fig. 9. Schematic diagram illustrating the definition geometric parameters used to describe the packing of secondary-structural elements. (*a*) Ω_o, the packing or orientation angle between two α helices, (*b*) $\Omega_{\alpha\beta}$, the orientation or horizontal-projected angle between an α helix and a β sheet (290), (*c*) $\Omega_{\beta\beta}$, the orientational or horizontal-projected angle between two β sheets (299). The signs of these angles are defined by looking from any one of the two secondary-structural elements toward the other: If the element closer to viewer must be rotated in a counterclockwise direction to bring the orientation angle to zero, the angle is given a negative sign.

packing (47), and this approach has led to numerous insights. The geometric approach is also consistent with energetic considerations, because good fits between complementary surfaces generally imply nonbonded interactions are favorable. It is necessary, however, to compute the energies of various packed structures to assess their relative stabilities, to determine whether a geometrically feasible structure is stable energetically, and to obtain more fine-grained information on the factors that influence packing. In some cases, energy calculations have predicted structures not found by geometrical analysis (286, 287, 289, 290, 297, 298). Thus, consideration of the packing energy provides more detailed information on the stability of folded structures.

D. HELIX–HELIX PACKING

Previous studies of the packing of two α helices based on a number of empirical models, viz., knobs into holes (300), strings and sausage (301, 305), ridges into grooves (295, 296), close-packed sphere (303), and polar and apolar spheres (302), have suggested that there exists only a small number of preferred orientations of the helix axes. Some of the models also include a requirement for hydrophobic interactions (302, 303). Each of these models has some advantageous features for describing a given packing arrangement (30). Actually, geometrical models generally emphasize the unevenness of the surface of an α helix caused by the presence of regularly spaced sidechains. In most of these arrangements, the side chains of one helix intercalate into spaces lying between the side chains of the other helix. For example, according to the ridges into grooves model, on a given helix, a set of side chains with a mutual separation of j residues along the sequence (with $j = 1$, 3, or 4), forms a ridge, with a continuous groove between them. The complementary packing of the ridges and grooves of two adjacent helices results in particular preferred orientations of the two helices. Because of the great variation of the size and geometries of side chains in actual α helices, it is possible to describe contacts by quite different idealized geometrical models. However, geometrical models cannot tell us which packing arrangement is more stable. Rather, it is the energetics of interaction between the helices that determines packing. The geometrical feature of surface complementarity describes only part of the total interaction energy (vide supra). Actually, some packing arrangements found in proteins were missed by the geometrical models but were successfully predicted by energy calculations [see, e.g. Chou et al. (290)]. It is therefore necessary to study packing of helices in terms of their interaction energies.

Energy calculations designed to investigate the packing of two (L-Ala)$_{10}$ α helices showed that (only) 10 stable packing arrangements occur within an energy range of 5.3 kcal/mol (Table II). Several distinct values of the orientation angles have been found, but the three lowest-energy packing arrangements are nearly antiparallel (occurring within ±35° of $\Omega_o = \pm 180°$). The orientation angle for the energetically most favorable packing is $\Omega_o = -154°$ (Fig. 10), which is the arrangement found in most proteins. The second most frequent

TABLE II.

Energy Parameters and Orientations of Two Packed Poly(L-Ala) α Helices[a]

Orientation Angle Ω_o (deg)	Total Relative Energy[b] (kcal/mol)	Interchain Energy (kcal/mol)			Classification of Packing[c] (ij)
		Total Energy E	Electrostatic	Nonbonded	
−154	0.00	−17.23	−2.20	−15.02	34
170	0.76	−16.42	−2.24	−14.28	None[d]
146	1.74	−15.43	−1.45	−13.98	13
−36	3.35	−13.82	0.54	−14.36	13
127	3.86	−13.31	−1.20	−12.11	44
30	4.35	−12.84	0.95	−13.79	34
79	4.57	−12.54	0.15	−12.70	14
−136	4.92	−12.26	−1.29	−10.98	None[d]
−155	5.17	−12.05	−1.58	−10.47	34
−87	5.30	−11.89	−0.14	−11.74	33

[a] From Chou et al. (297). Minimum-energy packing arrangements computed for two CH_3CO-$(L\text{-Ala})_{10}$-$NHCH_3$ α helices.
[b] $\Delta E = E - E_0$, where E_0 is the energy of the structure in line 1.
[c] According to a geometrical ridges into glooves model (295). The indices i and j refer to the interacting ridges on the two helices (295). Two packings with the same value of ij correspond to the reversal of orientation of one of the helices.
[d] There is no ridges into grooves arrangement corresponding to this computed packing.

observed helix packing found in proteins (47, 295) occurs in the range $-50°$ $< \Omega_o < -30°$, and corresponds to the computed low-energy structure found at $\Omega_o = 36°$.

Both nonbonded and electrostatic contributions influence the packing of a α helices, as shown in Table II. The magnitude of the interchain nonbonded energy exceeds that of the electrostatic energy, implying that nonbonded interactions dominate the overall energy of stabilization of packed helices. On the other hand, the calculated contributions of these two terms to the energy difference among the various packing geometries are comparable in magnitude and both are significant for packing (297, 298).

The computations summarized above have been carried out in the absence of solvation. Solvent effects give rise to several additional energy contributions. The presence of water or, in general, a solvent with a high dielectric constant has two effects on electrostatic interactions (306). It reduces the magnitude of the interhelix dipole interactions, thereby decreasing their contribution to the energy differences between various arrangements, and it destabilizes packed structures because of the need to desolvate the helical dipoles before helix association can take place. On the other hand, the presence of water stabilizes the packing of structures containing nonpolar side chains, as a result of hydrophobic interactions. Both sets of analyses (297, 298, 306) indicate that electrostatic interactions play a minor role in helix packing, in contrast to earlier estimates based on considerations of the dipole moment of the α helix alone (307, 308).

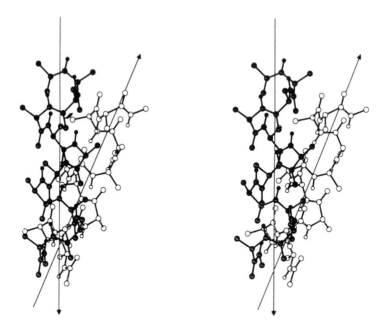

Fig. 10. Stereoscopic diagram of two (L-Ala) $_{10}$ α helices in their lowest-energy packing geometry, with $\Omega_o = -154°$. The helix axes are indicated by arrows directed from the N- to C-terminal residues. The front helix is indicated by shaded atoms and bonds. Hydrogens on the alanyl and terminal methyl groups have been omitted. The interdigitation of the two helices can be seen in the region of contact between them: Each turn of one helix fits into the space between adjacent turns of the other helix ('ridges into grooves' model). [Reproduced from (297).]

It can be seen from Table II that large differences exist in the relative stabilities of various geometrically feasible packing arrangements according to the geometrical ridges into grooves model (296). Furthermore, some low-energy packings have not been found by such geometrical modeling.

The packing of α helices may be influenced by the size and shape of the side chains in contact (as well as possible specific interactions between them, such as hydrogen bonds). A computation of the packing of a poly (L-Leu) with a poly (L-Ala) α helix (298) indicated that the lowest-energy near-antiparallel packings were not affected significantly by the substitution of Leu for Ala. The relative energies and the geometry of the packing in other, higher-energy arrangements, however, depend sensitively on side-chain inter-actions. Sequence-specific interactions between α helices may restrict the number of probable ways of packing even further (303).

The mutual influence of several α helices on packing was investigated for the A, G, and H helices in sperm whale myoglobin. Two nearly antiparallel tightly packed helices (G and H) are not affected by the introduction of a third, neighboring helix. On the other hand, a pair of nearly perpendicular helices with weak interactions (A and H) can pack in a variety of ways, but, in this case, the presence or absence of a third helix (G) alters the relative energies of the A/H packings (304). Small shifts in the packing of neighboring α helices can have important effects on the mechanisms of biologically important conformational changes (309).

An important application of the energetic approach to helix packing is found in studies of four-α-helix bundle proteins. The main features of this structural motif, which is found in many proteins (30, 310, 311), can be explained in terms of nonbonded interactions among the constituent helices (312). In almost all observed four-α-helical bundle proteins, neighboring helices are oriented in a nearly antiparallel fashion, an orientation that is favored by electrostatic interactions between the macrodipoles of the helices (308). It has been pointed out, however, that the presence of an aqueous solvent environment decreases the role of electrostatic interactions (306).

The energy-minimized structure of the four unlinked helices (312) shown in Fig. 11 also possesses the characteristic left-handed twist of the bundle as observed in essentially all proteins in this class. The calculated distances of closest approach between two pairs of diagonally related helices are 7.8 and 13.7 Å, respectively, indicating, however, that the bundle structure has a cross section that is more diamond shaped than square, as is generally observed for such proteins. If the effects of the connecting chains (loops) between the helices are taken into account, the computed lowest-energy structure shown in Fig. 12 is obtained (313).

The structure possesses the following features: (1) the four α helices again exhibit the proper left-handed twist and an approximately square cross section; (2) the distances of closest approach between two adjacent interhelix axes are 7.9 ± 0.3 Å, and those between two diagonal interhelix axes are 11.3 ± 0.2 Å; (3) the adjacent interhelix angles are $-165 \pm 2°$; and the diagonal interhelix angles are $21 \pm 4°$. All of these values are quite similar to those found on average for four-α-helical bundle proteins. A detailed discussion concerning the role of the connecting loops in the folding of four-α-helix bundle proteins was given recently by Carlacci and Chou (313). As pointed out by DeGrado et al. (314), the design and synthesis of a model four-α-helical bundle polypeptide must include considerations of α-helix packing and the effect of the connecting loops on the overall bundle structure.

E. HELIX—SHEET PACKING

Alpha helices are often associated with β sheets in globular proteins. The frequently observed $\beta\alpha\beta$ or $\beta\alpha\beta\alpha\beta$ structural motif known as Rossmann fold (30, 49) involves the connection of two neighboring parallel β strands by an α-helical segment. There are, however, many other helix—sheet packing arrangements where the helices and sheets come from widely separated parts of the sequence. The packing geometries of complexes of α helices and β sheets have been studied extensively by several laboratories (315, 316). These investigators sought to characterize the preferred chain orientations and residue

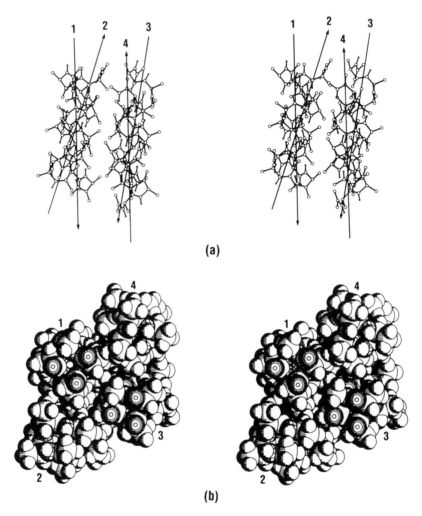

Fig. 11. Stereoscopic diagram of a minimum-energy idealized four-α-helical bundle protein formed from four unlinked CH_3CO-$(L\text{-}Ala)_{10}$-$NHCH_3$ chains. In (*a*) the view is from the side of the bundle. Only heavy atoms and amide hydrogens are shown, and the arrows denoting the helix axes are directed from the N- to C-terminal residues of each helix. Note the antiparallel packing of adjacent helices. (*b*) A space-filling representation of the bundle viewed from the "bottom" of (*a*), namely the C-terminus of helix 1. All atoms are included; nitrogen atoms are cross-hatched, oxygen atoms are denoted by concentric circles, and all atoms are half shaded. The bundle possesses a left-handed twist with an orientation angle, $\Omega_o = -168 \pm 7°$. The distances of closest approach between two helices located across the diagonal from one another are 7.8 and 13.7 Å, respectively, indicating that the bundle does not possess the square cross section observed in x-ray crystallographically determined structures. [Reproduced from (312).]

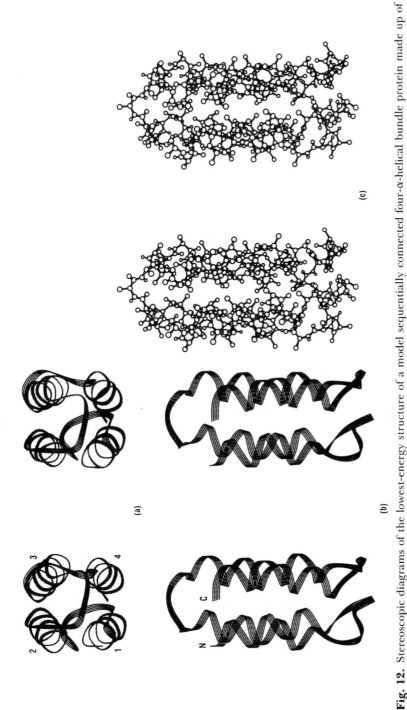

Fig. 12. Stereoscopic diagrams of the lowest-energy structure of a model sequentially connected four-α-helical bundle protein made up of polyalanine residues. The helix and loop portions are made up of (poly-Ala)$_{12}$ and (poly-Ala)$_{10}$ segments, respectively. (a) Ribbon diagram viewed along the central axis of the bundle, (b) ribbon diagram viewed from the side of the bundle, and (c) atomic model including all heavy atoms and amide hydrogen atoms, as viewed from the side of the bundle. The bundle shows the expected left-handed twist with Ω_o = −165 ± 2°, and an approximate fourfold symmetry about the bundle axis as observed in most four-α-helical bundle proteins of known structure. [Reproduced from (313).]

47

contact patterns typically found in helix—sheet packing. The observed preferences were interpreted in terms of the complementarity between the surface
topologies of the two regular structures (295, 315), or in terms of the interdigitation of side chains on the two surfaces (55, 316).

Energy calculations on a poly (L-Ala) α helix packed against a five-
stranded poly (L-Val) β-sheet provide useful data on the relative stabilities
of various packing geometries (290). Four distinct classes of low-energy
packing arrangements were found, each differing in the orientation of the helix
axis with respect to the direction of the strands, as illustrated in Fig. 13.

In the lowest-energy set structures (class 1), the helix lies nearly parallel or
antiparallel to the strands such that $-10° < \Omega_{\alpha\beta} < 10°$. This is the class that
occurs with greatest frequency in proteins, and has been proposed as the most
favorable helix—sheet packing arrangement based on purely geometric considerations (295, 315). In the second-lowest-energy set of structures (class 2),
the helix runs nearly perpendicular to the strands such that $80° < |\Omega_{\alpha\beta}| < 100°$.
While Class 2 packing structures were found by energy minimization, they
were not obtained by geometric means as was the case in class 1 structures.
Numerous class 2 structures are also found in proteins.

Both classes are stable because the helix lies along a tangent of the curved
surface of the β sheet, such that residues that lie along the entire helix can
interact with those in the β sheet. In another set of relatively low-energy
structures (class 3), the helix lies along a diagonal of the β sheet in such a
manner (recall the overall shape of the β sheet is saddle-shaped as a result of
the right-handed twisting) that its central part packs tightly against the
saddle-shaped surface of the sheet such that $-60° < \Omega_{\alpha\beta} < -40°$. As was
true for class 1 structures, class 3 structures occur frequently in proteins, and
have also been obtained from a purely geometrical analysis (295, 315). Class 4
structures are energetically even less favorable. In these structures, the helix
lies along the other diagonal ($\Omega_{\alpha\beta} \sim 60°$) greatly reducing the extent of
interresidue contact. Classes 1—3 correspond to the frequently observed α/β
packings in proteins, but a few examples of class 4 exist. The frequency
distribution of observed orientation angles in proteins, $\Omega_{\alpha\beta}$, has maxima
corresponding closely to these four classes of computed structures (cf. Fig. 12
of (290)). Stereodrawings of the three computed lowest-energy α/β packings
are shown in Fig. 14.

Electrostatic (dipole) interactions are unimportant in α/β packing. Most of
the interaction energy arises from nonbonded interactions. Consequently, the
orientation of the helix axis is unimportant in this case. The detailed analysis
[see, e.g., Chou et al. (284, 290)] of the interaction energies and of the shape
of the β sheet in the computed α/β structures has demonstrated that an
antiparalled β sheet is more flexible than a parallel β sheet, because the former
can be deformed more easily in response to packing interactions.

F.   SHEET—SHEET PACKING

Two distinct classes of computed low-energy arrangements exist for the packing

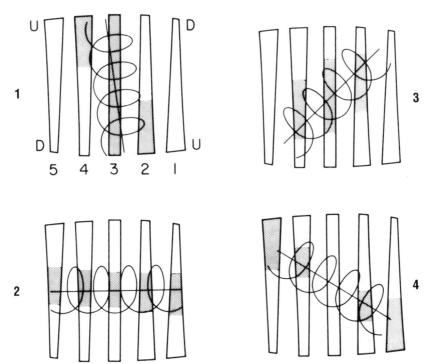

Fig. 13. Schematic representation of the various classes of α-helix/β-sheet packing geometries. The strands of the twisted β sheet are shown schematically as rectangles drawn in perspective to indicate their tilts relative to the plane of the drawing. The letters U and D denote the corners of the sheet located above and below the plane of the page, respectively. Shading indicates the region of the strands in contact with the α helix. [Reproduced from (290).]

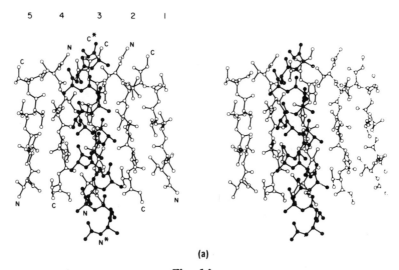

(a)

Fig. 14 a

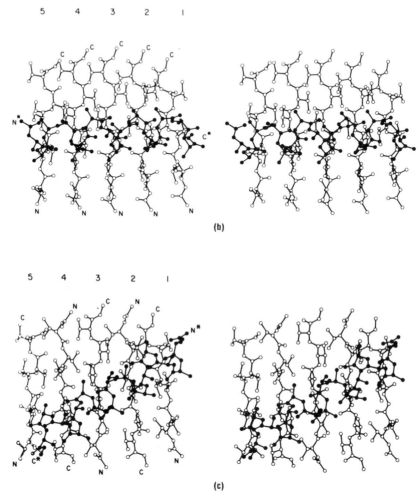

Fig. 14. Stereoscopic diagram of a CH_3CO-(L-Ala)$_{16}$-$NHCH_3$ α helix (filled atoms and bonds) and a five-stranded antiparallel CH_3CO-(L-Val)$_6$-$NHCH_3$ β sheet. The numbering of the strands of the β sheet is indicated at the top of each figure, and the N- and C-termini are labeled to indicate strand direction. All hydrogen atoms are omitted, except for amide hydrogens. (a) The lowest-energy α/β packing geometry, which corresponds to class 1 of Fig. 13. Geometric models also predicted this geometry. (b) The next lowest-energy α/β packing geometry, which corresponds to class 2 of Fig. 13. Geometric models failed to predict this geometry, although it has been observed in numerous proteins. (c) The second next lowest-energy α/β packing geometry, which corresponds to class 3 of Fig. 13. [Reproduced from (290).]

of two β sheets (299). In the lowest energy class, the strands of the two sheets are aligned nearly parallel or antiparallel to each other, resulting in complementary packing of the two saddle-shaped surfaces, as shown in Fig. 15. The computed horizontal projected orientation angle lies in the range $-26° < \Omega_{\beta\beta} < 5°$. Observed values (295, 317, 318) fall into the range $-20°$ to $-45°$. In the other class, the strands are nearly perpendicular to each other, with the computed $\Omega_{\beta\beta}$ between $93°$ and $107°$ and observed values ranging from $71°$ to $99°$ (295, 319). While the saddle-shaped surfaces are not complementary in this arrangement, their interactions are favorable because there is good packing between the corner of one sheet and the interior part of the other sheet. The intersheet energy is $1-4$ kcal/mol higher than in the first class.

These two packing classes have been termed *aligned* and *orthogonal*. Chothia and Janin (319) have proposed that the orientational angle $\Omega_{\beta\beta}$ for aligned structures is related in a simple manner to the difference of twists of the two sheets. In fact, the value of $\Omega_{\beta\beta}$ can be predicted (299, 315) from the twists to within $\pm 4°$. Orthogonal packings are usually formed by two β sheets folded back on themselves, in such a manner that a strand passes from one sheet to the other sheet near their corner, forming a near $90°$ bend (315). In such cases, a covalent connection is usually present, but energy calculations show that its presence is not required to stabilize the orthogonally packed structure.

Insights gained from studies of the energetics of packing are important for the prediction of protein folding. Information about preferred packing arrangements makes it possible to select for a given protein a small set of probable conformations, which then can be used as starting points for detailed energy computations. This potential application is significant because it helps to alleviate the multiple-minimum problem, one of the greatest difficulties that remains in the prediction of three-dimensional structures of proteins (vide infra). The number of probable conformations can be lowered considerably by selecting likely structures at a level between those of local conformational preferences and the overall folding of the molecule. Thus, the analysis of packing constitutes an important link between conformational analysis of small peptides and the solution of the protein-folding problem.

5. TERTIARY STRUCTURE

Currently, only about 400 structures reside in the PDB (4), and although this number is increasing each year, it is doing so very slowly, necessitated by the well-recognized difficulty of determining protein structures crystallographically. In contrast, more than 10 times this number of primary structures are available (9, 320) and, due to the relative ease with which proteins and their parent nucleic acids can be sequenced, this number is growing rapidly. Thus, reliable methods for predicting the tertiary structures of proteins from their primary structures, that is, methods for solving what has become known as the *folding problem*, are needed if the full potential of the available sequence data is to be realized.

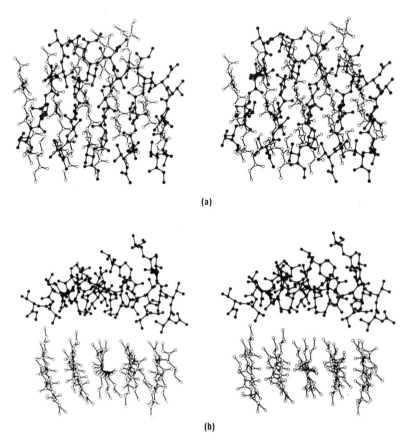

(a)

(b)

Fig. 15. Stereoscopic diagram of the lowest-energy packing structure of two β sheets: a parallel β sheet consisting of five CH_3CO-$(L$-$Ile)_6$-$NHCH_3$ strands (unfilled atoms) and an antiparallel β sheet consisting of five CH_3-$(L$-$Val)_6$-$NHCH_3$ strands (filled atoms) as viewed from above (a) and from the edge (b) of the sheets. All hydrogen atoms have been omitted for clarity. [Reproduced from (299).]

Practical a priori methods, that is, methods based solely on the sequence and molecular characteristics of the constituent amino acids and the physico-chemical laws governing the interactions of their atoms, do not exist. A serious impediment to the development of such methods is the existence of the multiple-minimum problem (see, e.g., Gibson and Scheraga (111)). This multiple-minimum problem is caused by the chain-like character of polypeptides and their associated large number of degrees of freedom, which give rise to numerous local minima. During energy minimization, if the variables charac-terizing the structure correspond to one of these minima, the minimization would stop, since there is no feasible algorithm that could distinguish this from the global minimum (110). Moreover, the minima encountered depend upon the initial starting geometry and, to some extent, on the method employed.

Thus, traditional energy minimization procedures will not suffice as a practical means for a priori 3D structure prediction of large polypeptides and proteins. And, although a number of clever schemes have been devised to circumvent the multiple-minimum problem, none have been totally successful when applied to systems as large as real proteins (see, e.g., the reviews of Gō (321) and Skolnick et al. (322), and the discussion in Section 5.3). Energy minimization procedures, especially simulated annealing, have been successfully employed in the refinement of structures generated crystallographically (154, 158, 323), by 2D- and 3D-NMR (6), and through model building, as discussed in Section 5.1. In all such cases the initial structures minimized are close to the native structure or, in situations where this is not the case, specific constraints are employed to insure that the starting structure moves toward the native one without being trapped in local minima.

Heuristic procedures represent another important means for solving the folding problem. These approaches are characterized by an adaptive process where one of several possible solutions is selected at each of a number of successive steps for use in the next step. Methods for generating solutions may involve knowledge- and/or energy-based procedures. At the present time heuristic methods dominate tertiary structure prediction efforts, especially since the totally energy-based methods are hampered by the multiple-minimum problem. Heuristic approaches in use today may be broadly classified into two categories, homology-based approaches and combinatoric ones.

Homology-based methods, which represent the most ubiquitous of the heuristic approaches in use today, are discussed in Section 5.1. These methods require that at least one protein of known 3D structure exist (the *template*) which is similar in some sense to the protein whose 3D structure is to be predicted (the *target*). The template protein then serves as a model or prototype from which to build, by analogy, the structure of the target protein. In addition, a number of knowledge-based and energy-based procedures may be used to deal with, for example, the generation of appropriate loop structures and refinement of the completed structure.

Heuristic approaches that employ rule-based procedures, while not as popular as homology-based ones, nevertheless have shown some promise as a means for solving the folding problem. In such procedures higher-level structures are built up from lower-level ones. The resulting combinatorial explosion of structures is then processed to manageable proportions using rules developed from the knowledge-base of protein structure information. Although only modest success on several classes of proteins has been obtained with rule-based procedures, they are useful at least for limited classes of proteins. A discussion of this approach is presented in Section 5.2.

True a priori methods, in contrast to heuristic ones, seek to predict a protein's tertiary structure without knowledge of structural information derived from known protein structures, related or otherwise. In many cases, however, information derived from protein structures is incorporated into the methodology. This is particularly true of the many simplified psuedo-potential energy functions employed in a priori folding studies (see Section 5.3). Moreover, it is not necessary that an a priori method model or simulate an actual or presumed folding process, although many a priori methods are based to some extent on specific folding paradigms derived, more or less, from an analysis of in vitro experimental data. One would (neglecting a number of fine points to be discussed later) simply need to find the conformation corresponding to the global minimum of the protein's potential (or psuedo-potential) energy function. Unfortunately, the multiple-minimum problem and the need to include environmental factors such as solvent represent serious challenges to the practical implementation of such an approach. Nevertheless, a priori approaches, which are discussed in Section 5.3, do show some promise, although considerable work remains to be done before the a priori prediction of tertiary structure becomes a practical reality. The following three subsections will provide a glimpse of the tertiary structure prediction activities ongoing in many laboratories throughout the world.

5.1. Homology-Based Approaches to Tertiary-Structure Prediction

Most homology-based approaches predominantly employ knowledge-based procedures (259, 324−329), although energy-based procedures are used for a number of tasks. Knowledge-based procedures rely heavily on structural information residing in the PDB. Currently, the PDB contains data on about 400 structures covering more than 100 unique, nonhomologous proteins (4). A critical limitation of all present knowledge-based methods is the limited availability of data on structurally diverse proteins. The most serious consequence of this limitation is that new structural features or motifs, i.e., those not already in the PDB, cannot be discovered with the use of knowledge-based procedures.

Energy-based procedures, however, hold some promise as a means for circumventing this difficulty, especially with regard to generation of loop structures (330−336). In addition, as discussed in Section 4.6, energy-based methods have also provided a number of useful insights on the packing arrangements of secondary-structural elements in super-secondary structures; and a priori folding methods are also showing some promise in the calculation of *qualitatively* correct tertiary structures (Section 5.3). Considerably more testing must be done, however, before the ability of energy-based methods to 'discover' new tertiary structure motifs (30, 40; cf. 9) can be assessed fully.

In essentially all practical procedures for protein structure prediction, a target protein is matched to that of a prototypical template protein. Prototypical, as it is used here, refers to the fact that the template protein may not be made up of a single structure. Rather, it may represent a consensus structure based on the structures of several related proteins, or on appropriate fragments of many proteins which may or may not be related functionally to the target protein. Although each protein must be handled somewhat independently, the steps generally followed in the model building process are given in Scheme 2.

Scheme 2

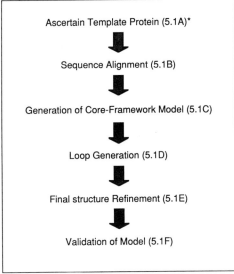

Ascertain Template Protein (5.1A)*

Sequence Alignment (5.1B)

Generation of Core-Framework Model (5.1C)

Loop Generation (5.1D)

Final structure Refinement (5.1E)

Validation of Model (5.1F)

* Numbers in paranthesis refer to the section in which the material
is discussed.

A. ASCERTAIN TEMPLATE PROTEIN

The first step is usually taken with a knowledge of the family or class to which the target protein belongs. If 3D structures of proteins in this family (class) are unavailable, then one must search the PDB for homologous proteins or proteins with similar functions.

B. SEQUENCE ALIGNMENT

This is one of the most crucial steps in the model-building process. In cases where sequence homology between the target and template proteins is high (>50%), sequence alignment is relatively straightforward, and can be accomplished by hand or by any one of a number of automated alignment procedures (10), in addition to the well-known Needleman and Wunsch (337) method. When the amount of sequence homology is moderate (~50%), the identification of structurally conserved regions can be useful in areas where sequence alignment is ambiguous.

Greer (324, 338) divides a protein into structurally conserved and variable regions, SCRs and VRs, respectively. SCRs correspond to what is commonly thought of as the core framework (339), and constitute the underlying structural architecture conserved throughout a protein family (43, 324, 338). Sequence homology is much higher in the SCRs. Note also, as discussed by Lesk and Chothia (339), that mutations occurring within the core of a protein generally do not significantly alter its chain topology; the main changes are primarily in the relative orientations of the secondary-structural elements. VRs are closely related to loop structures, where amino acid insertions, deletions, and mutations can be accommodated with considerably less effect on the general architecture.

Figures 16 and 17 illustrate this point. The nearly superimposed blue strands of the backbone chains depicted in Figure 16 (see color insert) represent β strands that correspond closely to the SCRs of six immunoglobulin domains, and the red strands represent the VRs.

Figure 17 shows the sequence alignment for the structures overlaid in Fig. 16. Note that regions of high sequence homology are confined mostly to the SCRs, the labelled β strands. The VRs, on the other hand, exhibit considerable variety in sequence; this variety is reflected in the structures.

Fortunately, structural homology persists for proteins in the same family, even when sequence homology may be quite low (20–30%) (259). Under such conditions it is difficult to produce reliable sequence alignments (cf. 12). However, several automated procedures have been devised to obtain maximally overlapped protein backbones (340–343), and figures similar to Fig. 16 have been presented in a number of other works (see Tang et al. (43)).

In cases where only one crystal (or NMR) structure exists, incorrect sequence alignment between the target and template proteins may introduce significant errors in the modeling process. This can be ameliorated somewhat if sequence information exists for a number of proteins closely related to the target protein. In such a situation, it may be possible to use sequence alignment techniques to ascertain which amino acid residues are conserved within the family, and by inference which residues may be of critical importance to the structure and function of proteins within the family. Alignment of the target and template proteins is then carried out giving these critical residues added weight in the sequence-alignment process. The alignment can be further enhanced if functionally important residues deduced from site-specific mutagenesis or chemical-modification experiments are also included. An example of this is the proposed correspondence between Lys-758 of the Klenow fragment (KF) of *Escherichia coli* DNA polymerase I, and Lys-263 of the retroviral reverse transcriptase (RT) of HIV-1. Reaction of the enzymes with pyridoxal phosphate, followed by reduction, attaches the pyridoxal group to Lys-758 in the case of KF (344) and Lys-263 in the case of RT (345), suggesting that the lysine residues in the two proteins are located in the same position relative to their nucleoside triphosphate binding site (vide infra).

The most difficult case arises when 3D structures of appropriate template proteins may not exist. Such is the case for the RT from HIV-1 (vide supra). Reverse transcriptase is a complex, multifunctional enzyme that, among other things, contains both RNA-directed and DNA-directed DNA polymerase activity. To date the only polymerase whose 3D structure is known is the KF from Pol I from *E. coli* (346). Currently, only the C_α coordinates are available in the PDB. Since the sequence homology between the template (KF) and target (RT) is low, reliable sequence alignment is a difficult task at best,

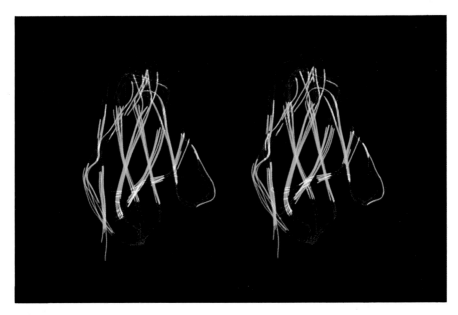

Fig. 16 Stereoscopic overlay of six immunoglobulin variable-light chain backbones represented as single strands. The blue segments represent β strands, which are related, although not identical, to the SCRs defined by Greer (324, 338). The red segments represent loop regions and correspond approximately to the VRs defined by Greer. The PDB codes for the immunoglobulins depicted in the figure are as follows: 1FBJ, 1IG2, 1MCP, 1REI, 2RHE, and 3FAB.

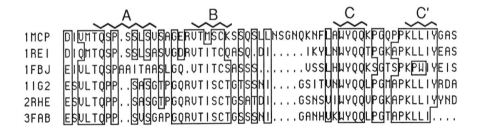

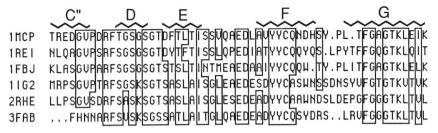

Fig. 17. Sequence alignment of six immunoglobulin variable-light chains, indicated by their PDB codes which correspond to those given in Fig. 16. Approximately the first 110 N-terminal residues of each protein are used. Identical or well-conserved matches are shown in boxes. Beta-strands, as identified by Williams [*Immunol. Today* **8**, 298–303, (1987)], are indicated above the sequences by zigzag lines.

and may not be possible at all unless site-specific mutagenesis or chemical-modification experiments can provide information on the location and nature of functionally or structurally equivalent amino acid residues (vide supra).

Generally, a clear-cut sequence alignment between the target and template proteins is very difficult to obtain except in cases of high sequence homology (>>50%). Previous homology model-building studies have shown that sequence alignment is a critical step in model generation, and errors in that step can adversely affect the accuracy of the resulting model (347, 348). Hence, when sequence homology is low (<25%), it may be hard to determine the best alternative alignment, and several models based on alternative alignments ought to be pursued. The best model is selected based upon its agreement with experimental data and with validation procedures (see Section 5.1.F).

C. GENERATION OF A CORE-FRAMEWORK MODEL

There are two stages in the generation of a core-framework model, the generation of a backbone, and the addition of side chains to the backbone. In cases of high-sequence homology, the backbone of the target protein is essentially the same as that of the template protein. When sequence homology is low, but coordinates of several structurally homologous proteins are available, a backbone model can be constructed through a cut-and-paste strategy, choosing backbone structures for different portions of the chain from different proteins, or a consensus template strategy where a ficticious new protein backbone is generated with minimum rms deviation from any of the known protein backbones (259). Identification of the core regions for this purpose can be accomplished either by visual inspection, or by using secondary-structure assignment methods such as that of Kabsch and Sander (203). Modeling the backbone can become difficult if there are insertions and deletions in the structurally conserved core-framework regions when the sequence of target and template proteins are aligned. Insertions and deletions, however, occur infrequently within core-framework regions (339, 349, 350), and if they do, the backbone can be reconstructed using limited local refinement procedures (see, e.g., Jones (351)).

Once the backbone of the core framework is available, side-chain modifications must be dealt with. Generally, this is done by replacing the appropriate side chains, keeping the X_1 angle fixed at the template value, and using the most probable values of the other side-chain angles observed in the PDB (328, 356, 359, 360). Energy-based structure refinement may then be used to adjust the mutated structure to the presence of the new side chains. In this procedure, special care must be taken to insure that hydrogen bonds stabilizing the framework are not disturbed significantly (132, 352; Granatir, unpublished results). Thus, amino acid substitutions are done in a stepwise fashion, where a small number of amino acids are replaced and the resulting structure relaxed at each step. Interestingly, Lesk and Chothia (339) showed that mutations in secondary structural regions of protein domains produced changes in 'orientation' of the secondary structure elements, but did not change the overall nature of the hydrophobic core of the protein (see also (349)).

D. LOOP GENERATION

Once the core framework of the protein has been constructed, loops, which give proteins within a given family their diversity, must be generated. This diversity is manifested in the variable length and sequence characteristics of loop regions. In some cases, especially those where high sequence homology exists, the loops of a target protein may be modeled by loops from other proteins within the same family. Greer (324, 338) has shown this to be the case for many serine proteinases, where he was able to model the loops of factor Xa from loops of serine proteinases of known structure. Chothia et al. (353, 354) have also shown that a significant portion of the loops in immuno-globulins can be described by a relatively small number of canonical loop structures. Given the limitations of the current PDB, however, it is unlikely that a sufficient number of canonical loop structures can be developed to treat all proteins of interest. Interestingly, Tramontano et al. (355) have shown that two basic structural motifs do describe the bulk of medium length loops. Nevertheless, the need for loop generation is encountered in most homology model-building studies carried out at the present time.

Two main approaches to loop generation are in use today: knowledge-based and energy-based. In knowledge-based approaches (259, 324−326, 338) loop structures are modeled by preexisting loops found in the PDB. Thus, if the environments of the target and template loops are similar, and if the loops contain similar amino acid residues, it is reasonable that the template loop will model the target loop with sufficient accuracy. A severe limitation of this approach is, of course, that the PDB may not contain a suitable variety of loop structures to adequately model the loops of many proteins. Nevertheless, for short loops, that is, loops of six or less residues, knowledge-based procedures are probably sufficient (254, 325, 326, 338, 356).

Energy-based approaches, on the other hand, circumvent many of the problems engendered by the limited size and diversity of the PDB, although they are considerably more demanding of computing resources. Energy-based approaches to loop generation typically involve two steps: (1) generation of elementary loops, (2) followed by addition and proper positioning of the sidechains. In elementary loops the side chains of all residues except glycine and proline are truncated to their C_β carbons, and thus are represented as alanine residues. Both cis and trans isomers of proline are customarily considered.

As is the case for protein folding in general, the multiple-minimum problem remains an impediment to the determination of energy-optimized loop structures. Moreover, the conformational space spanned by loops in proteins, while considerably smaller than that spanned by the whole protein, nonetheless is still large enough to frustrate any brute-force approach. It is possible, however, to sample the loop conformational space with highly optimized grid searches or biased random sampling procedures, and several promising approaches to loop generation have been developed over the last few years (330−333). While their ability to handle loops of seven or less residues appears to be adequate, efficient, and accurate, success with larger loops is questionable, except for perhaps the random tweak method (cf., 334). Martin et al. (335) have implemented an interesting hybrid approach that combines both knowledge- and energy-based procedures into an automated method that does not require user intervention. The method has been applied only to immunoglobulin loops to date, and thus its effectiveness in treating general loops in other types of proteins remains to be evaluated. Very recently, Dudek and Scheraga (336) have also addressed the loop generation problem with an energy-based approach that appears promising. But a broader range of systems (they studied BPTI) will have to be investigated before the usefulness and reliability of their approach can be determined.

The random tweak method differs significantly from those of Bruccoleri and Karplus and Moult and James in its approach to sampling backbone conformations. While the latter two methods essentially utilize polished grid search (systematic search) approaches, the former randomly samples conformational space by perturbing or tweaking all the backbone dihedral angles, and then optimally modifying these angles to insure that the geometric constraints at the ends of the loop are satisfied. It has been shown that the random tweak procedure generates, for short loops, at least, a uniform coverage of the reduced conformational space, and is considerably less demanding of computer resources than systematic search procedures.

Root-mean-square measures are typically used to assess the reliability of loop predictions. In addition, loop lengths are also generally quoted. Qualitatively, it is expected that longer loops, which possess a greater number of degrees of freedom than shorter ones, should be more flexible, and thus their geometries should be more difficult to predict accurately. However, there are exceptions to this expectation (334).

As is well known, three factors play major roles in determining loop geometries (357), namely, chain length (number of residues), chain separation at its initiation points (distance of separation of the N- and C-terminal loop residues), and chain orientation at the loop initiation sites. Thus, chain length is not a sufficient descriptor of chain geometry.

Recently, Shenkin et al. (330) discussed the concept of *fractional extension*, which may prove to be a useful chain descriptor. Fractional extension, as given by Shenkin et al., is the ratio of the $C_\alpha-C_\alpha$ distance of the terminal loop residues to its distance in a fully extended chain of the same number of residues. Thus, this descriptor contains conformational information on two of the structural features thought to be important in determining loop structure, namely the distance of chain separation measured at its initiation points and the orientation of the chain at these points (357). As the fractional extension approaches unity, the number of structurally allowable loop structures approaches one.

In the case of grid searches, which effectively uniformly sample *unconstrained* conformational space, the search is enhanced by choosing backbone dihedral angles based on allowed values derived from Ramachandran plots (333) or from analysis of crystal structure data (332). Systematic searching is actually a tree search, which, even under conditions of efficient pruning and careful selection of dihedral angle ranges, grows exponentially, and thus becomes computationally intractable for loops of more than about seven residues. Bruccoleri et al. (334) have applied the Bruccoleri−Karplus procedure CONGEN to the calculation of quite large loops with reasonably good results. However, the data reported covers only RMS deviations and loop lengths, which alone do not convey the range of conformations available to the loops, and thus the difficulty of predicting them correctly. In such cases, fractional extension would be an appropriate parameter to report since it provides a more complete picture of conformational flexibility.

Although all methods employ some type of estimation of van der Waals interactions to screen or filter out highly unlikely conformations obtained during the backbone generation phase (stage?) of the loop generation process, only the random tweak method utilizes energy minimization to further refine the bare loop backbone structures generated by the sampling procedure.

Even in determining loops without side chains it is important to account for the consequences of solvent effects on electrostatic interactions. For example, when solvent effects are inadequately treated, hydrogen bonding of the peptide moieties of loop residues with other loop residues or with residues located in the core framework of the protein will be overestimated. This will lead to greater numbers of low-energy structures stabilized by internal hydrogen bonds than is actually the case. Full treatment of solvent effects is, however, not necessary or desirable at this stage of the loop generation process. A simple remedy employed to circumvent such underestimation of solvent effects is to attenuate the electrostatic part of the potential energy function. This can be done through elimination of the electrostatic energy terms (including hydrogen-bonding terms if used), reduction of their contribution through the use of a dielectric constant close to that of water ($\varepsilon = 80$), or through the use of a distant-dependent dielectric constant (79).

Once an optimal set of backbone structures is available, appropriate side chains are added. Again, a number of approaches with many variations are possible and available. One can take a knowledge-based approach, using values of side-chain dihedral angles derived from an analysis of the PDB (328, 358, 359), being sure to account for unfavorable van der Waals contacts in the context of the model. A variation of this method is to utilize results from a statistical analysis of side-chain geometry as a function of main-chain geometries found in the PDB (Finzel, unpublished results). An energy-based approach has been used where selected low-energy conformations of side chains are stored, and suitable ones, based upon optimal van der Waals interactions for the loop in question, are retrieved (Moon and Granatir, unpublished results). One can also adopt a combination of energy-based and knowledge-based procedures (360). Whatever method is used, additional energy minimization would be needed to fine-tune the side-chain geometries.

An appropriate loop conformation must now be selected from among those generated by one of the three methods. In the Bruccoleri—Karplus approach the lowest-energy conformer of the set of conformers generated by their procedure is chosen, and additional energy minimization of the set or the chosen conformer is undertaken. The Moult—James approach eschews any structural refinement by energy minimization or related procedures, and selects the conformer based on considerations of solvent electrostatic energy, implemented within the framework of an image charge method developed by Friedman (361), tempered also by considerations of hydrophobic factors. The random tweak method (330, 331) allows for structure refinement by energy minimization methods before selecting the most appropriate loop conformer; the effect of solvent was estimated by a distance-dependent dielectric constant (79). Shenkin, Fine, and coworkers found that further structure optimization was important in longer loops, but had little real effect on shorter loops. All the above authors have stressed the need to consider solvent effects in more detail in future work, although for short loops this does not appear to be entirely necessary. The subject of loop generation remains one of great importance to homology model building, but a subject in which additional work must be done.

E. FINAL STRUCTURE REFINEMENT

Once an appropriate set of reasonable loop structures have been generated, the final structure of the protein may be further refined by energy minimization and/or molecular dynamics calculations. Again the multiple-minimum problem can be an impediment to finding the global energy minimum structure even when one starts from a structure that is quite close to it (362). Energy minimization methods tend to locate the minimum energy structure closest to that of the starting structure. For this reason, molecular dynamics calculations are usually employed to add thermal energy to the system, and thus to provide a means for it to overcome the small energy barriers of local minima. After a relatively short molecular dynamics run ($\sim 10-20$ ps), energy minimization may be used to locate the energy minimum of the energy basin, or catchment region (189), into which the system has passed. In all likelihood this will not be the global minimum. Comparison of the energy minimums of several such energy basins can be obtained by sampling points along the dynamical trajectory periodically, and minimizing each of the sample points. This methodology increases the chances of finding the global minimum. In such an approach, the longer the simulation, the higher is the probability of finding the global minimum. The Monte Carlo-based simulated annealing procedure overcomes local minima more effectively and allows a thorough exploration of conformational space. Despite this advantage, such procedures have not been tried due to the difficulty of treating polymers that are highly coupled systems. Molecular dynamics-based simulated annealing methods may be of some use in overcoming the difficulties associated with the highly coupled nature of polymer systems.

It is generally accepted that the treatment of solvent effects at some level of sophistication is an essential part of all these methods. However, solvent effect calculations tend to be computationally intensive, and thus are rarely done except for the cases where distance-dependent dielectric constants or related approaches are used.

F. VALIDATION OF THE MODEL

Validation of the model, or preferably the models, produced by the modeling process is essential if one is to have confidence in its reliability for further scientific applications, such as the design of ligands with specific binding properties. However, the model structure must be evaluated alongside a randomly generated one to insure that the evaluation criteria can distinguish between correct and incorrect structures. In cases where the predition is on a known structure, deviations of the predicted structure from the known one should be compared to the deviations from a randomly generated structure

(363). This is analogous to the situation discussed in Section 4.1 in the case of secondary structure prediction.

A model can be validated by examining it in terms of generally observed features of protein structure and in terms of derived properties such as those obtained from experimental IR, Raman, and CD measurements. A number of workers (339, 364, 365), have addressed the former, using such measures as solvent accessibility, residue packing densities, location of hydrophobic or charged/polar residues, and electrostatic and solvation free energies. At present, evaluation of predicted protein structures in terms of experimentally determined properties such as IR and CD spectra has not been investigated. Work by Manning and Woody (366) suggests that reliable CD spectra can be calculated for many proteins, which opens up the exciting possibility that CD spectroscopy may prove to be a useful tool for validating structure predictions. It is not clear, however, whether the theoretical methodology or the calculated CD spectrum itself is sensitive enough to small errors in structure to be of use.

5.2. Combinatoric Approaches to Tertiary-Structure Prediction

Combinatoric approaches involve the use of heuristics, and represent a middle ground between the homology-based approaches discussed in the previous section, which also employ heuristics, and the a priori prediction methods discussed in the following section. Basically, combinatoric approaches generate a vast number of possible tertiary folds; unreasonable structures are removed using rules based upon the relatively small number of structural patterns known to exist in proteins (30). Thus, an advantage of combinatoric approaches over homology-based ones is that knowledge of the 3D structure of a related protein or proteins is not required. On the other hand, information on the presence or lack of specific tertiary-fold motifs in proteins within a given class (e.g., α or α/β, see Section 2.1) is essential, and thus the need for protein-structure information remains. Consequently, combinatoric approaches cannot by their very nature be used to predict the structure of proteins possessing tertiary folds that have not been observed previously.

To date most of the work in combinatoric methods has been carried out by Cohen, Sternberg, and their coworkers (205, 229, 316, 367–372), although the first such study, which involved a prediction of helix packing in myoglobin, was carried out by Ptitsyn and Rashin (305) in 1975.

Basically, combinatoric methods, as described by Cohen et al. (e.g., 205), involve the following steps:

1. Determination of the class to which the protein belongs (see Section 2.1).
2. Determination of the location of loops.
3. Prediction of the location of secondary-structure elements within the backbone chain segments bounded by the loops.
4. Generation of a large number of possible packing arrangements of the secondary-structural elements.
5. Filtering out all unreasonable packing arrangements.

It should be noted that combinatoric methods can be applied only to individual domains and not to polypeptide chains possessing several domains. In the latter case, each domain must be handled separately (see also the discussion in Section 2.1).

Step 1 is crucial to the overall structure prediction process because it limits the scope of the problem to the smaller number of tertiary-fold motifs that are typically found within a given class. Furthermore, knowledge of the class to which a protein belongs considerably enhances the reliability of secondary structure prediction: For example, a protein in the β class would not be expected to possess any appreciable segments of α helix. Assignment of the class to which a protein belongs may be based on an analysis of its CD spectrum, which provides an estimate of percent of various secondary-structure types present (371, 372), and on biochemical data such as that obtained from proteolytic digestion experiments (372), or on computer-based predictions. As noted earlier in Section 4.4, computer-based methods of class prediction tend to be more reliable than secondary-structure predictions.

Step 2 continues the divide and conquer strategy begun in step 1 by locating all the loop structures on the backbone chain; Approximately 25% of all residues in globular proteins are involved in β turns (203). As was true for the prediction of a protein's class, the location of loops is generally predicted with greater reliability than other secondary-structural elements. Recently, Cohen et al. (267) developed a method based on pattern matching that appears to yield perhaps the most reliable loop predictions (ca. 90% in some instances). Steps 1 and 2 clearly lessen the difficulty of predicting the location of the remaining secondary-structural elements, step 3, by reducing the overall size of the problem (205).

Once all the secondary-structural elements have been located, the stage is set for steps 4 and 5, which constitute the most computationally intensive stages of the tertiary-structure prediction process. It should be noted that in all early studies using the combinatoric approach, both the class of the protein and the location of its secondary-structural elements were taken from known protein structures (305, 316, 369, 370, 373, 374). In those instances, the main purpose of the work was to validate the combinatoric procedures used rather than to predict tertiary structure de novo.

In step 4 a very large number of possible packing arrangements ($\sim 10^4$–10^8) are generated by docking the predicted α-helical and/or β-strand segments according to a small number of simple rules. For example, in the work of Cohen et al. (367) on the packing of α helices in myoglobin, a set of hydrophobic interaction sites were determined that were subsequently used as a basis for the generation of all possible helix pairings. Each helix pairing thus involved the matching of an interaction site on one helix with that on another: Eight geometrically different structures are associated with each site pairing.

Then in step 5 the set of packing arrangements thus generated is reduced to more manageable proportions through the use of additional structure-based rules, which may be obtained from "chemical intuition" or from information within the PDB. For example, in the case of helix packing Cohen et al. (367) used two simple rules: that the end-to-end distance of consecutive helices in the sequence be less than a maximum permissible value, and that only a limited number of close contacts be permitted in any acceptable structure. At this point in the tertiary-structure prediction, process data acquired from chemical modification, site-specific mutagenesis, fluorescence energy transfer, proteolytic digestion, or other biochemical experiments may be introduced to reduce the number of allowable structures further (372).

Up to the present, only a relatively small number of proteins have been studied by combinatoric approaches, and of that number most work has involved studies of proteins of known 3D structure. As noted earlier, the first such study was that of Ptitsyn and Rashin (305) on the folding of myoglobin, which was followed several years later by the work of Cohen et al. (367). In the latter work, the authors arrived at 20 possible myoglobin structures from an initial list of about 10^8. Shortly thereafter Cohen and Sternberg (368) employed information derived from chemical studies to further reduce this number to two structures with respective RMS deviations of 4.48 and 4.53 Å compared to the crystal structure.

Ptitsyn and coworkers (373, 374) also predicted the allowed topologies of β sandwiches using a combinatoric procedure. The most extensive work on this class of proteins, however, has been carried out by Cohen, Sternberg, and coworkers (317, 318, 369, cf. 375 on predicting the packing arrangement of β strands in the β sheets of globular proteins). These authors have also carried out a number of studies on α/β proteins (205, 316). To date no studies on α + β proteins have been undertaken.

Essentially all the work carried out prior to 1984 was directed toward establishing the feasibility of the basic combinatoric methodology (cf. 318). Subsequent to this work, very little in the way of actual tertiary-fold prediction has actually taken place. Of the predictions, two involved α proteins, namely human growth hormone (HGH) (371) and interleukin-2 (IL-2) (370), and the other involved an α/β protein, the α subunit of tryptophan synthase (α-TS) (372).

Both HGH and IL-2 were predicted to be four-α-helical bundle proteins (370, 371). As discussed by Cohen and Kuntz (371), the results of unpublished crystallographic work by Abdel-Meguid and his colleagues at Monsanto supports the general structural features predicted for HGH, although the chain topology of the predicted structure differed from that obtained crystallographically. At present a 3D structure has not been published for IL-2 (note that Don McKay has proposed one, and his work indicates that it is a modified 4-α-helical barrel-like protein), and thus a definitive comparison with experiment cannot at present be made.

The case of α-TS is an interesting one. The initial prediction by Hurle et al. was based on their assumption that α-TS possessed a β sheet rather than a TIM barrel tertiary-fold motif. Their assumption rested heavily on proteolytic digestion data, which proved to be misleading. Subsequent to their initial prediction, the results of preliminary x-ray crystallographic studies suggested that α-TS was, in fact, a TIM barrel, and this led them to modify their original prediction in light of the new evidence. This brings out the importance of reliable methods for distinguishing particular structural motifs within a given protein class, and, as alluded to by Hurle et al. (372), it also points to the difficulty of interpreting, in structural terms, many low-resolution experiments, such as proteolytic digestion.

As may be apparent from the foregoing discussion and papers cited therein, the combinatoric approach places emphasis on the prediction of the core framework of a protein at the expense of its loop structures. Loops are used mainly to aid in the prediction of secondary-structural elements, which are critical to the combinatoric folding process. At this point, energy minimization, such as described in Section 4.6, may be used to refine the packing geometry of the core framework further. Better loop structures can then be generated and refined using methods discussed in Section 5.1D. And finally the whole structure can be refined and validated as described in Sections 5.1E and 5.1F. However, such has not been the case for combinatoric experiments to date.

5.3. A Priori Approaches to Tertiary-Structure Prediction

As discussed above, heuristic approaches to the folding problem seek to determine a protein's tertiary structure by analogy to structures of related proteins and/or by the application of rules derived from the knowledge-base of protein structures and properties. In contrast, a priori approaches, in principle, seek to calculate a protein's structure on the basis of energetic principles (cf. 376, 377). It is not a requirement that a priori methods mimic any of the features of the actual folding process itself. Such mechanistic information, however, may be of use on the design of effective folding algorithms. In this regard, the respective roles of short-, medium-, and long-range interactions in the folding process have been the subject of numerous investigations (321, 378–382) that have been of value in the development and refinement of tertiary-structure prediction methods. Nevertheless, most studies carried out to date have been designed to provide mechanistic insights into the folding process, rather than to provide reliable predictions of tertiary structure (vide infra). Recent reviews by Gō (321) and by Skolnick and Kolinski (382) have extensively examined the literature on the folding problem.

Two basic computational procedures are employed in a priori approaches: function optimization and Monte Carlo (MC) sampling. In the first case, the energy of the system is generally represented as a continuous function of its geometry, while in the latter case the energy need not be continuous, and may, in fact, be given as a set of discrete values stored in some sort of table. Since potential and free energies are both used, sometimes even in the same energy expression, in this section we will use *energy function* to describe either possibility, or we will specify potential or free energy explicitly.

A. FUNCTION OPTIMIZATION METHODS

Function optimization methods, which generally involve minimization of some type of continuous energy function (cf. 376, 377), are plagued by the multiple-minimum problem. A number of strategies for overcoming this problem involving both MC and/or energy minimization procedures have been discussed in the recent review by Skolnick and Kolinski (382), but most are applicable only to oligopeptides and not to systems the size of proteins.

An interesting new procedure, the self-consistent electric field (SCEF) method proposed by Piela and Scheraga ((383); see also Ripoll and Scheraga (384) for more recent modifications) is based on the long-held notion that electrostatic interactions are crucial to the folding process. The procedure seeks to optimize the orientation of the dipole moment of each peptide group with respect to the electric field generated by the remainder of the protein. As each structural change affects the electric field, an iterative process is required to achieve self-consistency. The method has been applied to a number of relatively small systems with impressive success, but it is not yet clear how robust the method will be in dealing with the folding of real proteins. In addition, because the presence of solvent definitely influences the nature of a protein's electric field (e.g., 20, 21), the effect of neglecting solvent on the performance of the method is not clear.

Another, and perhaps the most obvious, approach to the multiple-minimum problem involves the use of simplified energy functions to reduce the number of local minima and, in some cases, the explicit use of biasing potentials to drive the folding process toward the native structure (115, 116, 385−389). Recently, the development of simulated annealing methods (112−114), which allow for the possibility of escaping local minima, holds promise for protein-folding studies ((115); cf. an interesting application by Robson and Osguthorpe (116) of simplex-based minimization in protein-structure prediction).

Simplified energy functions, in contrast to the more familiar and complex potential-energy functions described in Section 2.1, typically represent each amino acid residue as a single or a small number of interaction sites. The form of the site−site interaction energy is then obtained, and may be based upon an average interaction between residues (e.g., the time-averaged potentials of Levitt and Warshel (385, 386), or upon data derived from the distribution of amino acids in real proteins (115, 388). In addition, biasing potentials are often added to insure that the folding process converges to reasonably compact structures close to the native one. Levitt (386), and Levitt and Warshel (385), for example, employed holding and pushing potentials, while Robson and coworkers (28, 116, 390) employed compaction potentials, and Kuntz et al. (376) employed radial potentials (note that the Kuntz et al. work does not actually involve energy minimization but does involve minimization of an error function, which is related to energy in a very complicated and unknown way). Lastly, some account of solvent effects, or at least hydrophobic interactions, is usually included, albeit in an ad hoc fashion (385, 386, 391), to insure that proper folding occurs and that the resulting structure possesses an appropriately hydrophobic core surrounded by a hydrophilic shell (392). Energy functions derived directly from protein structure data (115, 388) implicitly contain information relating to solvent effects, and so additional solvent terms need not be included explicitly.

In their seminal work, Hagler and Honig (393) carried out folding calculations of PTI (pancreatic trypsin inhibitor) using a model polypeptide chain consisting of only alanine and glycine residues. The model polypeptide closely resembled the simplified representations of protein structure used by other workers in protein-folding studies. Hagler and Honig showed that folded structures could be generated that rivaled those produced by the simplified model potentials noted above (385, 386). These authors concluded that, aside from explicit biases introduced into protein-folding calculations of PTI, implicit biases (such as the proclivity of glycine residues for turn formation) were also present that tended to cause the model polypeptide to adopt reasonably compact structures (see also the discussion by Némethy and Scheraga (394) on this issue). Thus, they emphasized the need to examine the detailed structural features to assess the reliability of a prediction. For example, these authors pointed out that although compact structures were generated, they lacked certain topological features such as the threading of the 30−51 loop by the N-terminal sequence. Structural errors of this type cannot be overcome by further energy minimization using higher-level potential-energy functions (cf. the discussion of Levitt and Warshel (385)), thus these structures would not likely lead to the formation of a correctly folded polypeptide chain.

Due to the disappointing performance of these methods to produce reliable, unbiased predictions, interest in function-optimization energy-based approaches to tertiary structure prediction has waned considerably since the early flurry of activity that took place in the mid to late 1970s. Recently, however, Wilson and Doniach (115) have developed an approach to the folding problem that shows great promise, and should rekindle interest in the capability of a priori methods to provide useful and reliable tertiary structure predictions.

As is true with essentially all a priori folding methods, the Wilson and Doniach procedure employs a simplified model of protein structure to reduce the complexity and the number of local minima of the energy function (vide supra). In this model, each amino acid is represented by a single interaction site located at the centroid of an average side-chain geometry, and the protein's conformational degrees of freedom are restricted to the backbone Φ and Ψ angles. This approach is unlike that typically used in other folding studies based on simplified energy functions (see, e.g., 385, 386, 388; cf. 116). A distance-dependent free-energy function is derived from distributions of amino acid pair distances obtained from the PDB. Energy functions derived in this manner implicitly contain, among others, terms corresponding to electrostatic, H-bonding, van der Waals, and hydrophobic interactions, but no attempt is made to resolve the various energy components. It should also be noted that such energy functions may contain biases due to the limited set of protein structures available in the PDB. Structure refinement is carried out by simulated annealing. No attempt is made to build in any features of known or assumed folding pathways.

The Wilson—Doniach method was used by the authors to fold crambin, a small (46 residues) protein of well-determined structure (395), and the results of the study indicate a number of important features of their method (115):

1. Secondary structures are formed in appropriate regions of the polypeptide chain starting from a number of random starting geometries.

2. Secondary—structure formation is sequence specific and depends on long-range interactions.

3. Once assigned, the α helices and β strands associate as in the native structure.

4. The best folded structure obtained by the W-D method possesed an RMS deviation of 4.1 Å, and showed excellent agreement with the x-ray structure in terms of the mutual alignment of the α helices, although the overall packing of the β strands was poor.

5. Pairs of cysteines corresponding to native disulfides associated more frequently than other cysteine pairs.

The validity of the Wilson—Doniach approach will, of course, require considerable testing on a variety of proteins to establish its general applicability. Wilson and Doniach, in the same paper, also report on preliminary folding studies of BPTI. While the details were not presented, it appears that their method yields a distance matrix error (i.e., RMS error of all the corresponding $C_\alpha - C_\alpha$ distances between the native x-ray structure of the protein compared to the predicted structure (see, e.g., (28) for further discussion) of 4.5 Å, which is better than that obtained by any previous method. Moreover, considering that the Wilson—Doniach method starts from random structures, rather than the biased extended ones used in most other studies, their results are all the more interesting.

B. MONTE CARLO METHODS

The use of MC sampling bypasses the multiple-minimum problem, but runs into a combinatorial explosion of conformations for large proteins which necessitates the use of significantly reduced configuration spaces such as are found, for example, in lattices. Lattice-based methods have a certain appeal due to their simplicity. Nevertheless, they are not without pitfalls, perhaps the most serious being the severe restrictions placed on the types of structures, especially supersecondary structures, that can be generated. For example, current lattice models do not allow for the observed chain twists that give rise to the overall right-handed twists observed in β sheets, nor do they allow pairs of α helices to adopt their most preferred orientations (see, however, the recent work of Kolinski, Skolnick, and Sikorski noted in the review by Skolnick and Kolinski (382)). As was the case for optimization methods, MC-based calculations are facilitated by the use of simplified energy functions, generally even simpler than those employed in optimization approaches. MC methods also contain biases that induce the formation of particular structural features (e.g., β turns) or stabilize specific long-range interactions known to exist in the native structure of a protein (e.g, (379, 396); and review by Skolnick and Kolinski (382)). Some account of solvent/hydrophobic factors, which are crucial to the folding process, are also generally included (397). For example, ad hoc procedures have been employed in which solvent effects are accounted for by favorably weighting nearest-neighbor interactions (note that nearest-neighbor refers, in lattice calculations, to neighboring lattice sites) between like amino acid types, namely hydrophobic—hydrophobic or hydrophilic—hydrophilic residues, and penalizing interactions between amino acids of opposite types.

Two types of MC sampling have been implemented: sampling over Φ-Ψ space, and sampling over 2D or 3D lattice spaces (382, 394). Energy functions for use in Φ-Ψ space sampling calculations are generally obtained from experimental data on proteins of known structure (e.g., 394). Although energy functions derived in this manner may be represented as continuous functions, customarily when they are used in MC calculations they are not (394, 396). Energy functions employed in lattice-based calculations are generally of even simpler form than those used in methods based on Φ-Ψ space sampling. Moreover, the configuration space covered by lattice-based sampling is considerably more restricted than that covered by discrete Φ-Ψ space sampling.

Most folding studies using MC sampling have been more concerned with analyzing the factors critical to protein folding and the physicochemical features of the folding process, and less with predictions of tertiary structure (321, 382). Recently, work by Skolnick and coworkers (382, 397, 398) applied dynamic MC lattice-based methods, using less biased lattice-type energy functions, to predict the structure of several interesting structural motifs, namely the four-α-helix bundle (397, 398), the β barrel (399), and the six-stranded Greek key (322, 382). The structures they obtained, subject of course to the inherent structural limitations of the carbon lattice used in their calculations, were quite reasonable and hold future promise that with improved, more flexible lattices (see Kolinski, Skolnick, and Sikorski cited in (382)) and with more realistic energy functions it may be possible to predict tertiary structure reliably with this method. As was the case in the Wilson—Doniach approach, considerable further work must be done on a wide variety of systems before the usefulness of the method, or a close relative thereof, for tertiary structure prediction will be established.

C. MISCELLANEOUS METHODS

Recently, a novel approach to protein folding based on an associative memory Hamiltonian was described by Friedrichs and Wolynes (400). Construction of the folding Hamiltonian was based on work related to the statistical mechanics of associative memories and spin glasses (see Bryngelson and Wolynes, (401); see also Ref. 7 in (400)). Sequence-specific information was incorporated in terms of amino acid features, referred to as charges in the paper, such as hydrophobicity, atomic charge, volume, etc., and the system's memory was based on structural data obtained from the PDB. While some bias was introduced into the folding procedure, the results of their study did illustrate the potential for such an approach. Although all current approaches to the folding problem suffer limitations, considerable progress has been made during the last decade. Moreover, it appears that some of these approaches, their variants, or some unexpected breakthrough leading to fundamental changes in our view of the problem may very well occur within the decade of the nineties. Thus, there is some reason for optimism with regard to the future of a priori tertiary structure prediction, and hopefully the renewed interest in these methods will lead to a robust solution to this important problem in the not too distant future.

6. QUATERNARY STRUCTURE

Protein—protein interactions play essential roles in biochemical reactions and cellular functions. These may be divided into two classes. The first includes phenomena such as allosteric interactions in, for example, hemoglobin, aspartate transcarbamylase and phosphofructokinase, where the enzymatic activity of a multimeric enzyme is regulated by intersubunit structure and interactions. The second class includes regulation of enzymatic or other biochemical activities such as the activation of phospholipase A2, signal transduction by G-proteins, and model systems for electron transport in respiratory proteins; in these systems, molecular recognition is the requisite step in initiating subsequent, more extensive protein—protein interactions. The following discussion on quaternary interactions will focus on structural aspects of several studies in these two areas.

6.1. Allosteric Proteins

Allostery was first introduced conceptually and studied for hemoglobin (402, 403). Similar allosteric regulation of enzymes has since been found in many multimeric enzyme systems such as E. coli aspartate transcarbamylase (404), muscle glycogen phosphorylase (405, 406), and bacterial phosphofructokinase (407, 408). Structural studies of allostery are represented by a number of different theoretical investigations of the cooperativity of ligand binding in hemoglobin. Warshel (409) studied the energy−structure correlation in metalloporphyrins with a simulated heme binding site by semiempirical quantummechanical calculation of the π electrons in the porphyrin system. The energy surface of the system was shown to provide a quantitative explanation of the control of ligand binding as a function of hemoglobin conformations. Statistical-mechanical models for hemoglobin cooperativity have also been formulated to account for thermodynamic and spectroscopic data (410, 411). Details of the structural changes related to ligand binding in hemoglobin have been known in crystallographic studies and allosteric mechanisms have been proposed based on such studies (e.g., 412). Tertiary structural changes in the ligand binding site of hemoglobin upon ligand binding can be calculated in a theoretical reaction path constructed for the deoxy- to oxyhemoglobin transition (412, 413); it was inferred that ligand-induced tertiary-structure changes within the allosteric core were coupled to the displacement of the core in the quaternary structure transition and that this coupling is an essential element in the cooperative mechanism. It would be interesting to study quaternary-structure changes directly by molecular mechanics or dynamics methods, although a study of such a large multimeric system is somewhat beyond the scope of current methods. Nevertheless, it is possible to represent a quaternary complex in a simplified model and to focus on the electrostatic aspect of protein−protein interactions, as shown in a recent study of aspartate transcarbamylase (414).

6.2. Protein Molecular Recognition

Protein−protein interactions are also important in nonallosteric enzyme systems. In this category, quaternary complexes are either nonexistent or only loosely formed and a molecular recognition step is required before intermolecular interactions can take place. Moreover, the complex must then turn over upon deactivation or completion of functional transactions. Examples include respiratory electron transfer complexes (vide infra), dimerization and activation of porcine pancreatic phospholipase A2 (415), and signal transduction systems (416).

A number of physiological and nonphysiological cytochrome c complexes have been studied experimentally as a means of investigating biological electron transfer mechanisms (e.g., 417). Structural models of some of these cytochrome c complexes have also been constructed as a means for studying the characteristics of protein−protein interactions in electron transfer, namely, cytochrome c and cytochrome b5 (418, 419), cytochrome c and flavodoxin (420, 421), cytochrome b5 and methemoglobin (422), and cytochrome c and cytochrome c peroxidase (423).

The complex of cytochrome c and cytochrome b5 (418) was the first of such model complexes to be proposed. For each of the two molecules in the binary complex, an irregular polyhedron model is constructed using crystallographically determined positions of the most exposed surface atoms as the vertices. Based on surface charge distributions, a front-to-front orientation was identified for the bimolecular complex. In a least-squares fitting procedure, the polyhedron model of cytochrome b5 is then rotated into a position in which complementary intermolecular charge pairs have ca. 3.0-Å separations. Rotations of some of the lysine side-chains of the cytochromes then allowed further steric, nonionic complementarity of the complex. Subsequent to this work, crystallographically determined coordinate sets have been directly manipulated to construct bimolecular complexes based on charge and structural complementarity requirements similar to those in the polyhedron model (420, 422, 423). Intermolecular contacts are usually relieved by energy minimization calculations, and electrostatic potential surfaces (424, 425) are also useful in the construction of bimolecular complexes. In molecular complexes for which the initial orientation of the two molecules are less obvious than in cytochrome c complexes, a computational six-dimensional mapping method based on electrostatic and steric interactions may be employed (426).

Based on the early static model of the complex of cytochromes c and b5, a more recent study investigated intermolecular interactions in the complex by molecular dynamics simulation (419). Conformational transitions were observed in the intermolecular interface and some of the molecular dynamics-generated structures are substantially different and potentially more favorable for electron transfer reaction. The heme—heme distance was found to be shortened by several angstroms.

In addition to the cytochrome c—cytochrome b5 complex, the cytochrome c—cytochrome c peroxidase complex (427, 428) represents another nonphysiological model system for studying electron-transfer reactions, although there is evidence that cytochrome c may in fact act as a natural substrate for cytochrome c peroxidase in yeast (429). As described above, a model structure of the bimolecular complex has been constructed (423), and crystallographic structure determination has also been attempted (430). Similar to other electron-transfer protein complexes, electrostatic interactions make significant contributions to the stability of this bimolecular complex, as well as to the process of association and molecular recognition between these two molecules. As demonstrated recently in the Brownian dynamics simulation of diffusional association (431, 432), the association rate constant obtained in the simulation is smaller by an order of magnitude in the absence of the electrostatic forces. The Brownian dynamics simulation also provided additional details of the protein—protein interactions that took place during the diffusional encounter. For example, potentially productive association of cytochrome c with cytochrome c peroxidase can occur on more than one surface domain of cytochrome c peroxidase; an alignment of heme planes may also be required in the actual electron-transfer reactions. While a number of refinements of the diffusional encounter model can be made (432), the Brownian dynamics simulation has already revealed essential features of the molecular recognition and protein−protein interactions of the process.

7. FUTURE PROSPECTS

Considerable progress has been made in the prediction and analysis of protein structure. Nevertheless, much remains to be done. In particular, the protein−folding problem continues to be refractory to all attempts at its solution, due primarily to the intractability of the multiple-minimum problem. A number of new methods have been developed in an effort to overcome or circumvent this problem. Unfortunately, many of the methods, while applicable to oligopeptides, do not scale well and thus cannot be applied to realistic polypeptide systems. A priori methods appear to hold some promise, at least with regard to the generation of realistic, albeit incompletely folded structures. It is just too early to tell whether they can be improved sufficiently to play an important role in future work on the folding problem. It appears that we must better understand the rules that govern folding before we will be able to design more robust and reliable folding procedures. The rapid increase in computer power that continues to grow, while impressive, will not be sufficient to support a brute-force solution to the folding problem.

At the present time heuristic approaches continue to be our best hope. Heuristic approaches are, however, plagued by the need for structural data on proteins related to those of interest. Although x-ray crystallography and 2D/3D-NMR methods are providing 3D structural data at an accelerated rate, it is not fast enough. Moreover, very little structural data are available for some important protein families; for example, for DNA and RNA polymerases presently only one structure, a C_a trace of the Klenow frament of E. coli DNA pol I (346), is available in the PDB. The lack of a protein database of sufficient size even impacts the ability to obtain secondary-structure predictions of enhanced accuracy. It has been estimated that about 1500 3D structures will be needed to provide an adequate database (9). Even as the size of the PDB grows, the need for new ways to view protein structure is critical if we are to develop a deeper understanding of the many subtle relationships that underlie structure. For example, Rachovsky and Scheraga (433) have developed an interesting way, based on differential geometry, of describing the curve-like character of polypeptide chains. Recent topological characterizations of protein structure (24, 25) have also provided a number of interesting insights, as have investigations into the chiral features of helix packing (26, 61). The discovery of new and novel ways of representing protein structural features awaits the curious investigator, and the dividends for such discoveries may be significant breakthroughs in the way in which we think about protein structure and its relationship to protein function.

The ratio of primary- to tertiary-structure information continues to grow at a very rapid rate, and an important source of 3D structural information lies untapped. As the history of biochemical research well documents, the relationship of a protein's structure to its function has been an important theme — a theme that will not diminish in importance in the future, but rather will gain in importance as more structural information becomes available. Thus, as methods for predicting and analyzing all levels of protein structure improve, their role in biochemical research will be ultimately enhanced.

ACKNOWLEDGMENTS

The authors express their special thanks to C.A. Granatir for reading and patiently editing the complete manuscript, an effort that significantly improved its overall quality. We also thank J.B. Moon and J.R. Blinn for helpful discussions and for help with several figures, and J.K. Hammond for her help in preparing the many drafts and the final manuscript.

References

1. L. Pauling and R.B. Corey, *Proc. Natl. Acad. Sci. USA* **37**, 729−740 (1951).

2. D.W. Green, V.M. Ingram, and M.F. Perutz, *Proc. R. Soc. London Ser. A* **225**, 287−307 (1954).

3. K.U. Linderstrom-Lang and J.A. Schellman, in P.D. Boyer, Ed., *The Enzymes*, 2d ed., Vol. 1, Academic, New York, 1959.

4. R. Bernstein, T.F. Koetzle, G.J.B. Williams, E.F. Meyer, Jr., M.D. Brice, J.R. Rodgers, O. Kennard, T. Shimanouchi, and M. Tasumi, *J. Mol. Biol.* **112**, 535−542 (1977).

5. J.L. Markley and E.L. Ulrich, *Annu. Rev. Biophys. Bioeng.* **13**, 493−521 (1984).

6. K. Wüthrich, *NMR of Proteins and Nucleid Acids*, Wiley-Interscience, New York, 1986.

7. A.M. Gornenborn and G. Marius Clore, *Protein Seq. Data Anal.* **2**, 23−37 (1989).

8. E.L. Ulrich, U.L. Markley, and Y. Kyogoku, *Protein Seq. Data Anal.* **2**, 23−37 (1989).

9. J.M. Thornton and S.P. Gardner, *TIBS* **14**, 300−304 (1989).

10. J.F. Collins and A.F.W. Coulson, in M.J. Bishop and C.J. Rawlings, Eds., *Nucleic Acid and Protein Sequence Analysis: A Practical Approach*, IRL Press, Oxford, 1987, Chpt. 13.

11. R.H. Lathrop, T.A. Webster, and T.F. Smith, *Commun. ACM* **30**, 909−921 (1987).

12. W.R. Taylor, *Protein Eng.* **2**, 77−86 (1988).

13. P. Dean, *Molecular Foundations of Drug-Receptor Interaction*, Cambridge University Press, Cambridge, 1987.

14. J.A. McCammon and S.C. Harvey, *Dynamics of Proteins and Nucleic Acids*, Cambridge University Press, Cambridge, 1987.

15. C.L. Brooks III, M. Karplus, and B.M. Pettitt, *Proteins: A Theoretical Perspective of Dynamics, Structure, and Thermodynamics*, Wiley, New York, 1988.

16. K.C. Chou, *Biophys. Chem.* **30**, 3−48 (1988).

17. A. Warshel and S.T. Russell, *Quart. Rev. Biophys.* **17**, 283−422 (1984).

18. J.B. Matthew, *Annu. Rev. Biophys. Biophys. Chem.* **14**, 387−417 (1985).

19. B. Honig, W.L. Hubbell, and R.F. Flewelling, *Annu. Rev. Biophys. Biophys. Chem.* **15**, 163−193 (1986).

20. S.C. Harvey, *Proteins* **5**, 78−92 (1989).

21. K.A. Sharp and B. Honig, *Annu. Rev. Biophys. Biophys. Chem.*, in press.

22. P.Y. Chou and G.D. Fasman, *Adv. Enzymol.* **47**, 45−148 (1978).

23. J. Garnier, D.J. Osguthorpe, and B. Robson, *J. Mol. Biol.* **120**, 97−120 (1978).

24. B. Mao, *J. Am. Chem. Soc.* **111**, 6132−6136 (1989).

25. B. Mao, K.C. Chou, and G.M. Maggiora, *Eur. J. Biochem.*, **188**, 361−365 (1990).

26. G.M. Maggiora, B. Mao, and K.C. Chou, in P.G. Mezey, Ed., *New Developments in Molecular Chirality*, Reidel, in press.

27. G.E. Schulz and R.H. Schirmer, *Principles of Protein Structure*, Springer-Verlag, New York, 1979.

28. B. Robson and J. Garnier, *Introduction to Proteins and Protein Engineering*, Elsevier, Amsterdam, 1986.

29. D.E. Wetlaufer, *Proc. Natl. Acad. Sci. USA* **70**, 697−701 (1973).

30. J.S. Richardson, *Adv. Protein Chem.* **34**, 167−339 (1981).

31. J. Janin and C. Chothia, *Methods Enzymol.* **115**, 420−430 (1985).

32. R.R. Porter, *Science* **180**, 713−717 (1973).

33. C. Ghelis, M. Tempete-Gaillourdet, and J.M. Yon, *Biochem. Biophys. Res. Commun.* **84**, 31−36 (1978).

34. W.S. Bennet, Jr., and T.A. Steitz, *Proc. Natl. Acad. Sci. USA* **75**, 4848−4852 (1978).

35. G.M. Crippen, J. Mol. Biol. **126**, 315−332 (1978).

36. G.D. Rose, *J. Mol. Biol.* **134**, 447−470 (1979).

37. S.J. Wodak and J. Janin, *Biochemistry* **20**, 6544−6552 (1981).

38. G.M. Edleman and W.E. Gall, *Annu. Rev. Biochem.* **38**, 415−466 (1969).

39. G.M. Edelman, *Science* **180**, 830−840 (1973).

40. M.G. Rossmann, D. Moras, and K.W. Olsen, *Nature* **250**, 194−199 (1974).

41. H. Sakano, J.H. Rogers, K. Hüppi, C. Brack, A. Traunecker, R. Maki, R. Maki, R. Wall, and S. Tonegawa, *Nature* **277**, 627−633 (1979).

42. M. Levine, H. Muirhead, D.K. Stammers, and D.I. Stuart, *Nature* **271**, 626−630 (1978).

43. J. Tang, M.N.G. James, I.N. Hsu, J.A. Jenkins, and T.L. Blundell, *Nature* **271**, 618−621 (1978).

44. A. Wlodower, M. Miller, M. Jaskolski, B.K. Sathyanarayana, E. Baldwin, I.T. Weber, L.M. Self, L. Clawson, J. Schneider, and S.B. Kent, *Science* **245**, 616−621 (1989).

45. D.E. Wetlaufer, *Adv. Protein Chem.* **34**, 167−339 (1981).

46. M. Levitt and C. Chothia, *Nature* **261**, 552−558 (1976).

47. C. Chothia, *Annu. Rev. Biochem.* **53**, 537−572 (1984).

48. J.S. Richardson, *Methods Enzymol.* **115**, 341−358 (1985).

49. S.T. Rao and M.G. Rossmann, *J. Mol. Biol.* **76**, 241−256 (1973).

50. J.S. Richardson, D.C. Richardson, K.A. Thomas, E.W. Silverton, and D.R. Davies, *J. Mol. Biol.* **102**, 221−235 (1976).

51. J.S. Richardson, *Proc. Natl. Acad. Sci. USA* **73**, 2619−2623 (1976).

52. J.S. Richardson, *Nature* **268**, 495−500 (1977).

53. M.J.E. Sternberg and J.M. Thornton, *J. Mol. Biol.* **105**, 367−382 (1976).

54. M.J.E. Sternberg and J.M. Thornton, *J. Mol. Biol.* **110**, 269−283 (1977).

55. K. Nagano, *J. Mol. Biol.* **109**, 236—250, 251—274 (1977).

56. A.V. Finkelstein, O.B. Ptitsyn, and P. Bendzko, *Biofizika* **24**, 21—26 (1979).

57. O.B. Ptitsyn and A.V. Finkelstein, *Biofizika* **24**, 27—31 (1979).

58. M.H. Klapper and I.Z. Klapper, *Biochim. Biophys. Acta* **626**, 97—105 (1980).

59. T. Kikuchi, G. Némethy, and H.A. Scheraga, *J. Comp. Chem.* **7**, 67—88 (1986).

60. T. Kikuchi, G. Némethy, and H.A. Scheraga, *J. Comp. Chem.* **10**, 287—294 (1989).

61. G.M. Maggiora, P.G. Mezey, B. Mao, and K.C. Chou, *Biopolymers*, **30**, 211—214 (1990).

62. A.G. Murzin and A.V. Finkelstein, *J. Mol. Biol.* **204**, 749—769 (1988).

63. C. Chothia, *Nature* **337**, 204—205 (1989).

64. D.M. Walba, *Tetrahedron* **41**, 3136—3212 (1985).

65. S.A. Wasserman and N.R. Cozzarelli, *Science* **232**, 951—960 (1986).

66. G.M. Crippen, *J. Theor. Biol.* **51**, 495—500 (1975).

67. M.L. Connolly, I.D. Kuntz, and G.M. Crippen, *Biopolymers* **19**, 1167—1182 (1980).

68. J.R. Maple, U. Dinur, and A.T. Hagler, *Proc. Natl. Acad. Sci. USA* **85**, 5350—5354 (1988).

69. M. Levitt, *J. Mol. Biol.* **145**, 251—263 (1981).

70. A.T. Hagler, J.R. Maple, T.S. Thacher, G.B. Fitzgerald, and U. Dinur, in W.F. van Gunsteren and P.K. Weiner, Eds., *Computer Simulation of Biomolecular Systems*, ESCOM, Leiden, 1989, pp. 149—167.

71. W. Jorgensen, *J. Am. Chem. Soc.* **103**, 335—340 (1981).

72. U. Burkert and N.L. Allinger, *Molecular Mechanics*, ACS Monograph Vol. 177, American Chemical Society, Washington, DC, 1982.

73. A.T. Hagler in S. Udenfriend and J. Meienhofer, Eds., *The Peptides*, Vol. 7, Academic, New York, 1985, pp. 213—299.

74. K.D. Gibson, and H.A. Scheraga, *Proc. Natl. Acad. Sci. USA* **58**, 420—427 (1967).

75. W.F. van Gunsteren and M. Karplus, *Nature* **293**, 677—678 (1981).

76. B.R. Gelin and M. Karplus, *J. Am. Chem. Soc.* **97**, 6996—7006 (1975).

77. W.F. van Gunsteren, in W.F. van Gunsteren and P.K. Weiner, Eds., *Computer Simulation of Biomolecular Systems*, ESCOM, Leiden, 1989, pp. 27—59.

78. L.G. Dunfield, A.W. Burgess, and H.A. Scheraga, *J. Phys. Chem.* **82**, 2609—2616 (1978).

79. B.R. Brooks, R.E. Bruccoleri, B.D. Olafson, D.J. States, S. Swaminathan, and M. Karplus, *J. Comp. Chem.* **4**, 187—214 (1983).

80. S.J. Weiner, P.A. Kollman, D.A. Case, U.C. Singh, C. Ghio, G. Alagona, S. Profeta, and P. Weiner, *J. Am. Chem. Soc.* **106**, 765—784 (1984).

81. S.J. Weiner, P.A. Kollman, D.T. Nguyen, and D.A. Case, *J. Comp. Chem.* **7**, 230—252 (1986).

82. A.T. Hagler, E. Huler, and S. Lifson, *J. Am. Chem. Soc.* **96**, 5319—5327 (1974).

83. T.P. Lybrand and P.A. Kollman, *J. Am. Chem. Phys.* **83**, 2923—2933 (1985).

84. C.J.F. Böttcher, *Theory of Electronic Polarization*, Vol. 1, Elsevier, Amsterdam, 1973.

85. G. Némethy, Z.I. Hodes, and H.A. Scheraga, *Proc. Natl. Acad. Sci. USA* **75**, 5760—5764 (1978).

86. K.D. Gibson and H.A. Scheraga, *J. Comp. Chem.* **8**, 826—834 (1987).

87. S.H. Northrup, M.R. Pear, and J.A. McCammon, *Nature* **286**, 304—305 (1980).

88. B. Mao and J.A. McCammon, *J. Biol. Chem.* **258**, 12543—12547 (1983).

89. F.H. Stillinger and A. Rahman, *J. Chem. Phys.* **60**, 1545—1557 (1974).

90. O. Matsuoka, E. Clementi, and M.J. Yoshimine, *J. Chem. Phys.* **64**, 1351—1361 (1976).

91. H.J.C. Berendsen, J.P.M. Postma, W.F. van Gunsteren, and J. Hermans, in B. Pullman, Ed., *Intermolecular Forces*, Reidel, Dordrecht, 1981, pp. 331—342.

92 W. Jorgensen, J. Chandrasekhar, J. Madura, M. Impey, and M.L. Klein, *J. Chem. Phys.* **79**, 926−935 (1983).

93. W.L. Jorgensen and J. Tirado-Rives, *J. Am. Chem. Soc.* **110**, 1657−1666 (1988).

94. W.F. van Gunsteren and W.J.C. Berendsen, *J. Mol. Biol.* **176**, 559−564 (1984).

95. C.L. Brooks and M. Karplus, *J. Mol. Biol.* **208**, 159−181 (1989).

96. S. Lifson and A. Warshel, *J. Chem. Phys.* **49**, 5116−5129 (1968).

97. S. Lifson, A.T. Hagler, and P. Dauber, *J. Am. Chem. Soc.* **101**, 5111−5121 (1979).

98. S.W. Harrison, S. Swaminathan, and D.L. Beveridge, *Int. J. Quantum Chem.* **24**, 319−332 (1978).

99. W. Jorgensen, *J. Am. Chem. Soc.* **101**, 2011−2016 (1979).

100. E. Clementi, *Computational Aspects for Very Large Chemical Systems*, Springer-Verlag, Berlin, 1980.

101. U. Dinur and A.T. Hagler, *J. Am. Chem. Soc.* **111**, 5149−5151 (1989).

102. U. Dinur and A.T. Hagler, *J. Chem. Phys.* **91**, 2949−2958 (1989).

103. D. Hall and N. Pavitt, *J. Comp. Chem.* **5**, 441−450 (1984).

104. I.K. Roterman, K.D. Gibson, and H.A. Scheraga, *J. Biomol. Struct. Dynam.* **7**, 391−419, (1989).

105. P.K. Weiner and P.A. Kollman, *J. Comp. Chem.* **2**, 287−303 (1981).

106. W.F. van Gunsteren and H.J.C. Berendsen, *GROMOS: Groningen Molecular Simulation Library*, BIOMOS B.V., Groningen, 1987.

107. A.J. Hopfinger and R.A. Pearlstein, *J. Comp. Chem.* **5**, 486−499 (1984).

108. B. Mao and J.A. McCammon, *J. Biol. Chem.* **259**, 4964−4970 (1984).

109. B.R. Gelin and M. Karplus, *Proc. Natl. Acad. Sci. USA* **72**, 2002−2006 (1975).

110. P.E. Gill, W. Murray, and M.H. Wright, *Practical Optimization*, Academic, London, 1981.

111. K.D. Gibson and H.A. Scheraga, in M.H. Sarma and R.H. Sarma, Eds., *Structure and Expression*, Vol. 1, *From Proteins to Ribosomes*, Adenine, Guilderland, New York, 1988.

112. S. Kirkpatrick, C.D. Gellatt, Jr., and M.P. Vecchi, *Science* **220**, 671−680 (1983).

113. R.A. Donnelly, *Chem. Phys. Lett.* **136**, 274−278 (1987).

114. S.R. Wilson and W. Cui, *Biopolymers* **29**, 225−235 (1990).

115. C. Wilson and S. Doniach, *Proteins* **6**, 193−209 (1989).

116. B. Robson and D.J. Osguthorpe, *J. Mol. Biol.* **132**, 19−51 (1979).

117. S.R. Niketic and K. Rasmusen, *The Consistent Force Field: A Documentation*, Springer-Verlag, Berlin, 1977.

118. K.B. Wiberg, *J. Am. Chem. Soc.* **87**, 1070−1078 (1965).

119. A.S. Householder, *Principles of Numerical Analysis*, McGraw-Hill, New York, 1953.

120. N.L. Allinger, *Adv. Phys. Org. Chem.* **13**, 1−78 (1976).

121. J.W. Ponder and F.M. Richards, *J. Comp. Chem.* **8**, 1016−1024 (1987).

122. R. Fletcher and C.M. Reeves, *Comput. J.* **7**, 149−154 (1964).

123. R.A. Scott and H.A. Scheraga, *J. Chem. Phys.* **44**, 3054−3069 (1966).

124. W.F. van Gunsteren and M. Karplus, *J. Comp. Chem.* **1**, 266−274 (1980).

125. D.N.J. White and O. Ermer, *Chem. Phys. Lett.* **31**, 111−112 (1975).

126. O. Ermer, *Struct. Bonding* **27**, 163−211 (1976).

127. J.A. McCammon, B.R. Gelin, M. Karplus, and P.G. Wolynes, *Nature* **262**, 325−326 (1977).

128. F. Colonna-Cesari, D. Perahia, M. Karplus, H. Eklund, C.I. Brändén, and O. Tapia, *J. Biol. Chem.* **261**, 15273−15280 (1986).

129. V.T. Oi, T.M. Vuong, R. Hardy, J. Reidler, J. Dangl, L. A. Herzenberg, and L. Stryer, *Nature* **307**, 136−140 (1984).

130. S.C. Harvey, and J.A. McCammon, *Comput. Chem.* **6**, 173−179 (1982).

131 C.S. Tung, S.C. Harvey, and J.A. McCammon, *Biopolymers* **23**, 2173−2193 (1984).

132. H.H.-L. Shih, J. Brady, and M. Karplus, *Proc. Natl. Acad. Sci. USA* **82**, 1697−1700 (1985).

133. A. Rahman, *Phys. Rev. A* **136**, 405−411 (1964).

134. M. Karplus and J.A. McCammon, *CRC Crit. Rev. Biochem.* **9**, 293−349 (1981).

135. M. Karplus and J.A. McCammon, *Annu. Rev. Biochem.* **52**, 263−300 (1983).

136. L. Verlet, *Phys. Rev.* **159**, 98−103 (1967).

137. C.W. Gear, *Numerical Initial Value Problems in Ordinary Differential Equations*, Prentice-Hall, Englewood Cliffs, NJ 1971.

138. J.A. McCammon, B.R. Gelin, and M. Karplus, *Nature* **267**, 585−590 (1977).

139. W.F. van Gunsteren and M. Karplus, *Biochemistry* **21**, 2259−2274 (1982).

140. M. Levitt and R. Sharon, *Proc. Natl. Acad. Sci. USA* **85**, 7557−7561 (1988).

141. W.F. van Gunsteren, H.J.C. Berendsen, J. Hermans, W.G.J. Hol, and J.P.M. Postma, *Proc. Natl. Acad. Sci. USA* **80**, 4315−4319 (1983).

142. D.H. Kitson and A.T. Hagler, *Biochemistry* **27**, 5246−5257 (1988).

143. C.K. Woodward and B.D. Hilton, *Annu. Rev. Biophys. Bioeng.* **8**, 99−127 (1979).

144. P.G. Debrunner and H. Frauenfelder, *Annu. Rev. Phys. Chem.* **33**, 283−299 (1982).

145. G. Wagner and K. Wüthrich, *Methods Enzymol.* **131**, 307−326 (1986).

146. J.A. McCammon and M. Karplus, *Annu. Rev. Phys. Chem.* **31**, 29−45 (1980).

147. M. Levitt, *Annu. Rev. Biophys. Bioeng.* **11**, 251−271 (1982).

148. S.H. Northrup, M.R. Pear, J.D. Morgan, and J.A. McCammon, **153**, 1087−1109 (1981).

149. C.B. Post, B.R. Brooks, C.M. Dobson, P. Artymiuk, J. Cheetham, D.C. Phillips, and M. Karplus, *J. Mol. Biol.* **190**, 455−479 (1986).

150. B. Mao, M.R. Pear, J.A. McCammon, and S.H. Northrup, *Biopolymers* **21**, 1979−1989 (1982).

151. T. Ichiye and M. Karplus, *Proteins: Struct. Funct. Gene.* **2**, 236−259 (1987).

152. R.M. Levy, D. Perahia, and M. Karplus, *Proc. Natl. Acad. Sci. USA* **79**, 1346−1350 (1982).

153. C.B. Post, C.M. Dobson, and M. Karplus, *Proteins: Struct. Funct. Gene.* **5**, 337−354 (1989).

154. A.T. Brünger, J. Kuriyan, and M. Karplus, *Science*, **235**, 458−460 (1987).

155. M. Fujinaga, P. Gros, and W.F. van Gunsteren, *J. Appl. Crystallogr.* **22**, 1−8 (1989).

156. R. Kaptein, E.R.P. Zuiderweg, R.M. Scheek, R. Boelens, and W.F. van Gunsteren, *J. Mol. Biol.* **182**, 179−182 (1985).

157. L. Nilsson, G.M. Clore, A.M. Gronenborn, A.T. Brünger, and M. Karplus, *J. Mol. Biol.* **188**, 455−475 (1986).

158. A.T. Brünger, G.M. Clore, A.M. Gronenborn, and M. Karplus, *Proc. Natl. Acad. Sci. USA* **83**, 3801−3805 (1986).

159. A. DiNola, H.J.C. Berendsen, and O. Edholm, *Macromolecules* **17**, 2044−2050 (1984).

160. H.C. Andersen, *J. Chem. Phys.* **72**, 2384−2393 (1980).

161. J.P. Rykaert and G. Ciccotti, *J. Chem. Phys.* **78**, 7368−7374 (1983).

162. H.J.C. Berendsen, J.P.M. Postma, W.F. van Gunsteren, A. DiNola, and J.R. Haak, *J. Chem. Phys.* **81**, 3684−3690 (1984).

163. C.L. Brooks, A.T. Brünger, and M. Karplus, *Biopolymers* **24**, 843−865 (1985).

164. A.T. Brünger, C.L. Brooks, and M. Karplus, *Proc. Natl. Acad. Sci. USA* **82**, 8458−8462 (1985).

165. A. Warshel and S. Russell, *J. Am. Soc. Chem.* **108**, 6569−6579 (1986).

166. S.H. Northrup, M.R. Pear, C.Y. Lee, J.A. McCammon, and M. Karplus, *Proc. Natl. Acad. Sci. USA* **79**, 4035−4039 (1982).

167. J.A. McCammon, C.Y. Lee, and S.H. Northrup, *J. Am. Chem. Soc.* **105**, 2232–2237 (1983).

168. M. Karplus and J.N. Kushick, *Macromolecules* **14**, 325–332 (1981).

169. W. Nadler, A.T. Brünger, K. Schulten, and M. Karplus, *Proc. Natl. Acad. Sci. USA* **84**, 7933–7937 (1987).

170. T. Thacher, S. Ganesan, A. Askar, and H. Rabitz, *J. Chem. Phys.* **85**, 3655–3673 (1986).

171. T. Thacher and H. Rabitz, *Biophys. J.* **54**, 695–704 (1988).

172. U.C. Singh and P.A. Kollman, *J. Comp. Chem.* **7**, 718–730 (1986).

173. W.L. Jorgensen, J. Chandrasekhar, J.K. Buckner, and J.D. Madura, *Ann. N. Y. Acad. Sci.* **482**, 198–209 (1986).

174. P.A. Bash, M.J. Field, and M. Karplus, *J. Am. Chem. Soc.* **109**, 8092–8094 (1987).

175. P. Cieplak, P. Bash, U.C. Singh, and P.A. Kollman, *J. Am. Chem. Soc.* **109**, 6283–6289 (1987).

176. J. Aqvist and A. Warshel, *Biochemistry* **28**, 4680–4689 (1989).

177. A. Warshel and S. Creighton, in W.F. van Gunsteren and P.K. Weiner, Eds., *Computer Simulation of Biomolecular Systems*, ESCOM, Leiden, 1989, pp. 120–135.

178. C. Zheng, C.F. Wong, J.A. McCammon, and P.G. Wolynes, *Nature* **334**, 726–728 (1988).

179. N. Metropolis, A.W. Rosenbluth, M.N. Rosenbluth, A.H. Teller, and E. Teller, *J. Chem. Phys.* **21**, 1087–1092 (1953).

180. P.J. Rossky, J.D. Doll, and H.L. Friedman, *J. Chem. Phys.* **69**, 4628–4633 (1978).

181. S.H. Northrup and J.A. McCammon, *Biopolymers* **19**, 1001–1016 (1980).

182. H. Wakana, H. Wako, and N. Saitô, *Int. J. Peptide Protein Res.* **23**, 315–323 (1984).

183. N. Gō and H.A. Scheraga, *Macromolecules* **3**, 178–187 (1970).

184. T. Noguti and N. Gō, *Biopolymers* **24**, 527–546 (1985).

185. Z. Li and H.A. Scheraga, *Proc. Natl. Acad. Sci USA* **84**, 6611–6615 (1985).

186. D.L. Beveridge and F.M. DiCapua, *Annu. Rev. Biophys. Biophys. Chem.* **18**, 431–492 (1989).

187. D.L. Beveridge and F.M. DiCapua, in W.F. van Gunsteren and P.K. Weiner, Eds., *Computer Simulation of Biomolecular Systems*, ESCOM, Leiden, 1989, pp. 1–26.

188. W.F. van Gunsteren and P.K. Weiner, Eds., *Computer Simulation of Biomolecular Systems*, ESCOM, Leiden, 1989.

189. M. Mezei and D.L. Beveridge, *Ann. N.Y. Acad. Sci.* **482**, 1–23 (1986).

190. B.L. Tembe and J.A. McCammon, *Comput. Chem.* **8**, 281–283 (1984).

191. T.P. Lybrand, I. Ghosh, and J.A. McCammon, *J. Am. Chem. Soc.* **107**, 7793–7794 (1985).

192. U.C. Singh, F.K. Brown, P.A. Bash, and P.A. Kollman, *J. Am. Chem. Soc.* **109**, 1607–1614 (1987).

193. A Warshel, F. Sussman, and G. King, *Biochemistry* **25**, 8368–8372 (1986).

194. C.F. Wong and J.A. McCammon, *J. Am. Chem. Soc.* **108**, 3830–3832 (1986).

195. S.N. Rao, U.C. Singh, P.A. Bash, and P.A. Kollman, *Nature* **328**, 551–554 (1987).

196. T.P. Lybrand and J.A. McCammon, *J. Comput.-Aided Mol. Design* **2**, 259–266 (1988).

197. W.F. van Gunsteren, *Protein Eng.* **2**, 5–13 (1988).

198. M.P. Allen and D.J. Tildesley, *Computer Simulation of Liquids*, Clarendon Press, Oxford, 1987.

199. J.C. Owicki and H.A. Scheraga, *J. Am. Chem. Soc.* **99**, 7403–7412, 7413–7418 (1977).

200. I.P. Crawford, T. Niermann, and K. Kirschner, *Proteins* **2**, 118–129 (1987).

201. W.R. Taylor, in M.J. Bishop and C.J. Rawlings, Eds., *Nucleic Acid and Protein Sequence Analysis: A Practical Approach*, IRL Press, Oxford, 1987, Chapter 1.

202. G.E. Schulz, *Annu. Rev. Biophys. Biophys. Chem.* **17**, 1–21 (1988).

203. W. Kabsch and C. Sander, *Biopolymers* **22**, 2577–2637 (1983).

204. B. Busetta and M. Hospital, *Biochim. Biophys. Acta* **701**, 111–118 (1982).

205. F.E. Cohen, R.M. Abarbanel, I.D. Kuntz, and R.J. Fletterick, *Biochemistry* **22**, 4894–4904 (1983).

206. W.R. Taylor and J.M. Thornton, *Nature* **301**, 540–542 (1983).

207. W.R. Taylor and J.M. Thornton, *J. Mol. Biol.* **173**, 487–514 (1984).

208. R.C. Garratt, W.R. Taylor, and J.M. Thornton, *FEBS Lett.* **188**, 59–62 (1985).

209. G. Deleage and B. Roux, *Protein Eng.* **1**, 289–294 (1987).

210. P.Y. Chou and G.D. Fasman, *Biochemistry* **13**, 211–222 (1974).

211. P.Y. Chou and G.D. Fasman, *Biochemistry* **13**, 222–245 (1974).

212. B. Robson, *Biochem. J.* **141**, 853–867 (1974).

213. B. Robson and R.H. Pain, *Biochem. J.* **141**, 869–882, 883–897, 899–904 (1974).

214. J.-F. Gibrat, J. Garnier, and B. Robson, *J. Mol. Biol.* **198**, 425–443 (1987).

215. A.J. Corrigan and P.C. Huang, *Comput. Prog. Biomed.* **15**, 163–168 (1982).

216. K. Nishikawa, *Biochim. Biophys. Acta* **748**, 285–299 (1983).

217. N. Rawlings, K. Ashman, and B. Wittman-Liebold, *Int. J. Peptide Protein Res.* **22**, 515–524 (1983).

218. P. Argos and J. Palau, *Int. J. Peptide Protein Res.* **19**, 380–393 (1982).

219. J. Palau, P. Argos, and P. Puigdomenech, *Int. J. Peptide Protein Res.* **19**, 394–401 (1982).

220. D. Kotelchuck and H.A. Scheraga, *Proc. Natl. Acad. Sci. USA* **62**, 14–21 (1969).

221. E.A. Kabat and T.T. Wu, *Biopolymers* **12**, 751–774 (1973).

222. K. Nagano, *J. Mol. Biol.* **75**, 401–420 (1973).

223. A.W. Burgess, P.K. Ponnuswamy, and H.A. Scheraga, *Isr. J. Chem.* **12**, 239–286 (1974).

224. P.F. Periti, *Boll. Chim. Farm.* **113**, 187–218 (1974).

225. F.R. Maxfield and H.A. Scheraga, *Biochemistry* **15**, 5138–5153 (1976).

226. F.R. Maxfield and H.A. Scheraga, *Biochemistry* **18**, 697–704 (1979).

227. T.T. Wu and E.A. Kabat, *Proc. Natl. Acad. Sci. USA* **68**, 1501–1506 (1971).

228. M. Manning, *J. Pharm. Biomed. Anal.* **7**, 1103–1119 (1990).

229. M.J.E. Sternberg, in M.J. Geisow and A.N. Barrett, Eds., *Computing in Biological Science*, Chapter 5, Elsevier Biomedical, Amsterdam, 1983.

230. W. Kabsch and C. Sander, *FEBS Lett.* **155**, 179–182 (1983).

231. W.R. Taylor, *J. Mol. Biol.* **173**, 512–514 (1984).

232. V.I. Lim, *J. Mol. Biol.* **88**, 873–894 (1974).

233. R.M. Sweet, *Biopolymers* **25**, 1565–1577 (1986).

234. J.M. Levin, B. Robson, and J. Garnier, *FEBS Lett.* **205**, 303–308 (1986).

235. K. Nishikawa and T. Ooi, *Biochim. Biophys. Acta* **871**, 45–54 (1986).

236. W.R. Taylor, *J. Theor. Biol.* **119**, 205–218 (1986).

237. W. Kabsch and C. Sander, *Proc. Natl. Acad. Sci. USA* **81**, 1075–1078 (1984).

238. M.J. Rooman and S.J. Wodak, *Nature* **335**, 45–49 (1988).

239. D. Poland and H.A. Scheraga, *Theory of Helix-Coil Transitions in Biopolymers*, Academic, New York, 1970.

240. B.H. Zimm and J.K. Bragg, *J. Chem. Phys.* **31**, 526–535 (1959).

241. W.L. Mattice, *Annu. Rev. Biophys. Biophys. Chem.* **18**, 93–111 (1989).

242. P.N. Lewis, N. Gō, D. Kotelchuck, and H.A. Scheraga, *Proc. Natl. Acad. Sci. USA* **65**, 810–815 (1970).

243. P.N. Lewis, F.A. Momany, and H.A. Scheraga, *Proc. Natl. Acad. Sci. USA* **68**, 2293–2297 (1971).

244. S. Tanaka and H.A. Scheraga, *Macromolecules* **9**, 142−159, 159−167, 168−182, 812−833 (1976).

245. S. Tanaka and H.A. Scheraga, *Macromolecules* **10**, 9−20, 305−316 (1977).

246. H. Wako, N. Saitô, and H.A. Scheraga, *J. Protein Chem.* **2**, 221−249 (1983).

247. H. Wako and N. Saitô, *J. Phys. Soc. (Japan)* **44**, 1931−1938, 1939−1945 (1978).

248. O.B. Ptitsyn and A.V. Finkelstein, Biopolymers **22**, 15−25 (1983).

249. G.E. Schulz, C.D. Barry, J. Friedman, P.Y. Chou, G.D. Fasman, A.V. Finkelstein, V.I. Lim, O.B. Ptitsyn, E.A. Kabat, T.T. Wu, M. Levitt, B. Robson, and K. Nagano, *Nature* **250**, 140−142 (1974).

250. P. Argos, J. Schwarz, and J. Schwarz, *Biochim. Biophys. Acta,* **439**, 261−273 (1976).

251. V. Biou, J.F. Gibrat, J.M. Levin, B. Robson, and J. Garnier, *Protein Eng.* **2**, 185−191 (1988).

252. K. Nishikawa and T. Ooi, *J. Biochem.* **91**, 1821−1824 (1982).

253. K. Nishikawa, Y. Kubota, and T. Ooi, *J. Biochem.* **94**, 981−995 (1982).

254. K. Nishikawa, Y. Kubota, and T. Ooi, *J. Biochem.* **94**, 997−1007 (1982).

255. B. Busetta and Y. Barrans, *Biochim. Biophys. Acta* **790**, 117−124 (1984).

256. P. Klein and C. DeLisi, *Biopolymers* **25**, 1659−1672 (1986).

257. H. Nakashima, K. Nishikawa, and T. Ooi, *J. Biochem.* **99**, 153−162 (1986).

258. M.J. Zvelebil, G.J. Barton, W.R. Taylor, and M.J.E. Sternberg, *J. Mol. Biol.* **195**, 957−961 (1987).

259. M.J. Sutcliffe, I. Haneef, D. Carney, and T.L. Blundell, *Protein Eng.* **1**, 377−384 (1987).

260. C.M. Venkatachalam, *Biopolymers* **6**, 1425−1436 (1968).

261. G.D. Rose, L.M. Gierasch, and J.A. Smith, *Adv. Protein Chem.* **37**, 1−109 (1985).

262. E.J. Milner-White and R. Poet, *TIBS* **12**, 189−192 (1987).

263. C.M. Wilmot and J.M. Thornton, *J. Mol. Biol.* **203**, 221−232 (1988).

264. I.D. Kuntz, *J. Am. Chem. Soc.* **94**, 4009−4012 (1972).

265. G.D. Rose, *Nature* **272**, 586−590 (1978).

266. H. Cid, M. Bunster, E. Arriagada, and M. Campos, *FEBS Lett.* **150**, 247−254 (1982).

267. F.E. Cohen, R.M. Abarbanel, I.D. Kuntz, and R.J. Fletterick, *Biochemistry* **22**, 266−275 (1986).

268. M.J. Rooman, S.J. Wodak, and J.M. Thornton, *Protein Eng.* **3**, 23−27 (1989).

269. E. Domany, *J. Sta. Phys.* **51**, 743−775 (1988).

270. N. Qian and T.J. Sejnowski, *J. Mol. Bio.* **202**, 865−884 (1988).

271. L.H. Holley and M. Karplus, *Proc. Natl. Acad. Sci. USA* **86**, 152−156 (1989).

272. H. Bohr, J, Bohr, S. Brunak, R.M.J. Cotterill, B. Lautrup, L. NOrskov, O.H. Olsen, and S.B. Petersen, *FEBS Lett.* **241**, 223−228 (1988).

273. P. Argos, J.K. Mohana Rao, and P.A. Hargrave, *Eur. J. Biochem.* **128**, 565−575 (1982).

274. M.J. McGregor, T.P. Flores, and M.J.E. Sternberg, *Protein Eng.* **2**, 521−526 (1989).

275. J.C. Kendrew, R.E. Dickerson, B.E. Strandberg, R.G. Hart, D.R. Davies, D.C. Phillips, and V.C. Shore, *Nature* **158**, 422−427 (1960).

276. T. Ooi, R.A. Scott, G. Vanderkooi, and H.A. Scheraga, *J. Chem. Phys.* **16**, 4410−4426 (1967).

277. M. Hashimoto and S. Arakawa, *Bull. Chem. Soc. Japan* **40**, 1698−1705 (1968).

278. E.H. Erenrich, R.H. Andreatta, and H.A. Scheraga, *J. Am. Chem. Soc.* **92**, 1116−1119 (1970).

279. C. Chothia, *J. Mol. Biol.* **75**, 295−302 (1973).

280. K. Raghavendra and V. Sasisekharan, *Int. J. Peptide Protein Res.* **14**, 326−338 (1979).

281. F.R. Salemme and D.W. Weatherford, *J. Mol. Biol.* **146**, 101−117 (1981).

282. F.R. Salemme and D.W. Weatherford, *J. Mol. Biol.* **146**, 119−141 (1981).

283. F.R. Salemme, *Prog. Biophys. Mol. Biol.* **42**, 95−133 (1983).

284. K.C. Chou, M. Pottle, G. Némethy, Y. Ueda, and H.A. Scheraga, *J. Mol. Biol.* **162**, 89−112 (1982).

285. K.C. Chou and H.A. Scheraga, *Proc. Natl. Acad. Sci. USA* **79**, 7047−7051 (1982).

286. K.C. Chou, G. Némethy, and H.A. Scheraga, *J. Mol. Biol.* **168**, 389−407 (1983).

287. K.C. Chou, G. Némethy, and H.A. Scheraga, *Biochemistry* **22**, 6213−6221 (1983).

288. S. Lifson and C. Sander, *Nature* **282**, 109−111 (1979).

289. K.C. Chou, G. Némethy, M. Pottle, H.A. Scheraga, *Biochemistry* **24**, 7948−7953 (1985).

290. K.C. Chou, G. Némethy, S. Rumsey, R.W. Tuttle, and H.A. Scheraga, *J. Mol. Biol.* **186**, 591−609 (1985).

291. M.S. Edwards, M.J.E. Sternberg, and J.M. Thornton, *Protein Eng.* **1**, 173−181 (1987).

292. K.C. Chou, G. Némethy, M. Pottle, and H.A. Scheraga, *J. Mol. Biol.* **205**, 241−249 (1989).

293. K.C. Chou, L. Carlacci, and G.M. Maggiora, *J. Mol. Biol.*, **213**, 315−326 (1990).

294. K.C. Chou, A. Heckel, G. Némethy, G. Rumsey, L. Carlacci, and H.A. Scheraga, *Proteins, 8*, 14−22 (1990).

295. C. Chothia, M. Levitt, and D. Richardson, *Proc. Natl. Acad. USA* **74**, 4130−4134 (1977).

296. C. Chothia, M. Levitt, and D. Richardson, *J. Mol. Biol.* **145**, 215−250 (1981).

297. K.C. Chou, G. Némethy, and H.A. Scheraga, *J. Phys. Chem.* **87**, 2869−2881 (1983).

298. K.C. Chou, G. Némethy, and H.A. Scheraga, *J. Am. Chem. Soc.* **106**, 3161−3170 (1984).

299. K.C. Chou, G. Némethy, S. Rumsey, R.W. Tuttle, and H.A. Scheraga, *J. Mol. Biol.* **188**, 641−649 (1986).

300. F.H.C. Crick, *Acta Crystallogr.* **6**, 689−697 (1953).

301. O.B. Ptitsyn and A.A. Rashin, *Dokl. Akad. Nauk. SSSR (Biochem. Ser.)* **213**, 473−475 (1973).

302. A.V. Efimov, *J. Mol. Biol.* **134**, 23−40 (1979).

303. T.J. Richmond and F.M. Richards, *J. Mol. Biol.* **119**, 537−555 (1978).

304. M. Gerritsen, K.C. Chou, G. Némethy, and H.A. Scheraga, *Biopolymers* **24**, 1271−1291 (1985).

305. O.B. Ptitsyn and A.A. Rashin, *Biophys. Chem.* **3**, 1−20 (1975).

306. M.K. Gilson and B. Honig, *Proc. Natl. Acad. Sci. USA* **86**, 1524−1528 (1989).

307. W.G.J. Hol, L.M. Halie, and C. Sander, *Nature* **294**, 532−536 (1981).

308. R.P. Sheridan, R.M. Levy, and F.R. Salemme, *Proc. Natl. Acad. Sci. USA* **79**, 4545−4549 (1982).

309. A.M. Lesk and C. Chothia, *J. Mol. Biol.* **174**, 175−191 (1984).

310. P. Argos, M.G. Rossman, and J.E. Johnson, *Biochem. Biophys. Res. Commun.* **75**, 83−86 (1977).

311. P.C. Weber and F.R. Salemme, *Nature* **287**, 82−83 (1980).

312. K.C. Chou, G.M. Maggiora, G. Némethy, and H.A. Scheraga, *Proc. Natl. Acad. Sci. USA* **85**, 4295−4299 (1988).

313. L. Carlacci and K.C. Chou, *Protein Eng.*, **3**, 509−514 (1990).

314. W.F. DeGrado, Z.R. Wasserman, and J.D. Lear, *Science* **243**, 622−628 (1989).

315. J. Janin and C. Chothia, *J. Mol. Biol.* **143**, 95−128 (1980).

316. F.E. Cohen, M.J.E. Sternberg, and W.R. Taylor, *J. Mol. Biol.* **156**, 821−862 (1982).

317. F.E. Cohen, M.J.E. Sternberg, and W.R. Taylor, *J. Mol. Biol.* **148**, 253−272 (1981).

318. F.E. Cohen, J. Novotny, M.J.E. Sternberg, D.G. Campbell, and A.F. Williams, *Biochem. J.* **195**, 31−40 (1981).

319. C. Chothia and J. Janin, *Biochemistry* **21**, 3955−3965 (1982).

320. B.C. Orcutt, D.G. George, M.O. Dayhoff, *Annu. Rev. Biophys. Bioeng.* **12**, 419−441 (1983).

321. N. Gō, *Annu. Rev. Biophys. Bioeng.* **12**, 183−210 (1983).

322. J. Skolnick, A. Kolinski, and R. Yaris, *Proc. Natl. Acad. Sci. USA* **86**, 1229−1233 (1989).

323. A.T. Brünger, R.L. Campbell, G. Marius Clore, A.M. Gronenborn, M. Karplus, G.A. Petsko, and M.M. Teeter, *Science* **235**, 1049−1053 (1986).

324. J. Greer, *J. Mol. Biol.* **153**, 1027−1042 (1981).

325. T.L. Blundell, B.L. Sibanda, M.J.E. Sternberg, and J.M. Thornton, *Nature* **326**, 347−352 (1987).

326. T. Blundell, D. Carney, S. Gardner, F. Hayes, B. Howlin, T. Hubbard, J. Overington, D.A. Singh, B.L. Sibanda, and M. Sutcliffe, *Eur. J. Biochem.* **172**, 513−520 (1988).

327. B. Robson, E. Platt, R.V. Fishleigh, A. Marsden, and P. Millard, *J. Mol. Graphics* **5**, 8−17 (1987).

328. M.J. Sutcliffe, I. Haneef, D. Carney, and T.L. Blundell, *Protein Eng.* **1**, 385−392 (1987).

329. N. Summers, *Chem. Des. Autom. News* **4**(12), 1, 12−17 (1989).

330. P.S. Shenkin, D.L. Yarmush, R.M. Fine, H. Wang, and C. Levinthal, *Biopolymers* **26**, 2053−2085 (1987).

331. R.H. Fine, H. Wang, P.S. Shenkin, D.L. Yarmush, and C. Levinthal, *Proteins* **1**, 342−362 (1986).

332. J. Moult and M.N.G. James, *Proteins* **1**, 142−163 (1986).

333. R.E. Bruccoleri and M. Karplus, *Biopolymers* **26**, 137−168 (1987).

334. R.E. Bruccoleri, E. Haber, and J. Novotny, *Nature* **335**, 564−568 (1988).

335. A.C.R. Martin, J.C. Cheetham, and A.R. Rees, *Proc. Natl. Acad. Sci. USA* 9268−9272 (1989).

336. M.J. Dudek and H.A. Scheraga, *J. Comp. Chem.* **11**, 121−151 (1990).

337. S.B. Needleman and C.D. Wunsch, *J. Mol. Biol.* **48**, 443−453 (1970).

338. J. Greer, *Ann, N.Y. Acad. Sci.* **439**, 44−63 (1985).

339. A.M. Lesk and C.H. Chothia, *Philos. Trans. R. Soc. London Ser. A* **317**, 345−356 (1986).

340. W.R. Taylor, *J. Mol. Graphics* **1**, 5−8 (1983).

341. B.W. Matthews and M.G. Rossman, *Methods Enzymol.* **115**, 397−420 (1985).

342. W.R. Taylor and C.A. Orengo, *Protein Eng.* **2**, 505−519 (1989).

343. W.R. Taylor and C.A. Orengo, *J. Mol. Biol.* **208**, 1−22 (1989).

344. A. Basu and M.J. Modak, *Biochemistry* **26**, 1704−1709 (1987).

345. A. Basu, R.S. Tirumalai, and M.J. Modak, *J. Biol. Chem.* **264**, 8746−8752 (1989).

346. D.L. Ollis, P. Brick, R. Hamlin, N.G. Xuang, and T.A. Steitz, *Nature* **313**, 762−766 (1985).

347. R.J. Read, G.D. Brayer, L. Jurasek, and M.N.G. James, *Biochemistry* **23**, 6570−6575 (1984).

348. M.E.P. Murphy, J. Moult, R.C. Bleackley, H. Gershenfeld, I.L. Weissman, and M.N.G. James, *Proteins* **4**, 190−204 (1988).

349. M. Bajaj and T.L. Blundell, *Annu Rev. Biophys. Bioeng.* **13**, 453−492 (1984).

350. G.E. Schulz, *Angew. Chem. Int. Ed. Engl.* **20**, 143−151 (1981).

351. T.A. Jones, *Methods Enzymol.* **115**, 157−170 (1985).

352. M.E. Snow and L.M. Amzel, *Proteins* **1**, 267−279 (1986).

353. C. Chothia and A.M. Lesk, *J. Mol. Biol.* **196**, 901−917 (1987).

354. C. Chothia, A.M. Lesk, A. Tramontano, et al., *Nature* **342**, 877−883 (1989).

355. A. Tramontano, C. Chothia, and A.M. Lesk, *Proteins* **6**, 382−394 (1989).

356. T.A. Jones and S. Thirup, *EMBO J.* **5**, 819−822 (1986).

357. J.M. Thorton, *Nature* **343**, 411−412 (1990).

358. J.W. Ponder and F.M. Richards, *J. Mol. Biol.* **193**, 775−791 (1987).

359. M.J. McGregor, S.A. Islam, and M.J.E. Sternberg, *J. Mol. Biol.* **198**, 295–310 (1987).

360. N.L. Summers and M. Karplus, *J. Mol. Biol.* **210**, 785–811 (1989).

361. H.L. Friedman, *Mol. Phys.* **29**, 1533–1543 (1975).

362. J. Novotny, R.E. Bruccoleri, and M. Karplus, *J. Mol. Biol.* **177**, 787–818 (1984).

363. F.E. Cohen and M.J.E. Sternberg, *J. Mol. Biol.* **138**, 321–333 (1980).

364. L.S. Reid and J.M. Thornton, *Proteins* **5**, 170–182 (1989).

365. J. Novotny, A.A. Rashin, and R.E. Bruccoleri, *Proteins* **4**, 19–30 (1988).

366. M. Manning and R.W. Woody, *Biochemistry* **28**, 8609–8613 (1989).

367. F.E. Cohen, T.J. Richmond, and F.M. Richards, *J. Mol. Biol.* **132**, 275–288 (1979).

368. F.E. Cohen and M.J.E. Sternberg, *J. Mol. Biol.* **137**, 9–22 (1980).

369. F.E. Cohen, M.J.E. Sternberg, and W.R. Taylor, *Nature* **285**, 378–382 (1980).

370. F.E. Cohen, P.A. Kosen, I.D. Kuntz, L.B. Epstein, T.L. Ciardelli, and K.A. Smith, *Science* **234**, 349–352 (1986).

371. F.E. Cohen and I.D. Kuntz, *Proteins* **2**, 162–166 (1987).

372. M.R. Hurle, R. Matthews, F.E. Cohen, I.D. Kuntz, A. Toumadje, and W.C. Johnson, Jr., *Proteins* **2**, 210–224 (1987).

373. O.B. Ptitsyn, A.V. Finkelstein, and P. Falk-Bendzko, *FEBS Lett.* **101**, 1–5 (1979).

374. O.B. Ptitsyn and A.V. Finkelstein, *Quart. Rev. Biophys.* **13**, 339–386 (1980).

375. T. Kikuchi, G. Némethy, and H.A. Scheraga, *J. Protein Chem.* **7**, 473–490 (1988).

376. I.D. Kuntz, G.M. Crippen, P.A. Kollman, and D. Kimelman, *J. Mol. Biol.* **106**, 983–994 (1976).

377. I.D. Kuntz, G.M. Crippen, and P.A. Kollman, *Biopolymers* **18**, 939–957 (1979).

378. S. Tanaka and H.A. Scheraga, *Proc. Natl. Acad. Sci. USA* **72**, 3802–3806 (1975).

379. N. Gō and H. Taketomi, *Proc. Natl. Acad. Sci. USA* **75**, 559–563 (1978).

380. H. Abe and N. Gō, *Biopolymers* **20**, 1013–1031 (1981).

381. H. Taketomi, F. Kano, and N. Gō, *Biopolymers* **27**, 527–559 (1988).

382. J. Skolnick and A. Kolinski, *Annu. Rev. Phys. Chem.* **40**, 207–235 (1989).

383. L. Piela and H.A. Scheraga, *Biopolymers* **26**, S33–S58 (1987).

384. D.R. Ripoll and H.A. Scheraga, *Biopolymers* **27**, 1283–1303 (1988).

385. M. Levitt and A. Warshel, *Nature* **253**, 694–698 (1975).

386. M. Levitt, *J. Mol. Biol.* **104**, 59–107 (1976).

387. J.A. McCammon, S.H. Northrup, M. Karplus, and R.M. Levy, *Biopolymers* **19**, 2033–2045 (1980).

388. G.M. Crippen and V.N. Viswanadhan, *Int. J. Peptide Protein Res.* **25**, 487–509 (1985).

389. G.M. Crippen and P.K. Ponnuswamy, *J. Comp. Chem.* **8**, 972–981 (1987).

390. E. Platt and B. Robson, in M.J. Geisow and A.N. Barrett, Eds., *Computing in Biological Science*, Elsevier Biomedical, Amsterdam, 1983.

391. N. Saitô, T. Shigaki, Y. Kobayashi, and M. Yamamoto, *Proteins* **3**, 199–207 (1988).

392. F.M. Richards, *Annu. Rev. Biophys. Bioeng.* **6**, 151–176 (1977).

393. A.T. Hagler and B. Honig, *Proc. Natl. Acad. Sci. USA* **75**, 554–558 (1978).

394. G. Némethy and H.A. Scheraga, *Quart. Rev. Biophys.* **10**, 239–252 (1977).

395. W.A. Hendrickson and M.M. Teeter, *Nature* **290**, 107–113 (1981).

396. S. Miyazawa and R.L. Jernigan, *Biopolymers* **21**, 1333–1363 (1982).

397. A. Sikorski and J. Skolnick, *Biopolymers* **28**, 1097–1113 (1989).

398. A. Sikorski and J. Skolnick, *Proc. Natl. Acad. Sci. USA* **86**, 2668–2672 (1989).

399. J. Skolnick, A. Kolinski, and R. Yaris, *Proc. Natl. Acad. Sci. USA* **85**, 5057–5061 (1988).

400. M.S. Friedrichs and P.G. Wolynes, *Science* **246**, 371–373 (1989).

401. J.D. Bryngelson and P.G. Wolynes, *Proc. Natl. Acad. Sci USA* **84**, 7524–7528 (1987).

402. J. Monod, J.P. Changeux, and F. Jacob, *J. Mol. Biol.* **6**, 306–329 (1963).

403. J. Monod, J. Wyman, and J.P. Changeux, *J. Mol. Biol.* **12**, 88–118 (1965).

404. E.R. Kantrowitz and W.N. Lipscomb, *Science* **241**, 669–674 (1988).

405. D. Barford and L.N. Johnson, *Nature* **340**, 609–616 (1989).

406. E.J. Goldsmith, S.R. Sprang, R. Hamlin, N.H. Xuong, and R.J. Fletterick, *Science* **245**, 528–532 (1989).

407. Y. Shirakihara and P.R. Evans, *J. Mol. Biol.* **204**, 973–994 (1988).

408. T. Schirmer and P.R. Evans, *Nature* **343**, 140–145 (1990).

409. A. Warshel, *Proc. Natl. Acad. Sci. USA* **74**, 1789–1793 (1977).

410. A. Szabo and M. Karplus, *J. Mol. Biol.* **72**, 163–197 (1972).

411. A.W.M. Lee and M. Karplus, *Proc. Natl. Acad. Sci. USA* **80**, 7055–7059 (1983).

412. B.R. Gelin and M. Karplus, *Proc. Natl. Acad. Sci. USA* **74**, 801–805 (1977).

412. J. Baldwin and C. Chothia, *J. Mol. Biol.* **129**, 175–220 (1979).

413. B.R. Gelin, A.W.M. Lee, and M. Karplus, *J. Mol. Biol.* **171**, 489–559 (1983).

414. M.P. Glackin, M.P. McCarthy, D. Mallikarachchi, J.B. Matthew, and N.M. Allewell, *Proteins: Struct. Funct. Gene.* **5**, 66–77 (1989).

415. A.G. Tomasselli, J. Hui, J. Fisher, H. Zurcher-Neely, I.M. Reardon, E. Oriaku, F.J. Kézdy, and R.L. Heinrikson, *J. Biol. Chem.* **264**, 10041–10047 (1989).

416. A.G. Gilman, *Annu. Rev. Biochem.* **56**, 615–649 (1987).

417. L.M. Peerey and N.M. Kostic, *Biochemistry* **28**, 1861–1868 (1989).

418. F.R. Salemme, *J. Mol. Biol.* **102**, 563–568 (1976).

419. J.J. Wendoloski, J.B. Matthew, P.C. Weber, and F.R. Salemme, *Science* **238**, 794–797 (1987).

420. P.C. Weber and G. Tollin, *J. Biol. Chem.* **260**, 5568–5573 (1985).

421. J.T. Hazzard, M.A. Cusanovich, J.A. Tainer, E.D. Getzoff, and G. Tollin, *Biochemistry* **25**, 3318–3328 (1986).

422. T.L. Poulos and A.G. Mauk, *J. Biol. Chem.* **258**, 7369–7373 (1983).

423. T.L. Poulos and J. Kraut, *J. Biol. Chem.* **225**, 10322–10330 (1980).

424. S.J. Shire. G.I.H. Hamania, and F.R.N. Gurd, *Biochemistry* **13**, 2967–2974 (1974).

425. J.B. Matthew and F.M. Richards, *Biochemistry* **21**, 4989–4999 (1982).

426. J. Warwicker, *J. Mol. Biol.* **206**, 381–396 (1989).

427. B. Waldmeyer, R. Bechtold, H.R. Bosshard, and T.L. Poulos, *J. Biol. Chem.* **257**, 6073–6076 (1982).

428. J.B. Mauro, L.A. Fishel, J.T. Hazzard, T.E. Meyer, G. Tollin, M.A. Cusanovich, and J. Kraut, *Biochemistry* **27**, 6243–5256 (1988).

429. R. Rieder and H.R. Bosshard, *J. Biol. Chem.* **255**, 4732–4739 (1980).

430. T.L. Poulos, S. Sheriff, and A.J. Howard, *J. Biol. Chem.* **262**, 13881–13884 (1987).

431. S.H. Northrup, J.O. Boles, and J.C.L. Reynolds, *Science* **241**, 67–70 (1987).

432. S.H. Northrup, J.O. Boles, and J.C.L. Reynolds, *J. Phys. Chem.* **91**, 5991–5998 (1987).

433. S. Rachovsky and H.A. Scheraga, *Acc. Chem. Res.* **17**, 209–214 (1984).

Protein–Ligand Interaction as a Method to Study Surface Properties of Proteins

Tsutomu Arakawa, Yoshiko A. Kita, and Linda O. Narhi, *Amgen Inc., Thousand Oaks, California*

1. INTRODUCTION

Protein molecules are folded into a distinct three-dimensional conformation, resulting in a unique distribution of amino acid residues for a particular

Methods of Biochemical Analysis, Volume 35: Protein Structure Determination, Edited by Clarence H. Suelter.
ISBN 0–471–51326–1 © 1991 John Wiley & Sons, Inc.

protein. Those amino acids buried in the interior of the molecules and hence forming the core structure affect the stability and dynamics of the protein. On the other hand, those residues located on the protein surface play important roles both in the biological functions and in the biophysical and biochemical properties of the molecules. For biologically active proteins, the surface is involved in receptor binding, antigen—antibody recognition, enzyme reactions, and protein—protein interactions. The protein surface also determines such physical and chemical properties as solubility, stability, and self-association. Small differences in the protein surface could affect these properties dramatically.

Recent developments in recombinant protein engineering techniques make it possible to create a variety of mutant proteins with single or multiple amino acid substitutions. These substitutions can result in changes in the overall folding of the molecule, or in changes in the local conformation of the protein, both of which can consequently alter the surface properties of the mutant protein. Alternatively, these substitutions can simply result in changes in the chemical properties of the protein surface without affecting the folding. Therefore, these mutant proteins might display altered biological and physicochemical properties; for example, properties such as solubility, stability, and flexibility may be altered by the mutations introduced into the natural proteins. Understanding the overall surface properties of the various forms of the proteins would provide important information.

The protein surface, both in the native and unfolded states, is chemically heterogeneous, carrying positive and negative charges and hydrophobic and hydrophilic regions. These regions interact with solvent components differently, causing them to redistribute over the surface of the protein. A variety of methods can be used to characterize the properties of a protein surface; ligand binding measurement technique is one of these. For example, a fluorescent ligand such as 1-anilino-8-naphthalensulfonate (ANS) has been used to probe the exposure of hydrophobic surfaces in proteins (1—6). For strong ligand binding no complication arises from solvent hydration of proteins, since equilibrium measurements are performed at low ligand concentrations. Alternatively, it may be possible to use a ligand with weak binding affinity to proteins. With these weakly interacting ligands it is necessary to perform the interaction measurements at high ligand concentrations. Using high ligand concentrations requires extremely precise techniques to determine the concentration of ligand in the protein solution in equilibrium with the dialyzing solvent. Development of the high-precision densimeter made such measurements possible for a variety of ligands (7, 8). When ligand concentrations are high in equilibrium dialysis experiments, the binding parameters reflect all of the interactions that occur between the protein surface and the ligand or the water in the hydration shell. To distinguish these from strong binding, these parameters are referred to as a preferential (or selective) interactions (9—12). Differences between strong and weak ligand interactions and, hence, between absolute and preferential interaction are described in Section 2.

Preferential interaction parameters have been determined for a number of ligands to explain the effects of the ligands on the various properties of proteins, such as solubility, stability, and self-association (reviewed in references 13−17). These studies show that the preferential interaction of compounds that strongly stabilize or denature proteins is primarily determined by the surface area of the proteins; the interactions are relatively independent of their chemical natures. However, the interactions of some compounds are strongly affected by the chemical nature of the proteins. These interactions could be explained using the overall properties of the proteins involved. Although preferential ligand interaction measurements are not readily accessible to a simple experimental technique, these measurements can provide information about overall properties of the protein molecules which are difficult to obtain from other techniques.

Another technique that can be used to characterize the properties of protein surfaces is hydrophobic interaction chromatography. This technique is used to compare the relative hydrophobicity of protein surfaces by monitoring the avidity of protein binding to hydrophobic ligands attached to inert gel matrices. The protein−ligand interaction can be enhanced by manipulating solvent conditions, or by the type of hydrophobic ligand attached to the gel matrix. Hydrophobic interaction chromatography is done under nondenaturing conditions, with ligand type, ligand concentration, support matrix, and solvent conditions chosen so as not to effect the native protein conformation. This method also provides information on the surface characteristics of proteins that is not readily available using other techniques; however, the total hydrophobic content of the protein cannot be determined.

2. THEORY

2.1. Ligand Binding

Ligand binding to macromolecules can be measured upon equilibration as the difference in ligand concentration between protein solution and solvent. Fig. 1 displays examples of ligand distribution inside and outside the dialysis bag (18). Ligand binding is expressed with a parameter using the notation of Scatchard (19) and Stockmeyer (20); i.e., water = component 1, protein = component 2, and ligand = component 3,

$$\left(\frac{\partial g_3}{\partial g_2}\right)_{T,\mu_1,\mu_3}$$

where g_i is the concentration of component i in grams per gram water, T is the absolute (Kelvin) temperature, and μ_i is the chemical potential of component i. This parameter is related to the total binding of ligand (component 3) and water (component 1) in a three-component solution as (21)

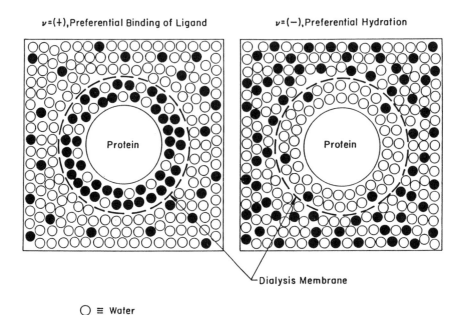

Fig. 1. Schematic diagram of preferential interaction. [Taken from Na and Timasheff (18). Reproduced with permission from Academic Press.]

$$\left(\frac{\partial g_3}{\partial g_2}\right)_{T,\mu_1,\mu_3} = A_3 - g_3 A_1 \tag{1}$$

where A_1 and A_3 are the total bindings of water and ligand, respectively. For strong ligand binding where binding saturates at low ligand concentration (g_3), the second term becomes negligible and the binding parameter is equal to the total ligand binding. For ligands that bind weakly to proteins and hence require higher ligand concentrations, the second term becomes significant. Therefore, ligand binding obtained at high concentrations reflects all of the interactions that occur between protein molecules and water or ligand.

The parameters, $(\partial g_3/\partial g_2)_{T,\mu_1,\mu_3}$, could be either positive or negative. Figure 1 illustrates these two cases. The left side of the figure illustrates the usual case in which the ligand concentration is greater in the protein solution than in the dialyzing solvent at equilibrium; that is, the protein preferentially binds the ligand. This gives a positive value of $(\partial g_3/\partial g_2)_{T,\mu_1,\mu_3}$. On the right side, the ligand concentration is lower in the protein solution; that is, the ligand is deficient in the protein domain relative to its concentration in the bulk. This gives a negative value of $(\partial g_3/\partial g_2)_{T,\mu_1,\mu_3}$. In the latter situation, water is in excess in the protein domain over its concentration in the bulk, in other words, the protein is preferentially hydrated. The preferential ligand interaction parameter, $(\partial g_3/\partial g_2)_{T,\mu_1,\mu_3}$ can be converted to the preferential hydration parameter (22) by

$$\left(\frac{\partial g_1}{\partial g_2}\right)_{T,\mu_1,\mu_3} = -\frac{1}{g_3}\left(\frac{\partial g_3}{\partial g_2}\right)_{T,\mu_1,\mu_3} \tag{2}$$

It is evident from this relation that when ligand binding is small and over-whelmed by hydration, $(\partial g_3/\partial g_2)_{T,\mu_1,\mu_3}$ becomes negative and $(\partial g_1/\partial g_2)_{T,\mu_1,\mu_3}$ becomes positive. Since $(\partial g_3/\partial g_2)_{T,\mu_1,\mu_3}$ or $(\partial g_1/\partial g_2)_{T,\mu_1,\mu_3}$ reflect various types of weak as well as strong interactions between the ligand and the proteins or water, it is impossible to quantitatively correlate the property of the ligand with its binding to the protein. However, the preferential interaction can indicate qualitatively the average property of the protein surface, which should determine the overall interaction between the ligand and the protein.

2.2. Hydrophobic Interaction Chromatography

Hydrophobic interaction chromatography (HIC) has been developed as a powerful technique for protein purification (23−26). This technique utilizes the hydrophobic nature of proteins. Binding of the proteins occurs between the hydrophobic groups on the protein and a hydrophobic ligand attached to a gel matrix such as agarose or synthetic polymers. Although the ligands used in this chromatography are relatively hydrophobic, they do not usually induce gross conformational changes in the proteins. This is due to a relatively low ligand density covalently linked to a stationary phase, which inhibits cooperative binding of the ligand to the proteins. Due to a limited number of contacts between ligand and protein, and to the weak interaction between the hydro-phobic region of the protein and the hydrophobic ligand, it is often necessary to enhance the binding between the protein and the ligand to achieve protein binding to the column. Binding is enhanced by incorporating a salting-out salt, usually $(NH_4)_2SO_4$, at high concentrations; the bound protein is sub-sequently eluted by decreasing the salt concentration. Although the system contains many variables, the salt concentration required to elute the proteins can be used to examine the hydrophobic character of the native proteins. Since the proteins are in the native state, the parameter reflects only the hydrophobic nature of the native protein surface, and has little correlation with the overall hydrophobicity which can be estimated from the amino acid compositions, as done by Bigelow (27). The salts such as $(NH_4)_2SO_4$ do not alter the intrinsic binding between the protein and the ligand, but simply add binding energy as briefly described below (28).

Let us describe the binding of the protein to the column by

$$L + P \overset{K}{=} LP \tag{3}$$

where L is the ligand of the column, P is the protein, and LP is the protein bound to the column. The equilibrium constant, K_w, in the absence of salt, may be given by

$$-RT \ln K_w = \Delta G_w \qquad (4)$$

where ΔG_w is the intrinsic free-energy change resulting from the protein binding to the column and is determined by the affinity between the protein and the ligand in the absence of salts. In the presence of salts, K is related to the free-energy change of the binding, ΔG, by

$$-RT \ln K = \Delta G = \Delta G_w + \Delta G_{sol} \qquad (5)$$

ΔG_{sol} is the free-energy change resulting from the preferential interaction of the salts with the proteins and ligand. Equation 5 indicates that protein binding to a hydrophobic column is a function of the intrinsic affinity of proteins with the column and of the solvent interactions with the proteins and the column. However, the preferential interaction and hence ΔG_{sol} are independent of the proteins involved for a strong salting-out salt such as $(NH_4)_2SO_4$. In other words, the binding constant K is determined by the intrinsic binding free energy ΔG_w dependent on the protein and by a constant, protein-independent value of ΔG_{sol}. This provides the basis for HIC. The value of ΔG_{sol} can be manipulated by using different salting-out salts; $(NH_4)_2SO_4$ and Na_2SO_4 result in larger values, strongly enhancing the binding; $MgSO_4$ and $NaCH_3COO$ result in intermediate values and NaCl results in a lower value, only weakly enhancing the binding.

3.   APPLICATION

3.1.   Techniques

Preferential protein−solvent interactions can be determined by measuring the concentration of ligand in the protein solution that is in equilibrium with that in the bulk solvent. This equilibration is conventionally attained by dialysis or gel filtration (7, 29). Since small differences in the ligand concentration must be measured against a high background, extremely sensitive techniques are required. Density measurements using an Anton-Parr precision density meter meet this requirement for those ligands that have partial specific volumes far different from that of water (e.g., <0.7 mL/g). When ligands have a different refractive index from water, as is true of many organic solvents, differential refractometry can be used for direct determination of the difference in refractive index between the protein solution and dialyzing solution. Detailed descriptions for these two techniques have been given elsewhere (7, 30, 31).

Much less accurate techniques can be employed for strong binding ligands, since the binding occurs at low ligand concentrations (less background).

Techniques employed include radioactivity, absorbance, fluorescence intensity, or colorimetric measurements. When the quantum yield of ligands is extremely different in the bound state from the free state, the fluorescence intensity directly reflects the amounts of ligand bound to the proteins. It should be emphasized here that these ligands, which bind strongly to proteins, may also bind nonspecifically to many column matrices and hence make it impossible to use gel filtration to attain equilibration.

Hydrophobic interaction chromatography is a very simple technique. A hydrophobic column is first brought to equilibrium in a high salt solution, and the protein is then applied to the column in the presence of the salt. The bound protein is subsequently eluted either with a gradient decreasing in salt concentration or isocratically.

3.2.   Strong Ligand Interaction

The most frequently used molecules in this category are fluorescent compounds, which bind to hydrophobic regions of native proteins. The determination of the amount of binding of these ligands is usually based on the property that the fluorescence yield is negligible for unbound ligand relative to that of bound ligand; the emission wavelength reflects the hydrophobicity of the protein region to which it binds. As one can see in the following examples, these ligands can bind to proteins at concentrations of the order of 10^{-3} mol/L. Under these conditions the measured binding corresponds, within experimental error, to the actual binding; in other words, the water binding term A_1 in Eq. 1 is negligible relative to the ligand binding term, A_3.

The fluorescent molecules most frequently employed as described above are bis-ANS [bis(8-anilinonaphthalene-1-sulfonate)] and ANS(anilinonaphthalene-sulfonate). The experiments performed with fluorescent ligands can be divided into two categories. In the first type of experiment, they are used to follow changes in the protein surface resulting from perturbations of the native protein structure due to changes in its environment. In the second type of experiment, they are used to characterize the protein surface of the native molecule.

A.   CONFORMATIONAL CHANGES

An example of the first type of experiment involves using bis-ANS as a probe for the effect of time and temperature on tubulin decay (32). When increasing concentrations of bis-ANS are added to tubulin, the fluorescence intensity increases hyperbolically between zero and 20 μM ANS and then increases sigmoidally at higher concentrations. Scatchard analysis of the hyperbolic portion is consistent with the presence of a single, high-affinity binding site (K_d = 2 μM); this site is responsible for inhibition of microtubule assembly. Using the double titration method (33), the binding at higher ANS concentrations can be attributed to six low affinity sites (K_d = 16 μM).

The binding parameters themselves do not change with time. However, when tubulin is incubated at 37 °C in the presence of bis-ANS the fluorescence increases gradually, reaching a maximum in about 2 h. The tubulin can be incubated for various lengths of time at 37 °C, and then mixed with bis-ANS with similar results, indicating that the bis-ANS itself is not responsible for tubulin decay, and that the increased fluorescence intensity is due to changes in the tubulin resulting in increased bis-ANS binding.

These experiments are done under conditions where the secondary structure remains constant, as indicated by circular dichroism spectroscopy, and no changes can be detected in the ultraviolet spectrum. Thus, bis-ANS binding is one of the few techniques that can be used to follow tubulin decay. The effect of antimitotic drugs on the binding of bis-ANS is shown in Fig. 2. The primary binding site remains unaffected as indicated by the initial increase in fluorescence in all samples. However, the effects of the four drugs on the binding of bis-ANS to the secondary binding sites of tubulin parallel their effects on the alkylation of tubulin, and also correlate with the ability of the drugs to induce polymerization of tubulin into nonmicrotubule structures. These experiments demonstrate that the decay of tubulin's ability to form microtubules is associated with the exposure of hydrophobic areas on the tubulin molecule. The tubulin−tubulin interaction sites lose their specificity, but increase their hydrophobicity, resulting in slow aggregation, and bis-ANS can be used to detect this change in the protein surface.

The fluorescent compounds can also be used to characterize changes in protein surfaces during thermal denaturation. An example of this is the use of ANS to probe the thermal denaturation of antithrombin III, and its stabilization by lyotropic anions (34).

When antithrombin III is heated in the presence of ANS, the weak fluorescence initially decreases with increasing temperature to 50 °C, at which temperature the fluorescence increases dramatically, reaches a maximum, and then again decreases with increasing temperature. As shown in Fig. 3, this allows one to define T_d. The increase in fluorescence correlates well with the thermally induced loss of activity. Duplicates heated in the absence of ANS and then mixed with ANS after reaching temperature give the same T_d, demonstrating that ANS itself is not effecting the denaturation reaction. The stabilizing effect of citrate concentration on antithrombin III during thermal denaturation can be followed by ANS fluorescence, yielding a linear relationship between citrate concentration and T_d. Heparin similarly stabilizes antithrombin III to heating, but in this case the relationship between T_d and stabilizer concentration is saturable, indicating that heparin stabilizes antithrombin III by binding to the protein, a result that is in agreement with the change in intrinsic antithrombin III fluorescence that accompanies heparin addition. A variety of neutral salts were also tested for their ability to stabilize antithrombin III during thermal denaturation. The results indicate that the effects of these molecules on antithrombin III correlate well with their position in the lyotropic series: Citrate, phosphate, and sulfate are strong stabilizers, while iodide and thiocyanate are potent destabilizers of the protein. Ethylenediaminetetracetic

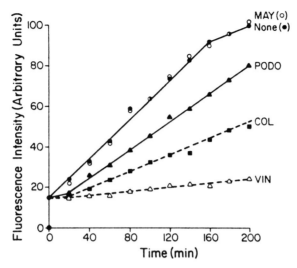

Fig. 2. Effect of drugs on the binding of bis-ANS to tubulin. May, maytasin; PODO, podophylotoxin; COL, colchicine; VIN, vinblastine. [Taken from Prasad et al. (32). Reproduced with permission from the American Chemical Society.]

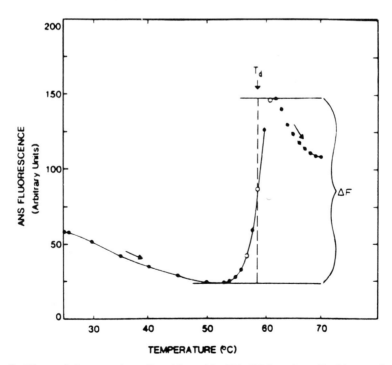

Fig. 3. Thermal denaturation of antithrombin III. [Taken from Bushby et al. (34). Reproduced with permission from the American Society of Biological Chemists.]

acid is also a potent stabilizer. While aggregation did occur during heating, it lags behind inactivation, and appears to be a consequence rather than a cause of thermally induced denaturation. In these experiments, ANS provides a convenient and sensitive probe into subtle conformational changes, and permits the identification of conditions that stabilize the protein to pasteurization conditions, increasing the safety of its use as a clinical therapeutic agent.

Changes in ANS fluorescence were also used to follow changes in the structure of the myosin−ANS complex during thermal denaturation (6). The role of hydrophobic interactions in the aggregation, filament formation, and ATPase activity of myosin has been the subject of speculation for some time. When the fish myosin−ANS complex is heated at 1 °C/min the results are very similar to those obtained with antithrombin III; that is, there is a large increase in fluorescence intensity with no change in emission wavelength between 27 and 41 °C, which is unique compared to the behavior of other protein−ANS complexes (albumin and lysozyme) and to ANS in organic solvents. At low ANS concentrations, myosin contains a single high-affinity binding site, and no thermal transition is seen. At higher concentrations of ANS (32 μM), where the thermal transition is seen, Scatchard analysis reveals that myosin contains numerous (at least 200) low-affinity binding sites with a multiplicity of binding. The number of sites and affinity for ANS remain unchanged with increasing temperature. The myosin−ANS complex was then heated in several different salts, as shown in Fig. 4. The results were analyzed using the procedure of Schrier and Schrier (35), and revealed that the change in enthalpy per residue during the thermal transition is approximately 500 cal/mol-deg. This suggests that only a small number of residues are contributing to the change in fluorescence, and that the change involved is primarily one of exposure of hydrophobic residues. From these experiments, and previously obtained analysis, it was postulated that the myosin molecule undergoes a thermally induced conformational change that exposes hydrophobic surfaces. This increase in hydrophobicity facilitates increased polymerization of myosin oligomers, and results in the increase in fluorescence. In this experiment the use of ANS results in a better understanding of the thermodynamics of the thermal denaturation of myosin, as well as a possible explanation of the reaction itself.

The fluorescent dyes can also be used to follow changes in protein structure with reduction. This experiment was done with an engineered form of interleukin-2, IL-2(Ala-125) (5). When this protein is reduced with 100 mM DTT (dithiothreitol) at pH 4.0 a large increase in fluorescence intensity is seen, along with a shift in the emission maximum from 467 to 462 nm. By following the change in fluorescence with time, after DTT addition, it is possible to monitor the time course of reduction of IL-2(Ala-125). The presence of ANS during incubation does not affect the rate of reduction by DTT, which is complete by 6 h. When these results are coupled with the near- and far-UV circular dichroism spectra, it is possible to conclude that reduction increases

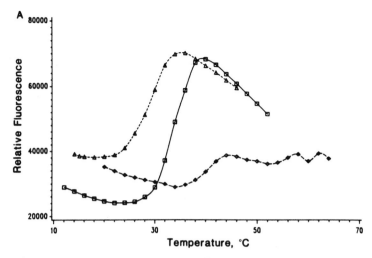

Fig. 4. Thermal denaturation of myosin ANS complex. □, in 0.6 M KCl, 50 mM potassium phosphate pH 6.5; ◊, in 0.5 M sodium acetate, 0.6 M KCl, 50 mM potassium phosphate pH 6.5; △, in 0.5 M sodium bromide, 0.6 M KCl, 50 mM potassium phosphate pH 6.5. [Taken from Wicker and Knopp (6). Reproduced with permission from Academic Press.]

the hydrophobicity of the protein and removes the apparent tertiary structure, but does not effect the secondary structure of the protein.

Studying the effect of pH on the ANS-IL-2(Ala-125) fluorescence spectra shows that ANS binding increases with decreasing pH from 4.0 to 2.0. Combining fluorescence spectral data with circular dichroic spectra in the same solvents allows Arakawa and Kenney (5) to conclude that the secondary structure is maintained in all solvents, while the IL-2(Ala-125) unfolds in acid pH; the dye binding reveals that it unfolds more extensively as the pH is lowered.

We have recently examined the effect of pH on the structure of human recombinant G-CSF (granulocyte-colony stimulating factor) using ANS binding. While the secondary structure remains constant from pH 7.5 to pH 2.5, with perhaps some enhancement of α-helix at the lowest pH values, the tertiary structure appears to change. ANS binding indicates that the maximum number of hydrophobic sites are exposed at pH 2.5, with ANS binding in the order of pH 2.5 > pH 3.1 > pH 4.5 > pH 6.1 = pH 7.5. In both of these cases ANS can be used in conjunction with other techniques to monitor protein surface changes due to solvent conditions. The interpretation of the binding of ANS with changes in pH, however, is qualitative, due to the charges of the dye and the protein, the latter changing with changing pH.

B. HYDROPHOBICITY OF NATIVE PROTEINS

Ligands such as the fluorescence dyes can also be used to characterize the surface of proteins under nondenaturing conditions. This has proven particularly useful when comparing related proteins such as analogous proteins from different species, engineered proteins involving point mutations or deletions, and in studying the association of subunits into proteins.

The use of ANS to study the protein surface of related proteins from different species is illustrated in the following example. Interferon-γ is a species-specific cytokine, as expected from the low degree of sequence homology between species. Human interferon-γ and murine interferon-γ, while performing similar functions, have little sequence homology. Murine interferon-γ contains a greater number of hydrophobic residues. That this results in very different surface hydrophobicity is demonstrated by their different ANS binding characteristics, shown in Fig. 5. Murine interferon-γ is a much more hydrophobic molecule; the fluorescence intensity of the ANS in the presence of the murine protein is much greater than the fluorescence intensity of ANS when mixed with the human protein, indicating that there are many more sites for ANS binding on the murine protein. The emission maximum of the murine interferon-γ—ANS complex was 465 nm, while that of the human interferon-γ—ANS complex was 480 nm, demonstrating that the binding sites for ANS on the murine protein are also in more hydrophobic regions (L.O. Narhi and T. Arakawa, unpublished). Thus, ANS can indicate surface differences between proteins that perform the same functions in different species.

ANS can also be used to examine differences in proteins that have been engineered to differ in specific regions. Recombinant Pst (porcine somatotropin) was cloned in three forms, the intact 22K protein, which is analogous to human somatotropin, a 21K form involving the deletion of amino acids 32–38, and a 20K form involving the deletion of amino acids 32–46 (T. Arakawa and T.D. Bartley, unpublished). The recombinant 22K Pst and

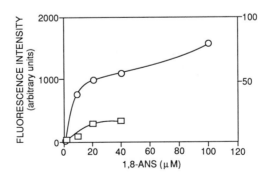

Fig. 5. Binding of ANS to human (○) and murine (□) interferon-γ. Result for human interferon-γ refers to the right-side ordinate and those for murine protein refer to the left-side ordinate.

natural pituitary Pst have equivalent activities in vivo, the 21K Pst has enhanced activity, and the 20K Pst has no activity at all. ANS binding (Fig. 6) reveals that while the recombinant 22K Pst has essentially identical fluorescence enhancement as the natural protein upon mixing with ANS, much stronger enhancement is observed for the 21K Pst. The emission maximum is also shifted, from 467 nm for the natural protein to 464—465 nm for the 21K protein. This shift indicates that the deletion mutation results in proteins with more ANS binding sites, and the hydrophobicity of the binding sites is also increased. This increased hydrophobicity of the 21K form could be due to different folding or simply altered protein surface by deletion, or both. The 20K protein appears to be even more hydrophobic than the 21K Pst. This is due to its inability to fold into a distinct tertiary structure as demonstrated by circular dichroism.

ANS can also occasionally detect differences in the surface properties of proteins differing only in a few amino acids. An example of this is shown in Fig. 7. Several different analogs of subtilisin, differing by a few point mutations, were mixed with ANS at identical concentrations (L.O. Narhi and T. Arakawa, unpublished). Subtilisin 124 and 127 differ only in a single Val to Cys point mutation, yet this single change results in increased ANS binding. Subtilisin 130 and 131 differ only in a single Asn, which has been changed to Asp in 131,

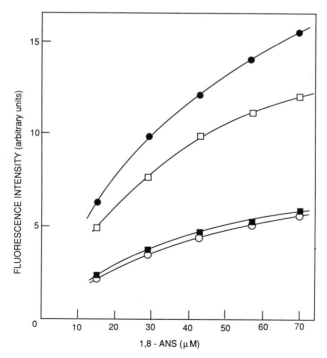

Fig. 6. Binding of ANS to recombinant porcine somatotropin. ■, pituitary Pst; ○, 22K Pst; □, 21K Pst; ●, 20K Pst.

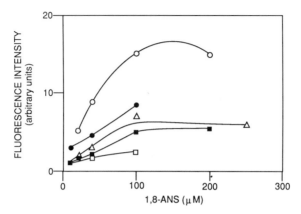

Fig. 7. Binding of ANS to subtilisins. ○, Novo; △, subtilisin 131; ●, subtilisin 124; □, subtilisin 127; ■, subtilisin 130.

and in this case the hydrophobicity of the protein surface has not been effected. Subtilisin Novo is 75% homologous to aprA-subtilisin, the parent subtilisin of these analogs, and appears to have an identical structure from molecular modeling. Yet this protein shows very different ANS binding behavior, and also a slightly lower emission wavelength maximum, indicating that it has a more hydrophobic surface. In this case as well, ANS binding can be affected by both altered surface and folding by mutations, although their enzymatic activity suggests that the folding has been maintained.

ANS is also useful for dissecting the association of the α and β subunits of several human glycoprotein hormones. While the monomeric subunits do not bind ANS, the intact hormones do. This makes ANS a useful tool in detecting the presence of the native hormone, and in studying the association–dissociation kinetics of the system (e.g., reference 36). In the presence of ANS, the native hormones form dimers and tetramers, enhancing ANS fluorescence. The rate of recombination of subunits as determined from ANS fluorescence and recovery of activity in bioassay (in the absence of ANS) are exactly the same, indicating ANS itself does not influence the rate of recombination of the subunits. The recombination reaction occurs independent of ANS concentration as well, further proof that ANS is not accelerating the rate of recombination.

Having established that the enhancement of ANS fluorescence is a valid indication of the recombination of HCG (human chorionic gonadotropin), this procedure was then used to examine the effect of increasing subunit concentrations on the rates of recombination (36). When subunit concentrations are varied, the final fluorescence values versus subunit concentration are linear, indicating that the proportionality between ANS fluorescence and hormone concentration is maintained. Ingham et al. (36) find that the initial rates of recombination are also proportional to the subunit concentration at low concentrations, while at the highest concentration ranges a threefold increase in concentration causes only a twofold change in initial rate. These data suggest that the free α and β subunits are in equilibrium with an $\alpha\beta$ oligomer which then undergoes a conformational change to form the active $\alpha\beta$ hormone. At low concentrations the first step is rate limiting, while at higher concentrations the second step becomes rate limiting. Because ANS does not effect the rate of recombination, the second step must not be an equilibrium; ANS must only interact with the active conformation of the $\alpha\beta$ oligomer. In addition to helping to define the mechanism of recombination of subunits, these studies indicate that the native $\alpha\beta$ conformation is more hydrophobic than either the individual subunits or the postulated $\alpha\beta$ intermediate(s). In this case, ANS is not only used to compare the surface characteristics of native proteins, but proves to be a powerful tool for detecting the presence of active hormone. In many cases, it is used as a quick assay for active $\alpha\beta$ hormone, when the time involved for bioassay is prohibitive.

The above examples demonstrate the use of strong binding ligands to characterize protein surfaces. The most common ligands used for this purpose, the fluorescent probes ANS and bis-ANS, have been used to follow changes in the hydrophobicity of the protein surface with changes in solvent conditions, temperature, and so on, or they can be used to compare the hydrophobic characteristics of the surfaces of related proteins. The degree of fluorescence enhancement reflects the number of hydrophobic regions, or ANS-binding sites, while the emission wavelength reflects the degree of hydrophobicity of these sites. As described, these techniques are elegantly employed not only to qualitatively analyze the types of change that occur but also to quantitatively determine the thermodynamics involved and the types of reaction mechanisms that are possible. This technique has the advantage that it is rapid, easy, and involves instrumentation that is widely available.

3.3. Preferential Ligand Interaction

Different ligands interact with proteins in widely different manners at high concentrations. Table I lists values of preferential interaction parameters obtained for BSA (bovine serum albumin) in various solvent systems. They range from a large negative to a large positive value. A positive value means that the first term A_3 is greater than the second term g_3A_1, that is, the proteins preferentially bind ligands. A negative value means deficiency of the ligand in the protein domain relative to the bulk solvent. This table shows that the sign and magnitude of the interaction varies widely with the type of compounds used. However, the values for these compounds depend little on the actual protein tested.

A.   GLYCINE

Preferential interactions for glycine with several proteins have been extensively studied (42−44). Glycine stabilizes protein in solution (13, 42−45) and during

TABLE I.

Preferential Interaction Parameters of BSA with Various Ligands[a]

Ligand	$\left(\dfrac{\partial g_3}{\partial g_2}\right)_{T,\mu_1,\mu_3}$ (g/g)	$\left(\dfrac{\partial g_1}{\partial g_2}\right)_{T,\mu_1,\mu_3}$ (g/g)	$\left(\dfrac{\partial m_3}{\partial m_2}\right)_{T,\mu_1,\mu_3}$ (mol/mol)	Reference
2-Chloroethanol, 20%	0.433		366	10
2-Chloroethanol, 40%	0.619		522	10
Guanidine HCl, 6 M	0.06		44	31
Ethylene glycol, 40%	−0.097	0.130		37
Glycerol, 40%	−0.154	0.185		37
Mannitol, 15%	−0.034	0.204		37
Glucose, 1 M	−0.099	0.212	−37.4	38
Na glutamate, 1 M	−0.088	0.477	−35.5	39
Lysine HCl, 1 M	−0.047	0.222	−17.4	39
NaCl, 1 M	−0.0145	0.243	−16.8	40
Na$_2$SO$_4$, 1 M	−0.074	0.524	−35.4	40
MgSO$_4$, 1 M	−0.047	0.388	−26.5	41
MgCl$_2$, 1 M (pH 4.5)	−0.004	0.041	−2.8	41

[a] $(\partial g_3/\partial g_2)_{T,\mu_1,\mu_3}$, preferential binding of ligand (component 3) with protein (component 2); $(\partial g_1/\partial g_2)_{T,\mu_1,\mu_3} = -\dfrac{1}{g_3}\left(\dfrac{\partial g_3}{\partial g_2}\right)_{T,\mu_1,\mu_3}$, preferential binding of water; $(\partial g_3/\partial g_2)_{T,\mu_1,\mu_3}$, preferential ligand binding expressed in molal unit.

freeze-thawing (46). It also accumulates at high concentrations in halophilic organisms to overcome the high osmolality of their surrounding media (47, 48). The data in Table II show different preferential interactions for β-LG (β-lactoglobulin) when compared with BSA and lysozyme.

The different interaction profile for β-LG is more clearly seen in Fig. 8, which is a plot of $(\partial g_1/\partial g_2)_{T,\mu_1,\mu_3}$ versus g_3, the concentration of ligand. While the preferential interaction values for BSA and lysozyme show little dependence on g_3, the value for β-LG is negative at low concentrations, then becomes positive, and approaches the same level as that observed for the other two proteins as the ligand concentration is increased. With α- and β-alanine, the values are much smaller for β-LG than for BSA and lysozyme. The constant positive values for BSA and lysozyme can be explained with

$$\left(\frac{\partial g_1}{\partial g_2}\right)_{T,\mu_1,\mu_3} = A_1 - \frac{A_3}{g_3} \qquad (6)$$

derived by combining Eqs. 1 and 2. If BSA and lysozyme do not bind glycine, then the second term is zero, and hence $(\partial g_1/\partial g_2)_{T,\mu_1,\mu_3}$ is equal to A_1, the hydration of the proteins. From this point of view, the observed difference for β-LG must be ascribed to its binding with glycine. At low glycine concentrations, the second term overwhelms the hydration, leading to negative values of

TABLE II.

Preferential Interaction Parameters of Glycine with Proteins

Concentration (M)	$\left(\dfrac{\partial g_3}{\partial g_2}\right)_{T,\mu_1,\mu_3}$ (g/g)	$\left(\dfrac{\partial g_1}{\partial g_2}\right)_{T,\mu_1,\mu_3}$ (g/g)	$\left(\dfrac{\partial m_3}{\partial m_2}\right)_{T,\mu_1,\mu_3}$ (mol/mol)	Reference
		Lysozyme		
0.7	−0.0353	0.646	−6.72	42
1.4	−0.0685	0.610	−13.0	42
2.0	−0.0784	0.474	−14.9	42
		Bovine Serum Albumin		
0.7	−0.0245	0.450	−22.2	42
1.4	−0.0463	0.412	−41.9	42
2.0	−0.0687	0.416	−62.2	42
		β-Lactoglobulin		
0.5	0.0025	−0.064	1.22	45
1	−0.0133	0.169	−6.52	45
2	−0.0485	0.294	−23.8	45

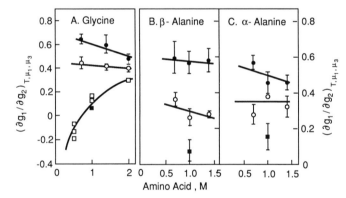

Fig. 8. Comparison of $(\partial g_1/\partial g_2)_{T,\mu_1,\mu_3}$ for β-lactoglobulin with its value for BSA and lysozyme. Upper curve, BSA; middle curve, lysozyme; bottom curve and solid triangle in *B* and *C*, β-LG. [Taken from Arakawa and Timasheff (44). Reproduced with permission from the American Chemical Society.]

$(\partial g_1/\partial g_2)_{T,\mu_1,\mu_3}$ and then becoming progressively smaller, with A_3 becoming saturated as the glycine concentration is increased. These results demonstrate that β-LG has a unique surface property that binds a neutral electrolyte, glycine.

The classical work by Cohn and Ferry (49) shows that glycine salts in β-LG even at relatively high concentrations; glycine usually acts as a salting-out agent. Cohn and Fery ascribe this salting-in effect of glycine on β-LG to the large dipole moment of β-LG. Arakawa and Timasheff (44) use the same argument to explain the specific glycine binding by the protein. Glycine, being a zwitterion, possesses a large dipole moment and interacts with β-LG through dipole–dipole interactions. Therefore, preferential interaction of proteins with glycine, a dipolar ligand, may be used to determine whether a protein molecule has a unique asymmetric charge distribution, resulting in a molecule with a dipole moment.

Knowledge of dipole moments of proteins may shed light on the nature of protein–protein interactions, and specific protein–ligand interactions. For example, β-LG aggregates in the absence of electrolytes. As the electrolyte concentration is increased, its solubility increases until the electrolytes exert their salting-out action. This phenomenon can be explained by the mutual interaction between β-LG molecules resulting in low solubilities and increasing interaction between β-LG and electrolytes, in favor of the dispersed monomeric form of the protein.

B. NACL

NaCl, being a salt universally found in living systems, also interacts with proteins. The values of $(\partial g_1/\partial g_2)_{T,\mu_1,\mu_3}$ range from 0.2 to 0.3 g/g, almost independent of which protein molecule is analyzed. This suggests that NaCl binds slightly to all proteins, and that protein hydration overwhelms the binding as indicated for the glycine systems with BSA and lysozyme. Therefore, a protein that has values of $(\partial g_1/\partial g_2)_{T,\mu_1,\mu_3}$ different from the above may have unique overall surface properties.

The data in Table III and Fig. 9 show that β-LG has unique surface properties when NaCl binding is studied. The preferential interaction parameter was determined for BSA, lysozyme, and β-LG as a function of pH. The values of $(\partial g_1/\partial g_2)_{T,\mu_1,\mu_3}$ for lysozyme and BSA are nearly independent of pH, being ~0.3 g/g for BSA and ~0.4 g/g for lysozyme. In contrast to these results, β-LG showed much smaller values above pH 4 and approached the level for the others below this pH. Arakawa and Timasheff (44) explained this peculiar result using the same mechanism as described for the glycine systems. That is, β-LG binds more NaCl than lysozyme and BSA, due to the large dipole moment of the protein via electrostatic interaction between dipole and ions. Below pH 4.0, the carboxyl groups of the protein are titrated and hence β-LG loses its unique charge distribution and is no longer dipolar. This should lead to decreased binding and as a consequence an increase in $(\partial g_1/\partial g_2)_{T,\mu_1,\mu_3}$. In other words, the observed peculiar interaction of β-LG with NaCl, relative to the interactions for BSA and lysozyme, is an indication that β-LG has a unique surface property, supporting the conclusion obtained from its interaction pattern with glycine.

The solubility of β-LG at low ionic strength is low around pH 4.5 and is increased by the addition of NaCl; that is, NaCl binds to the protein, neutralizes its dipole moment and reduces protein–protein interactions with a consequent

TABLE III.

Preferential Interactions of NaCl with Proteins in 1 M Salt Solution

pH	$\left(\dfrac{\partial g_3}{\partial g_2}\right)_{T,\mu_1,\mu_3}$ (g/g)	$\left(\dfrac{\partial g_1}{\partial g_2}\right)_{T,\mu_1,\mu_3}$ (g/g)	$\left(\dfrac{\partial m_3}{\partial m_2}\right)_{T,\mu_1,\mu_3}$ (mol/mol)
	Bovine Serum Albumin (MW 68 000)		
4.5	−0.0158	0.265	−18.4
5.6	−0.0145	0.243	−16.9
7.0	−0.0142	0.239	−16.5
9.0	−0.0158	0.266	−18.4
	Lysozyme (MW 14 000)		
3.0	−0.0269	0.451	−6.6
4.5	−0.0253	0.424	−6.2
7.0	−0.0237	0.398	−5.8
	β-Lactoglobulin (MW 36 800)		
2.0	−0.0174	0.292	−10.9
2.5	−0.0206	0.346	−13.0
3.0	−0.0206	0.346	−13.0
4.0	−0.0158	0.265	−9.9
5.1	−0.0079	0.133	−5.0
7.0	−0.0063	0.106	−4.0
10.0	−00.48	0.080	−3.0

Source. Arakawa and Timasheff (44).

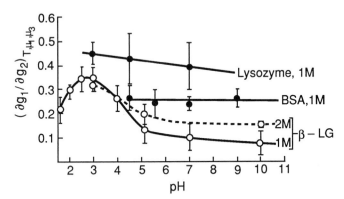

Fig. 9. Comparison of $(\partial g_1/\partial g_2)_{T,\mu_1,\mu_3}$ for β-LG with its value for lysozyme and BSA. [Taken from Arakawa and Timasheff (44). Reproduced with permission from the American Chemical Society.]

increase in solubility. On the other hand, the solubility of β-LG decreases with increasing NaCl concentrations at low pH.

Peculiar interaction patterns have also been observed with two enzymes prepared from an extreme halophilic bacteria (50). Figure 10 illustrates the preferential interaction parameter for malate and glutamate dehydrogenase purified from an extreme halophilic bacteria. For comparison, results for BSA in NaCl and guanidine hydrochloride and DNA in NaCl are given. The results for BSA in NaCl show nearly constant, positive values. Lysozyme shows a profile similar to BSA (44). This profile is reminiscent of β-LG in glycine, although the value of $(\partial g_1/\partial g_2)_{T,\mu_1,\mu_3}$ is much more negative. Analyzing the results in the same way as for β-LG shows an overwhelming binding of NaCl relative to hydration. The binding saturates with increasing NaCl concentration and, hence, the second term in Eq. 6 becomes the same magnitude as that of the first, the hydration term. Therefore, it appears that these enzymes have unique surface properties through which they bind a large number of NaCl molecules.

Salting-out salts, such as NaCl, usually stabilize proteins through the Debye−Hückel effect at low salt concentrations, in common with other salts, and through preferential exclusion at high salt concentration. However, the malate and glutamate dehydrogenase from halophilic bacteria respond differently to addition of salt than do most proteins. They require extremely high salt concentrations for their stability and activity. Figure 11 shows a plot of the α-helical content of malate dehydrogenase against NaCl concentration; the data show that the α-helical content reaches a maximal value at 2.0 M and becomes constant. The secondary structure of the enzyme is lost with decreasing

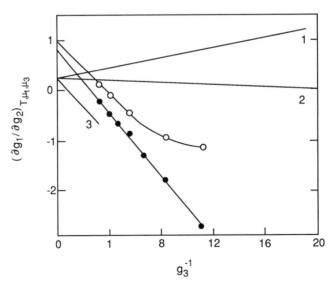

Fig. 10. Preferential interaction of halophilic malate and glutamate dehydrogenase. ●, Malate dehydrogenase; ○, glutamate dehydrogenase; 1, DNA in NaCl; 2, BSA in NaCl; 3, BSA in guanidine hydrochloride. [Replotted from Pundak and Eisenberg (50).]

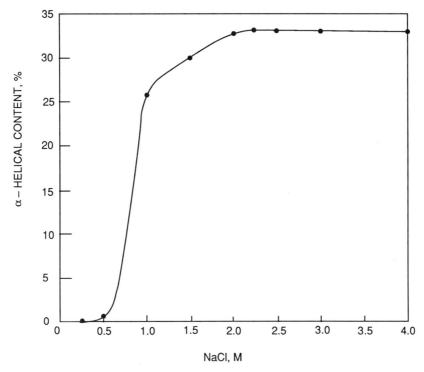

Fig. 11. Plot of α-helical content of malate dehydrogenase versus NaCl concentration, [Replotted from Pundak and Eisenberg (50).]

NaCl concentration. The tertiary structure and enzyme activity are maintained better at higher salt concentration, correlating with the observed change in the secondary structure. Therefore, the preferential interaction parameters for these enzymes, measured under what are physiological conditions for most proteins, reflect the surface properties of these proteins in the denatured state. Determination of the preferential interaction parameter as a function of NaCl concentration demonstrates that these enzymes bind a large number of NaCl molecules in the native state that can also be achieved at high salt concentrations. In other words, the native, stable form of the proteins has a unique surface property that is characterized by unusual binding capacity for NaCl. However, if one measures the interaction at, for example, 2 M NaCl, it would not be possible to draw the same conclusion, since the preferential hydration is slightly negative and might not appear to differ greatly from the expected, usual results.

C. GLYCEROL

Preferential interactions of glycerol with proteins have been studied to understand the effects of glycerol on protein stability (51−56), the self-assembly of tubulin into microtubules (18, 57, 58), and the stability of collagen (59). These studies show that most proteins are preferentially hydrated in aqueous glycerol solutions, an exception being calf skin collagen. Table IV shows the values of preferential interaction parameters obtained in 30% glycerol. These values are smaller than the usual hydration, $0.2-0.4$ g/g, indicating that glycerol binds to the proteins. A positive value of $(\partial m_3/\partial m_2)_{T,\mu_1,\mu_3}$ (this parameter is equal to $(\partial g_3/\partial g_2)_{T,\mu_1,\mu_3}$ multiplied by a molecular weight factor) indicates that the calf skin collagen binds glycerol more strongly than the other proteins.

Gekko and Timasheff (56) have attempted to find a correlation between the preferential interactions of glycerol and the property of the proteins analyzed. A priori, it was expected that, in the aqueous glycerol system, the preferential interactions with solvent components would be related to protein polarity (such as that calculated by Bigelow, see reference 27) rather than to the hydrophobicity, since glycerol is a structure-stabilizing cosolvent, while correlation with hydrophobicity is found only with strong denaturants that interact directly with hydrophobic residues of proteins. Figure 12 shows the plot of $(\partial g_3/\partial g_2)_{T,\mu_1,\mu_3}$ against the polarity parameter of the proteins. A clear correlation can be obtained: The preferential binding of glycerol increases with the increasing polarity of the proteins. This means that proteins with higher polarities bind more glycerol.

Gekko and Timasheff (56) calculated the binding of glycerol to the proteins, assuming their hydration values, as plotted in Fig. 13. At both pH 2.0 and pH 5.8, glycerol binding increases with increasing polarity of the proteins. They suggested that glycerol can bind to proteins because it is capable of occupying a part of the solvation sheath in the polar regions of proteins through hydrogen bonds.

The polarity parameter can be calculated from the amino acid compositions of proteins and simply reflects the ratio of polar residues in the proteins. It most likely parallels the number of polar residues on the protein surface.

TABLE IV.
Preferential Interactions of Proteins with 30% Glycerol

Protein	$\left(\dfrac{\partial g_2}{\partial g_2}\right)_{T,\mu_1,\mu_3}$ (g/g)	$\left(\dfrac{\partial g_1}{\partial g_2}\right)_{T,\mu_1,\mu_3}$ (g/g)	References
Tubulin	−0.127	0.234	18
Ribonuclease A (pH 5.8)	−0.087	0.163	56
BSA (pH 2.0)	−0.020	0.037	56
BSA (pH 5.8)	−0.113	0.212	56
Collagen, native	493	—	59
(4.5 M glycerol)	(in mol/mol)		
Collagen, denatured	108	—	59
(4.5 M glycerol)	(in mol/mol)		

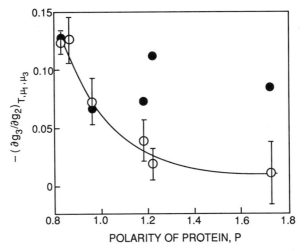

Fig. 12. Dependence of $(\partial g_3/\partial g_2)_{T,\mu_1,\mu_3}$ on protein polarity, P, in 30% glycerol solution at pH. 2.0 (○) and pH 5.8 (●). [Taken from Gekko and Timasheff (56). Reproduced with permission from the American Chemical Society.]

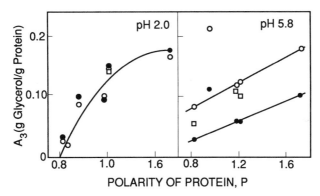

Fig. 13. Dependence of glycerol binding on protein polarity, P, in 30% glycerol solutions. Different symbols correspond to different methods of calculating hydration, A_1, from which the binding, A_3, was obtained with Eq. 2. Proteins used are insulin ($P = 0.86$), β-chymotrypsin (0.85), chymotrypsinogen A (0.83), lysozyme (1.18), BSA (1.22) and ribonuclease A (1.73). [Taken from Gekko and Timasheff (56). Reproduced with permission from the American Chemical Society.]

Therefore, the correlation suggests that the binding of higher amounts of glycerol indicates a more hydrophilic protein surface.

A preferential binding of glycerol is observed for the native, triple-helical collagen, meaning that the protein binds more glycerol than do the globular proteins. Since the polarity parameter is low for this protein, 0.89, the protein should bind less glycerol than those proteins shown in Fig. 12. However, as Na (59) explains, this result could be due to an unique sequence of this protein, as expressed by the repeating unit of Gly-X-Y, with hydroxyproline occurring exclusively at the Y position. These hydroxyl groups and the peptide bonds, along with the fact that the collagen triple helix has an extended structure, suggest that glycerol binds to the protein through multiple hydrogen bonds competing with water molecules. In other words, the observed extremely large binding of glycerol with collagen indicates a greater exposure of groups capable of forming hydrogen bonds.

D. OTHER LIGANDS

Preferential interaction measurements have been carried out for a variety of ligands. Certain compounds, such as sucrose, polyols (number of (OH) > 3) and strong salting-out salts (e.g., ammonium sulfate and sodium phosphate), appear to depend only on the surface area of the proteins; these compounds show a preferential hydration greater than 0.3 g/g. This means that these compounds are actually repelled from the protein surface. Such strong exclusion effects will obscure binding to proteins even if any is occurring.

Preferential interactions have been studied to explain how ligands at high concentrations can induce protein stabilization, self-association, and denaturation. Only a few examples, such as those described here, can be found in the literature that demonstrate the application of preferential interactions for analyzing the surface property of native proteins. However, other ligands with weak hydrophobic character, such as dimethyl sulfoxide, may be used for surface characterization when the concentrations are below those that induce protein denaturation.

The use of preferential interactions can provide information on the hydrophobicity of protein surfaces, on the charge distribution and dipole moment of protein surfaces, and on the importance of hydrogen bonds at the protein surface. This information is not easily obtained using any other technique presently available, and complements the information on surface hydrophobicity obtained using the strong binding ligands described in the previous section.

3.4. Hydrophobic Interaction Chromatography

Hydrophobic interaction chromatography (HIC) is another approach that uses protein−ligand interactions to probe the surface properties of proteins. This technique is used primarily in two ways: as a means of analyzing and following changes in the conformation of proteins and as a means of separating and characterizing proteins based on their native surface hydrophobicity.

A. CONFORMATIONAL CHANGE

The effectiveness of HIC to study conformational changes of proteins has been demonstrated numerous times. Sulkowski and colleagues (60) employed HIC on octyl-Sepharose CL-4B and phenyl-Sepharose CL-4B to follow the N-F transition (or "acid expansion") of HSA (human serum albumin). Quantitatively similar results are obtained with both resins, though the albumin binds more avidly to octyl-Sepharose than to phenyl-Sepharose. In this example, the protein is bound to the column in the presence of buffer (5.0 mM sodium phosphate pH 7.4, or 100 mM sodium acetate pH 4.0 to 5.5), but in the absence of any salting-out salts, and is then eluted with an ethylene glycol gradient. The albumin exhibits pH-dependent binding characteristics. On the phenyl-Sepharose, no binding is seen at pH 7.5 or 5.5, with the amount of protein bound increasing as the pH is decreased from 5 to 4. The quantitative binding correlates with the N-F isomerization. Thus, HIC can be used to follow acid-induced changes in the protein conformation, and indicates that the acid expansion results in an increase in surface hydrophobicity.

Zizkovsky et al. (61) employed HIC to study the accessibility of nonpolar areas of HSA and α-fetoprotein (AFP), and also examined the effect of pH changes on the behavior of these two proteins during HIC. These two proteins share some homology, and also the ability to bind many biologically important molecules, but little was known about the role or structure of AFP at the time of this study. Both proteins bind avidly to octyl-Sepharose CL-4B even at very low salt concentrations (20 mM phosphate, pH 7.4), and require *tert*-butanol for elution, eluting at between 15 and 20% alcohol. In contrast, neither binds to the rigid resin Spheron 300, which has very flat hydrophobic sites, even in the presence of 2 M NaCl. This behavior is quite different than that seen for trypsin, lysozyme, and chymotrypsin, which bind to both resins, and is taken as evidence that the large hydrophobic areas of HSA and AFP are located in crevices which are accessible only to the flexible octyl side chains of the octyl-Sepharose. The N-F transition of these proteins was then studied, using Spheron as the hydrophobic resin (Fig. 14). Both proteins undergo acid and alkaline conformational transitions that result in an exposure of hydrophobic sites, as expected from the results described above for albumin. A more gradual decrease in the hydrophobicity of AFP versus HSA is seen with increasing pH (from 3 to 7) and could be due to a multiplicity of forms (at least three) which are seen following gradient elution from the octyl-Sepharose. The midpoint of the acid transition (to increased hydrophobicity) of AFP is shifted to a slightly higher pH than that of HSA (5.3 vs. 4.65). This study indicates that the hydrophobic characteristics of the two protein surfaces are very similar, and that both proteins undergo both N-F and alkaline transition. It reveals some slight differences in behavior between the two proteins as well.

HIC can also be used to follow changes in the surface conformation of steroid and glucocorticoid receptors following hormone binding and activation. Increased binding of steroid hormone receptors to hydrophobic columns has been demonstrated for androgen receptors from rat prostatic nuclei (62) and for progesterone uterine cytosol receptors (63). The salt-induced DNA binding form of the uterine estrogen receptor binds strongly to phenyl-Sepharose (64).

For the glucocorticoid receptors, binding to DNA-cellulose, adsorption to

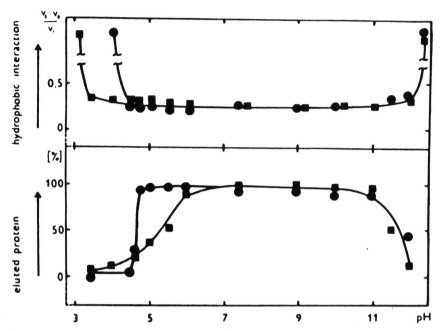

Fig. 14. Hydrophobic interaction retention parameter and amount of eluted protein (■, AFP; ●, HSA) as a function of the pH of the elution buffer. Elution buffers at pH 3.4−7.4; 0.05 M citrate−phosphate buffers with 0.15 M NaCl; at pH 7.4−12, 0.03 M borate−NaOH buffers with 0.15 M NaCl. [Taken from Zizkovsky et al. (61). Reproduced with permission from Elsevier Science Publishers, B.V.]

Whatman G/C glass fiber filters, and a decrease in the sedimentation coefficient from 9.25 to 3.85 have all been used to follow changes in protein conformation following activation, with different kinetics observed. Densmore et al. (65) then used HIC to investigate the hydrophobic characteristics of activated and unactivated glucocorticoid−type II receptor complex from mouse brain cytosol. Glucocorticoid receptor was labeled by binding with [³H]triamcinolone acetonide and activated by heating at 22 °C. The hydrophobicity of the activated and unactivated receptor complex was then assessed by HIC. Figure 15 shows the isocratic elution patterns of the glucocorticoid receptor from a series of hydrophobic columns of increasing alkyl chain length. The activated complex is more hydrophobic as demonstrated by its increased retention on the C_3 to C_6 and phenyl agarose columns. The protein binds to the octyl, decyl, and dodecyl agarose columns so avidly that it undergoes destabilization and loses its affinity for hormone. By choosing the appropriate salt conditions the activated and unactivated receptor complexes can be totally separated from each other on the pentyl-Sepharose column. The strong binding of the activated receptor complex to the phenyl column could be due in part to the structural similarity of this resin to DNA to which the activated receptor binds strongly.

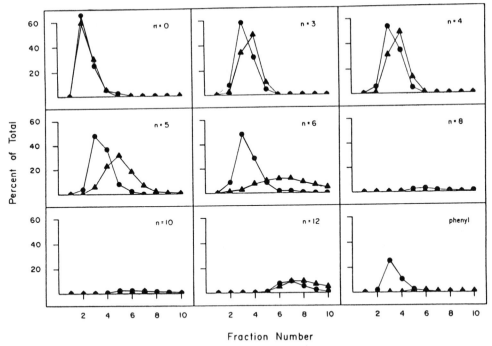

Fig. 15. Elution patterns of activated (▲) or unactivated (●) glucocorticoid type II receptor complexes from a series of alkyl Sepharose columns ($Cn:n$ = number of methylene groups) or a phenyl Sepharose column equilibrated and eluted isocratically under low-salt conditions (20 mM HEPES, 2 mM DTT, 20 mM sodium molybdate, pH 7.4). [Taken from Densmore et al. (65). Reproduced with permission from Raven Press Ltd.]

In this case, HIC was used to demonstrate unambiguously that the surface hydrophobicity of glucocorticoid receptors increases following activation of the receptor complex to the DNA binding form.

In yet another example, Wu et al. (66) used HIC at various temperatures and in the presence and absence of metal ions to follow changes in the stability of the calcium binding protein α-lactalbumin. Using C_1-ether-agarose as the hydrophobic resin, they showed that from 0 to 15 °C the addition of metal ion has little effect on retention. As the temperature is raised, the retention time increases most in the presence of Mg^{2+}, intermediate in the absence of metal ions, and least in the presence of 1 mM Ca^{2+}. These data indicate that the removal of Ca^{2+} results in a less stable protein with a more hydrophobic surface and that temperature-induced changes also result in increases in the hydrophobicity of the protein surface. This increase in hydrophobicity is further illustrated by the protein's behavior on the more hydrophobic support C_2-ether-agarose. At ambient temperatures, the calcium-depleted protein elutes as two peaks, with the second peak containing denatured protein, as indicated by spectral analysis. The percentage of protein eluting in the second peak increases with increasing temperature, and also with increasing resident time on the column. The addition of calcium again stabilizes the protein in the less hydrophobic conformation, and results in less protein in the second denatured peak. This work demonstrates the use of HIC to follow thermally induced changes in protein conformation, and changes resulting from metal-ion binding. However, care must be exercised for the former case, because temperature changes alone can affect hydrophobic interaction between the proteins and ligand, without reflecting conformational changes of the proteins.

The above examples illustrate the usefulness of HIC in following changes in the hydrophobic nature of the protein surfaces. In a modification of this application of HIC, Pivel et al. (67) used hydrophobic chromatography as a model for the membrane—membrane protein interactions of the F_1-ATPase from *Micrococcus lysodeiklicus*. F_1-ATPase is a complex oligomeric protein, which exhibits greatly increased ATPase activity following binding and elution from C_2- or C_4-agarose. The protein was applied to the columns in 30 mM Tris—HCl, pH 7.5 and activated fractions were then eluted with increasing LiCl. The activated fractions no longer contain the γ and δ subunits, and also show a molar decrease in the amount of α subunit. Following HIC, the active form requires LiCl to maintain solubility. The active form also shows an increase in fluorescence and a blue shift in its emission spectrum, with the tyrosine contribution increasing dramatically. The circular dichroic spectra reveals an increase in the α-helical character of the activated fractions. These data suggest that the activation of the ATPase is accompanied by changes in the quaternary structure, the tertiary structure, and the secondary structure of the protein.

In this study, HIC was used to mimic the extrinsic membrane surface because it allows the protein molecules to interact simultaneously with a hydrophobic phase, an aqueous phase, and the interface of the two. The interaction depends on the ligand arm used, with the protein binding more strongly to the C_4 than to the C_2 column, and with the protein binding so avidly to the C_6 ligand that only an inactive form could be eluted with SCN^-. It also depends on the ions used, with Li^+ being more effective as an eluent than Na^+, and SCN^- eluting the protein more efficiently than Cl^-. While this application of HIC resulted in interesting information, it also illustrates some of the difficulties that can be encountered when using this technique to purify or characterize proteins, particularly oligomeric proteins. The assumption that the technique itself is passive needs to be verified following use.

B.   HYDROPHOBICITY OF NATIVE PROTEINS

The most common application of HIC is in the purification and characterization of native proteins, and as a tool for comparing the surface hydrophobicity of different proteins. Memoli and Doellgast (68) employed columns of phenylalanine-Sepharose and aniline-Sepharose in Tris acetate, pH 8.0 at 4 °C, and various concentrations of ammonium sulfate to compare the surface hydrophobicity of serum albumin and hemoglobin. Both proteins are retained on aniline-Sepharose at lower ammonium sulfate concentrations than on phenylalanine-Sepharose, as expected from the increased hydrophobicity of this ligand; more protein is retained as the ammonium sulfate concentration is increased as well. Hemoglobin is retained at lower ammonium sulfate concentrations than serum albumin on both resins, demonstrating that it has a greater surface hydrophobicity.

Hrkal and Rejnkova (69) used HIC on phenyl-Sepharose at pH 7.0 and a linear gradient of decreasing ammonium sulfate to separate serum proteins from albumin depleted human serum, and found that the hydrophobicity of the molecules could be deduced to be IgG > IgA > transferrin > α_1-acid glycoprotein.

Nishikawa and Baillon (70) utilized caprylyl hydrazide-agarose gels at neutral pH and observed that ovalbumin, β-lactoglobulin, and BSA (bovine serum albumin) interact with this resin according to their hydrophobicities. That is, ovalbumin interacts very little, while β-lactoglobulin is intermediate, and BSA interacts very strongly. This order remains constant regardless of the concentration of ligand used, or the internal concentration of sodium sulfate. Narhi et al. (71) observed similar elution behavior when HIC is performed at alkaline pH on a phenyl superose column. Ribonuclease A, ovalbumin, and β-lactoglobulin elute in this order in the presence of monosodium glutamate, sodium sulfate, or guanidine sulfate. This agrees with the theory described in the introduction; the interaction of salt with the proteins and ligands simply adds binding energy to the intrinsic binding energy between the protein and the column ligand.

Pahlman et al. (72) studied the elution of human serum albumin, ovalbumin, and phycoerythrin from decyl- and pentyl-Sepharose at ambient temperature in the presence of different neutral salts. Ovalbumin has the shortest retention time on the columns, while human serum albumin elutes last, and phyco-erythrins display an intermediate hydrophobicity. This order of hydrophobic interaction remains constant regardless of what salt is used. The salts themselves are more or less effective in enhancing interactions with the hydrophobic resin depending on their position in the Hofmeister series, when compared at constant ionic strength. They detected changes in the tertiary structure of the proteins in both NaBr and NaSCN, indicating that care must be exercised when choosing the salt to be used, in order to cause as little denaturation of the proteins as possible.

Kato et al. (73) examined the binding of various proteins to Phenyl-5PW, a polymer based resin (Toyosoda, Tokyo, Japan) in the presence of different concentrations of ammonium sulfate. Figure 16 demonstrates that 10 proteins exhibit widely different degrees of retention on the resin as the ammonium sulfate concentration is increased. Once again, the order of binding affinity parallels the degree of hydrophobicity of the proteins. When these proteins are

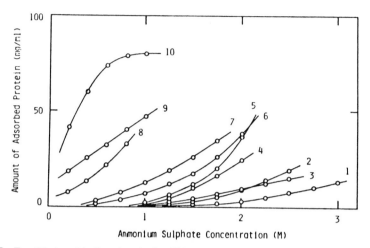

Fig. 16. Equilibrium binding (mg/mL of Phenyl-5PW) of various proteins with phenyl-5PW as a function of ammonium sulfate concentration. Proteins are 1, cytochrome c; 2, myoglobin; 3, ribonuclease; 4, lysozyme; 5, ovalbumin; 6, β-chymotrypsin; 7, β-chymotrypsinogen; 8, γ-globulin; 9, thyroglobulin; 10, ferritin. [Taken from Kato et al. (73). Reproduced with permission from Elsevier Science Publishers, B.V.]

bound to the phenyl-5PW column at 1.7 M ammonium sulfate and eluted with a linear gradient from 1.7 to 0 M, the order of elution (Fig. 17) is identical to the binding affinity determined by equilibrium binding. The elution of these same proteins from reverse phase HPLC occurred in a different order (Fig. 18), indicating that the denaturation that occurs on reverse phase HPLC results in a binding affinity that is quite different from that of the native proteins.

Three proteins, cytochrome c, conalbumin, and β-glucosidase, eluted from a number of columns with different hydrophobicities in the identical order; cytochrome c is always the most hydrophilic and β-glucosidase the most hydrophobic (74). This information indicates that the relative hydrophobicity of different proteins can be compared using different columns. An example of the elution profile is shown in Fig. 19 when the ligand is butyrate. Here again, cytochrome c has a very weak binding affinity. This order is essentially identical to those obtained by Kato using the phenyl-5PW column. They also show that lysozyme behaves as a very hydrophobic protein, although the hdyrophobic parameter estimated from the amino acid composition is relatively hydrophilic (75). Therefore, the hydrophobic character of the surface determined by hydrophobic interaction chromatography reflects that of the native state, and not that of the total protein sequence.

Strop (76) investigated the behavior of proteins on the semirigid, highly crosslinked gel Spheron 300. As previously mentioned, this gel has a non-polar backbone and interacts with the hydrophobic surface areas of proteins,

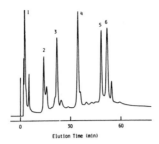

Fig. 17. Hydrophobic interaction chromatography of several proteins on phenyl-5PW eluted with a descending linear gradient from 1.7 to 0 M ammonium sulfate. Proteins are 1, cytochrome c; 2, myoglobin; 3, ribonuclease; 4, lysozyme; 5, α-chymotrypsinogen; 6, α-chymotrypsin. [Taken from Kato et al. (73). Reproduced with permission from Elsevier Science Publishers, B.V.]

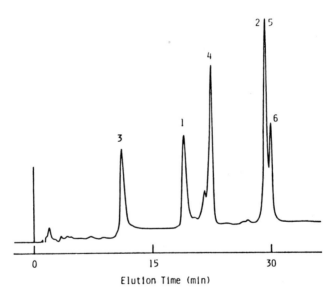

Fig. 18. Reverse-phase HPLC of several proteins on a C$_4$ column eluted with a linear gradient from 10 to 50% acetonitrile in 0.05% TFA (pH 2.2). Proteins are 1, cytochrome c; 2, myoglobin; 3, ribonuclease; 4, lysozyme; 5, α-chymotrypsinogen; 6, α-chymotrypsin. [Taken from Kato et al. (73). Reproduced with permission from Elsevier Science Publishers, B.V.]

but is unable to interact with clefts and crevices. Figure 20 shows the dependence of retention $[(V_e - V_o)/V_t]$ on ammonium sulfate concentration. When sodium sulfate is used, the order of retention is identical; with NaCl, it differs only for lysozyme and bovine pancreatic trypsin inhibitor. When KBr is used as the salt, the binding for some proteins is enhanced while it is

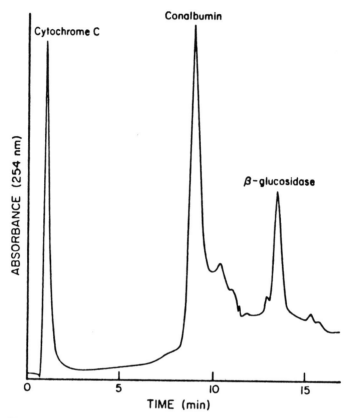

Fig. 19. Hydrophobic interaction chromatography of several proteins on a butyrate column with a linear gradient of 3.0 to 0 mM sodium sulfate. [Taken from Fausnaugh et al. (74). Reproduced with permission from Academic Press.]

decreased for others, and increased then decreased for lysozyme. The data suggest that KBr denatures the protein, or that the salt can affect the binding of a few proteins, due to the specific effects of these salts on the surfaces of these proteins. The order of effectiveness of the various salts in eluting proteins follows the Hofmeister series. Adding less polar agents, such as butanol, to the buffer decreases the binding of the proteins.

The above studies demonstrate that various HIC resins and different salts can be used to separate proteins based on their hydrophobicity, and that the absolute order of binding and elution of proteins remains constant in spite of the ligand or salt used, provided denaturation or other specific conformational changes do not occur under the chosen conditions. These experiments were all done using proteins with different structures and hydrophobicities.

Fausnaugh and Regnier (77) expanded on this, examining how the retention of lysozyme on phenyl-5PW is affected by amino acid substitutions in the

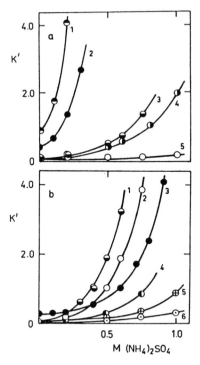

Fig. 20. Dependence of the distribution coefficient, K', on the concentration of ammonium sulfate during equilibrium hydrophobic interaction chromatography on Spheron P-300. Proteins are (*a*) 1, bull seminal plasma trypsin isoinhibitor, BUSI IA; 2, bull seminal plasma trypsin isoinhibitor BUSI IB; 3, basic pancreatic trypsin inhibitor; 4, cow colostrum trypsin inhibitor; 5, bull seminal plasma basic trypsin inhibitor BUSI IIA; (*b*) 1, chymotrypsinogen; 2, β-trypsin; 3, lysozyme; 4 α-chymotrypsin; 5, ribonuclease; 6, chymopapain. [Taken from Strop (76). Reproduced with permission from Elsevier Science Publishers, B.V.]

protein. They used the column in the isocratic mode, with ammonium sulfate as the salt. The degree of hydrophobicity of the proteins was analyzed by graphing log K' (the retention factor) versus m (the molal salt concentration). This plot should yield straight lines with slopes proportional to the contact area between the protein and the column. As shown in Fig. 21, when log K' versus m is plotted for each of seven lysozymes, a series of parallel lines is produced. The constant slope implies that the contact surface area is the same for all seven proteins. The retention factor varies widely between proteins at a given salt concentration, indicating that the intrinsic free energy of interaction is different for each protein. The differences in free energy of interaction between proteins correlates well with the transfer of free energy for the amino acid residues substituted. However, not all of the amino acid substitutions affected retention; an example of this is the two substitutions between Peking

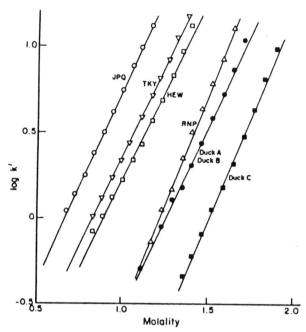

Fig. 21. Plot of retention parameter, log K', versus molality of ammonium sulfate during hydrophobic interaction chromatography on phenyl-5PW at pH 7.8 for several bird lysozymes. Proteins are JPQ, Japanese quail lysozyme; TKY, turkey lysozyme; HEW, hen egg white lysozyme; RWP, ring necked pheasant lysozyme; duck A, B, and C, Peking duck A, B, or C lysozyme. [Taken from Fausnaugh and Regnier (77). Reproduced with permission from Elsevier Science Publishers, B.V.]

duck A and Peking duck B lysozyme. The proteins show identical behavior on the column, which probably indicates that the substitutions are outside of the contact area. In this way, they could not only compare the hydrophobicity of the proteins' surface, but also define a contact area between the protein and the column. Changing the pH of the elution buffer changes the retention parameter, but not the slope, indicating that the contact surface itself is not affected, though the charge on the amino acids appears to affect the retention time.

A similar observation was made when acidic fibroblast growth factor was applied to a phenyl-Superose column. The analog molecule, in which histidine 93 was replaced with glycine, elutes at lower ammonium sulfate concentrations than the natural sequence molecule does. These data suggest that residue 93 in the protein is exposed to the protein surface and is located in the area of the protein involved in interactions with the column (T. Arakawa and T. Horan, unpublished).

Osthoff et al. (78b) employed HIC to characterize several elapid cardio-toxins. Table V gives the elution order and retention times of various snake

TABLE V.

Retention Times and Elution Orders of Snake Venom Cardiotoxins by RP-HPLC and HIC-HPLC

| | RP-HPLC | | HIC-HPLC | | |
Toxin	Retention Time (min)[a]	Elution Order	Retention Time (min)[a]	Elution Order	LD$_{50}$ (µg/g mouse)
N. m. mossambica V^{II}4	22.44 ± 0.17	1	18.01 ± 0.06	1	2.0
N. n. kaouthia CM-7 + 7A	23.16 ± 0.15	2	21.72 ± 0.23	2	1.2
N. h. annulifera V^{II}I	23.22 ± 0.02	3	23.65 ± 0.03	4	1.4
N. m. mossambica V^{II}2	23.67 ± 0.22	4	23.54 ± 0.15	3	1.1
N. nivea V^{II}2	24.82 ± 0.44	5	28.25 ± 0.08	6	1.5
N. m. mossambica V^{II}I	26.00 ± 0.33	6	26.22 ± 0.36	5	0.8

[a] Means ± SE for three corrected retention times of three runs.

Source. Osthoff et al. (78b).

venom cardiotoxins following reverse phase HPLC on a C-18 column with a linear gradient of 40 to 60% acetonitrile containing 0.1% TFA (trifluoroacetic acid) and HIC-HPLC on a TSK phenyl 5PW column with a linear gradient of decreasing ammonium sulfate in 0.1 M sodium phosphate at pH 7.0. When the structure of the cardiotoxins eluting from both columns was assessed using circular dichroism, it was obvious that the structure of all but one of the toxins was perturbed by the acetonitrile/TFA buffer system, while the buffer system used for HIC did not perturb the protein spectra. Thus, the hydrophobicity observed with HIC under these conditions is a reflection of the surface hydrophobicity of the native proteins, while that seen with reverse phase results from the hydrophobicity of the unfolded protein.

Hydrophobic interaction chromatography is also widely used to separate and characterize proteins from the same organism. Mevarech et al. (78a) demonstrated that Sepharose 4B itself was hydrophobic enough to adsorb several enzymes from a halophilic bacteria at 2.5 M ammonium sulfate, which then eluted based on their hydrophobic character as the salt concentration was decreased. Of these halophilic enzymes, malate dehydrogenase is less hydrophobic than aspartate transcarbamylase, which is less hydrophobic than glutamate dehydrogenase.

This technique was also used to characterize the surface hydrophobic properties of wheat gliadins and glutenines; in this case the alcohols were required for elution from the phenyl-Sepharose column (79). Proteins from wheat gluten were solubilized with acetic acid, and fractionated on the column using a linear gradient of increasing ethanol in 0.02 M ammonia. HIC showed that the surface hydrophobicity of these proteins increases in the order of ω-gliadin < β-gliadin < α-gliadin < γ-gliadin.

Raymond et al. (80) used HIC to separate and characterize saline-soluble sunflower proteins. Using octyl-Sepharose as the hydrophobic matrix the saline extract was absorbed to the column in 10% NaCl, 20% $(NH_4)_2SO_4$ and 0.02 M borate, pH 7.4. The first protein fraction was eluted by decreasing the ionic strength to 20% $(NH_4)_2SO_4$ and 0.02 M borate pH 7.4, the second by decreasing the ionic strength further to 10% NaCl in buffer, the third with 0.02 M borate alone, the fourth with 0.001 M in borate, and the fifth with water. Strongly hydrophobic fractions were then eluted by the addition of chaotropic agents, first with 0.1 M $MgCl_2$, and finally with 50% 2-methoxyethanol.

In a final example of the use of HIC to characterize the surface conformation of related proteins, Smyth et al. (81) compared the behavior of porcine entero-pathogenic *E. coli* with or without K88 antigen on various agarose resins. Many different strains of *E. coli* were applied to phenyl, octyl, naphthoyl, and palmitoyl agarose in 4 M NaCl, and in every case those expressing K88 antigen bound to the columns, while those without the antigen failed to bind. An example of this behavior is shown in Fig. 22. Binding to the hydrophobic columns does not correlate with serogroup or other characteristics of the cells, and when K88-positive strains were grown at 18 °C, where K88 is not expressed, the cells also failed to bind to the hydrophobic resins. Preliminary screening of K99-positive strains from pigs, lambs, and cows, and of colonization-factor positive or negative strains from infantile diarrhea also exhibited the same

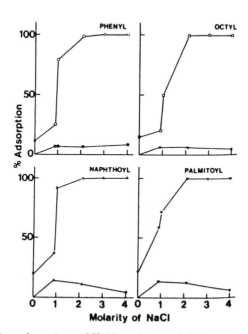

Fig. 22. The relative adsorptions of K-88 positive and K88-negative strains of *E. coli* to the hydrophobic ligands indicated with increasing ionic strength at 20 °C. △, ○, □, strain K-12 (K88-ab); ▲, ●, ■, strain K-12 (K88⁻). [Taken from Smyth et al. (81). Reproduced with permission from the American Society for Microbiology.]

behavior. This report demonstrates that HIC can be used to characterize the surface properties of bacteria as well, and from these results it appears that the hydrophobic character of the K88 antigen is at least partly responsible for the illi-mediated adhesion by enteropathogenic *E. coli*.

As illustrated above, HIC can provide important information about the surface hydrophobicity of proteins in the native state. Care must be exercised in choosing the conditions used, such as solvent, ligand type and concentration, and temperature, to ensure that the protein maintains its native conformation. When this is done HIC can be used to obtain information about contact sites, the nature of the hydrophobic areas on the protein surface, the effect of mutations on the protein surface, and the effect of solvent conditions on protein conformation. While more qualitative than the other techniques we have described, this technique is rapid, and can be used without any special instrumentation, and as such is readily available to everyone. It can provide information that complements that obtained using the ligand binding techniques.

4. CONCLUSION

Recent developments in recombinant cloning technology have made it possible to produce large quantities of a protein and related mutant proteins with a point or multiple substitutions. Analysis of these proteins by conventional biophysical techniques such as those described here will shed light on structure—function relationships when used in conjunction with the analysis of biological activities. Other techniques available for characterizing the native and unfolded forms of proteins are described in other chapters of this volume. Combinations of these techniques will reveal clues to the molecular architecture of proteins, providing important information about structure and function.

ACKNOWLEDGMENT

We thank Joan Bennett for typing the manuscript and Julie Heuston for artwork.

References

1. L. Stryer, *J. Mol. Biol.* **13**, 482–495 (1965).

2. D.C. Turner and L. Brand, *Biochemistry* **7**, 3381–3390 (1968).

3. W.O. McClure and G.M. Edelman, *Biochemistry* **5**, 1908–1919 (1966).

4. S. Damodaran, *Biochim. Biophys. Acta* **914**, 114–121 (1987).

5. T. Arakawa and W.C. Kenney, *Int. J. Peptide Protein Res.* **31**, 468–473 (1988).

6. L. Wicker and J.A. Knopp, *Arch. Biochem. Biophys.* **266**, 452–461 (1988).

7. J.C. Lee, K. Gekko, and S.N. Timasheff, *Methods Enzymol.* **61**, 26–29 (1979).

8. O. Kratky, H. Leopold, and H. Stabinger, *Methods Enzymol.* **27**, 98−110 (1973).

9. H. Inoue and S.N. Timasheff, *J. Am. Chem. Soc.* **90**, 1890−1897 (1968).

10. S.N. Timasheff and H. Inoue, *Biochemistry* **7**, 2501−2513 (1968).

11. J.A. Schellman, *Annu. Rev. Biophys. Chem.* **16**, 115−137 (1987).

12. J.A. Schellman, *Biopolymers* **26**, 549−559 (1987).

13. T. Arakawa and S.N. Timasheff, *Methods Enzymol.* **114**, 49−77 (1985).

14. S.N. Timasheff and T. Arakawa, *J. Crystal Growth* **90**, 39−46 (1988).

15. S.N. Timasheff and T. Arakawa, in T.E. Creighton, Ed., *Protein Structure, Practical Approach*, IRL Press, Oxford, 1988, pp. 331−345.

16. T. Arakawa and S.N. Timasheff, *Volume and Osmolality Control in Animal Cells*, in press.

17. T. Arakawa, J.F. Carpenter, Y.A. Kita, and J.H. Crowe, *Cryobiology*, **27**, 401−415 (1990).

18. G.C. Na and S.N. Timasheff, *J. Mol. Biol.* **151**, 165−178 (1981).

19. G.J. Scatchard, *J. Am. Chem. Soc.* **68**, 2315−2319 (1946).

20. W.H. Stockmeyer, *J. Chem. Phys.* **18**, 58−61 (1950).

21. H. Inoue and S.N. Timasheff, *Biopolymers* **11**, 737−743 (1972).

22. S.N. Timasheff and M.J. Kronman, *Arch. Biochem. Biophys.* **83**, 60−75 (1959).

23. Z. Er-el, Y. Zaidenzaig, and S. Shaltiel, *Biochem. Biophys. Res. Commun.* **49**, 383−390 (1972).

24. S. Shaltiel and Z. Er-el, *Proc. Natl. Acad. Sci. USA* **70**, 778−781 (1973).

25. S. Hjerten, *J. Chromatogr.* **87**, 325−331 (1973).

26. B.H.J. Hofstee, *Biochem. Biophys. Res. Commun.* **70**, 778−781 (1973).

27. C.C. Bigelow, *J. Theor. Biol.* **16**, 187−211 (1967).

28. T. Arakawa, *Arch. Biochem. Biophys.* **248**, 101−105 (1986).

29. J.P. Hummel and W.J. Dreyer, *Biochim. Biophys. Acta* **63**, 530−532 (1962).

30. E.P. Pittz, J.C. Lee, B. Bablowzian, R. Townend, and S.N. Timasheff, *Methods Enzymol.* **27**, 209−256 (1975).

31. J.C. Lee and S.N. Timasheff, *Biochemistry* **13**, 257−265 (1974).

32. A.R.S. Prasad, R.F. Luduena, and P.M. Horowitz, *Biochemistry* **25**, 739−742 (1986).

33. P.M. Horowitz and N.L. Criscimagna, *Biochemistry* **24**, 2587−2593 (1985).

34. J.F. Busby, D.H. Atha, and K.C. Ingham, *J. Biol. Chem.* **256**, 12140−12147 (1981).

35. E.E. Schrier and E.B. Schrier, *J. Phys. Chem.* **71**, 1851−1860 (1967).

36. K.C. Ingham, B.D. Weintraub, and H. Edelhook, *Biochemistry* **15**, 1720−1726 (1976).

37. K. Gekko and T. Morikawa, *J. Biochem.* **90**, 39−50 (1981).

38. T. Arakawa and S.N. Timasheff, *Biochemistry* **21**, 6536−6544 (1982).

39. T. Arakawa and S.N. Timasheff, *J. Biol. Chem.* **259**, 4979−4986 (1984).

40. T. Arakawa and S.N. Timasheff, *Biochemistry* **21**, 6545−6552 (1982).

41. T. Arakawa and S.N. Timasheff, *Biochemistry* **23**, 5912−5923 (1984).

42. T. Arakawa and S.N. Timasheff, *Arch. Biochem. Biophys.* **224**, 169−177 (1983).

43. T. Arakawa and S.N. Timasheff, *Biophys. J.* **47**, 411−414 (1985).

44. T. Arakawa and S.N. Timasheff, *Biochemistry* **26**, 5147−5153 (1987).

45. T. Arakawa, *Biopolymers* **28**, 1397−1401 (1987).

46. J.F. Carpenter and J.H. Crowe, *Cryobiology* **25**, 244−255 (1988).

47. P.H. Yancey, M.E. Clark, S.C. Hand, R.D. Bowlus, and C.N. Somers, *Science* **217**, 1214−1222 (1982).

48. J.L. Milner, D.J. McClellan, and J.M. Wood, *J. Gen. Microbiol.* **133**, 1851−1860 (1987).

49. E.J. Cohn and J.D. Ferry, in E.J. Cohn and J.T. Edsall, Eds., *Proteins, Amino Acids and Peptides*, Reinhold, New York, 1943, pp. 588−622.

50. S. Pundak and H. Eisenberg, *Eur. J. Biochem.* **118**, 463−470 (1981).

51. S.Y. Gerlsma and E.R. Stuur, *Int. J. Peptide Protein Res.* **4**, 377−383 (1972).

52. S.Y. Gerlsma, *J. Biol. Chem.* **243**, 957−961 (1968).

53. S.Y. Gerlsma, *Eur. J. Biochem.* **14**, 150−153 (1970).

54. S.Y. Gerlsma and E.R. Stuur, *Int. J. Peptide Protein Res.* **6**, 65−74 (1974).

55. K. Gekko and S.N. Timasheff, *Biochemistry* **20**, 4677−4686 (1981).

56. K. Gekko and S.N. Timasheff, *Biochemistry* **20**, 4667−4676 (1981b).

57. M.L. Shelanski, F. Gaskin, and C.R. Cantor, *Proc. Natl. Acad. Sci. USA* **70**, 765−768 (1973).

58. J.C. Lee and S.N. Timasheff, *Biochemistry* **14**, 5183−5187 (1975).

59. G.C. Na, *Biochemistry* **25**, 967−973 (1986).

60. E. Sulkowski, M. Madajewicz, and D. Doyle, in I.M. Chaiken, Ed., *Affinity Chromatography and Biological Recognition*, Academic, New York, 1983, pp. 489−494.

61. V. Zizkovsky, P. Strop, S. Lukesova, J. Korcakorb, R. Dvorak, *Oncodev. Biol. Med.* **2**, 323−330 (1981).

62. N. Bruchovsky, P.S. Rennie, and T. Comeau, *Eur. J. Biochem.* **120**, 399−405 (1981).

63. D.J. Lamb and D.W. Bullock, *J. Steroid Biochem.* **19**, 1039−1045 (1983).

64. M. Gschwendt and W. Kittstem, *Mol. Cell. Endocrinol.* **29**, 251−260 (1980).

65. C.L. Densmore, Y.-C. Chou, and W.G. Luttys, *J. Neurochem.* **50**, 1263−1271 (1985).

66. S.L. Wu, A. Figueroa, and B.L. Karger, *J. Chromatogr.* **371**, 3−27 (1986).

67. J.P. Pivel, E. Munoz, and A. Marquet, *J. Biochem. Biophys. Methods* **10**, 211−219 (1984).

68. V.A. Memoli and G.J. Doellgast, *Biochem. Biophys. Res. Commun.* **66**, 1011−1016 (1975).

69. Z. Hrkal and J. Rejnkova, *J. Chromatogr.* **242**, 385−388 (1982).

70. A.H. Nishikawa and P. Baillon, *Anal. Biochem.* **68**, 274−280 (1975).

71. L.O. Narhi, Y.A. Kita, and T. Arakawa, *Anal. Biochem.*, **182**, 266−270 (1989).

72. S. Pahlman, J. Posengen, and S. Hjerten, *J. Chromatogr.* **131**, 99−108 (1977).

73. Y. Kato, T. Kitamara, and T. Hashimoto, *J. Chromatogr.* **282**, 418−426 (1984).

74. J.L. Fausnaugh, E. Pfannkoch, S. Gupta, and F.E. Regnier, *Anal. Biochem.* **137**, 464−472 (1984).

75. M.J. O'Hare and E.C. Nice, *J. Chromatogr.* **171**, 209−226 (1979).

76. P. Strop, *J. Chromatogr.* **294**, 213−221 (1984).

77. J.L. Fausnaugh and F.E. Regnier, *J. Chromatogr.* **359**, 131−146 (1986).

78a. M. Mevarech, W. Leicht, and M.M. Werber, *Biochemistry* **15**, 2383−2387 (1976).

78b. G. Osthoff, A.I. Louw, and L. Vissy, *Anal. Biochem.* **164**, 315−319 (1987).

79. Y. Popineau, *J. Cereal Sci.* **3**, 29−38 (1985).

80. J. Raymond, J.-L. Azenza, and M. Fotso, *J. Chem.* **212**, 199−209 (1981).

81. C.J. Smyth, P. Jensson, E. Olsson, O. Soderland, J. Rosengren, S. Hjerten, and L. Wadstrom, *Infect. Immun.* **22**, 462−472 (1978).

Fluorescence Techniques for Studying Protein Structure

Maurice R. Eftink, *Department of Chemistry University of Mississippi University, Mississippi*

Methods of Biochemical Analysis, Volume 35: Protein Structure Determination, Edited by Clarence H. Suelter.
ISBN 0−471−51326−1 © 1991 John Wiley & Sons, Inc.

Symbols used*

α_i	amplitude of fluorescence decay component i
β_i	amplitude associated with an individual rotational correlation time
γ	efficiency of quenching reaction
ϵ	molar extinction coefficient
η	viscosity
Θ	average angle between transition moments, i.e., between absorption vector and emission vector
κ^2	orientation factor for resonance energy transfer
λ_{max}	wavelength of fluorescence maximum
τ	fluorescence lifetime (in absence of quencher, τ_0)
$\bar{\tau}$	weighted average fluorescence lifetime, $= \Sigma_i \alpha_i \tau_i^2 / \Sigma_i \alpha_i \tau_i$
$<\tau>$	mean fluorescence lifetime, $= \Sigma_i \alpha_i \tau_i$
ϕ	rotational correlation time
Φ	quantum yield, i.e., for fluorescence (in absence of quencher, Φ_0)
A	acceptor for resonance energy transfer
A_∞	fraction of limiting anisotropy associated with rapid wobbling motion
D	diffusion coefficient
D	donor for resonance energy transfer
d_i	depolarization factor
E	efficiency for resonance energy transfer
f_i	fractional steady-state contribution of component i
I	fluorescence intensity (in absence of quencher, I_0)
J	overlap integral for resonance energy transfer
K_A	association constant for formation of a dark complex (static quenching)
k_b	Boltzmann's constant
k_{ix}	rate constant for intersystem crossing between singlet and triplet states
k_{nr}	rate constant for nonradiative transition from excited state to ground state
k_{pc}	rate constant for photochemical reaction of excited state
k_q	rate constant for quenching reaction
k_r	rate constant for radiative transition from excited state to ground state
k_{ret}	rate constant for resonance energy transfer
K_{sv}	dynamic quenching constant
N'	Avogadro's number per millimole
Q	quencher
R	molecular radius or gas constant
r	fluorescence anisotropy

* Abbreviations for various methods are given in Table V.

R_0 critical distance for 50% efficiency of resonance energy transfer

r_0 limiting anisotropy in the absence of motion

S order parameter

S_0, S_1, S_2 electronic ground state, first and second excited singlet states

T_1 first excited triplet state

V hydrodynamic molar or molecular volume, or static quenching constant

1. INTRODUCTION

Fluorescence spectroscopy is established as a very important and widely used method for studying the structure, dynamics, and interactions of proteins in solution (1−9). The usefulness of this method is due to the rich variety of molecular details that it can reveal about proteins, including the solvent exposure of amino acid side chains, the existence of protein conformers, the rate of rotational diffusion of a protein, and the distance between sites on a protein. This richness of information content, together with the sensitivity (i.e., only a few nanomoles of sample is required) of the measurements and the existence of intrinsic fluorescing groups (and/or the ability to introduce extrinsic groups), has made fluorescence spectroscopy a method of particular importance in studies with proteins.

Some of the spectral properties that can be used to characterize a fluorescing molecule are its excitation and emission spectral contours and maximum positions, quantum yield, fluorescence decay time (lifetime), and polarization properties. In addition, the response of these spectral properties to changes in solvent composition, pH, temperature, or the addition of quenching agents can further be used to characterize a sample. The above fluorescence spectral properties are very responsive to these changes in the environment of a fluorophore. This responsiveness of fluorescence spectroscopy is much greater than that observed in related types of spectral measurements, such as UV-visible absorption spectroscopy.

The key to the environmental responsiveness of fluorescence spectroscopy is that it is in essence a kinetic measurement. Regardless of whether one excites a sample with a continuous lamp, a train of pulses from a lamp or laser, or a sinusoidally modulated beam, the character of the measured fluorescence signal depends on the competition between (1) the rate of the radiative transition of the excited state down to the ground state, and (2) the rate of a variety of other excited-state reactions that may lead to quenching (i.e., nonradiative return to the ground state), entry into the triplet manifold, resonance transfer of the excitation energy to a second chromophore, rotation of the fluorophore, and relaxation of solvent molecules or other polar groups around the excited state dipole. Some of these competing, excited-state processes involve the motion of the fluorophore or the motion of molecules and protein side chains in the neighborhood of the fluorophore. This is the basis for the ability of fluorescence to reveal information on molecular dynamics. The magnitude of the fluorescence decay rate determines the observation time frame. This point was illustrated by Weber (10) through the use of Einstein's relationship for translational motion, $\Delta \times = (2D\tau)^{1/2}$, where $\Delta \times$ is the average distance of translational diffusion for a solute with diffusion coefficient, D, in time interval, τ (which here is the fluorescence lifetime). For a short time interval of 1 ns, a molecule having $D = 2.0 \times 10^{-5}$ cm^2/s will be able to diffuse about 20 Å in water at 20 °C. For a long-lived excited state of 100 ns, a molecule will be able to diffuse a greater average distance of 200 Å.

This chapter will begin with an overview of basic concepts, followed by a description of intrinsic and extrinsic probes that are studied in protein systems. Then we will discuss how fluorescence decay measurements and some of the above-mentioned excited-state processes (rotational diffusion, solute quenching, and resonance energy transfer) can provide information about protein structure and dynamics. We will then present some experimental strategies for separating fluorescence components. A brief overview of advances in experimental methods and analytical procedures will be given, followed by some more quantitative uses of fluorescence spectroscopy of proteins.

The attempt here is not to provide a comprehensive review of the literature or to present experimental details. Instead, the intent is to describe the range of possible fluorescence studies as they apply to the investigation of protein structure and dynamics.

2. BASIC PRINCIPLES

The relationship between the absorption and emission of photons by a substance is illustrated by the Jablonski diagram in Fig. 1. Here S_0, S_1, and S_2 are the electronic ground state and first and second excited singlet states. T_1, is the first excited triplet state. Vibrational levels for each state are shown. The absorption of light into S_0 results in a nearly instantaneous ($\sim 10^{-15}$ s) promotion to the S_1 or S_2 state. This is followed by internal conversion (i.e., $S_2 \rightarrow S_1$) and vibrational relaxation ($\sim 10^{-13}$ s) to the lowest S_1 level. From S_1 several processes can occur to dissipate the excess energy, including (1) radiative decay with rate constant k_r to produce fluorescence, (2) nonradiative interconversion to S_0 by various quenching mechanisms with the sum of rate constants Σk_{nr}, (3) photochemical reactions to produce protoproducts (k_{pc}), and (4) intersystem crossing to the T_1 state with rate constant k_{ix}. If solute quenchers, Q, or resonance energy transfer acceptors, A, are present, deactivation of S_1 will also occur with constants $k_q[Q]$ and k_{ret}, respectively. (The "forbidden" radiative transition from T_1 to S_0 will produce phosphorescence. This type of luminescence will be considered only briefly in this article.)

The quantum yield of fluorescence is defined as

$$\Phi = \frac{k_r}{k_r + \Sigma k_{nr}} \tag{1}$$

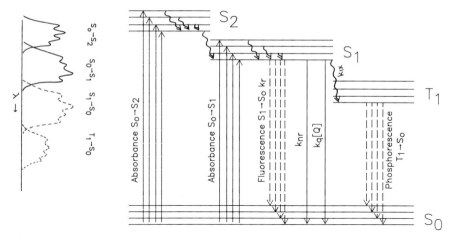

Fig. 1. Jablonski diagram for the absorption of photons from the S_0 ground state to excited singlet states S_1 and S_2. Spectra for the various transitions are shown on the left.

(unless specified, all other nonradiative rates, such as k_{ix} and $k_q[Q]$, are included in Σk_{nr}).

The fluorescence lifetime is defined as

$$\tau = \frac{1}{k_r + \Sigma k_{nr}} \tag{2}$$

These and other equations and plots are collected in Fig. 2 for quick reference.

The fluorescence intensity of a substance is expected to decay in an exponential manner with time. Following an infinitely brief excitation pulse, the impulse–response function is

$$I(t) = I_0 e^{-t/\tau} \tag{3}$$

where I_0 is the intensity at $t = 0$. Integration of the impulse–response function over time yields the steady-state fluorescence intensity, I, for continuous (nonsaturating) excitation of a large ensemble of molecules:

$$I = \int_0^\infty I(t)dt \tag{4}$$

If there is more than one type of excited state, which emits with distinguishable τ_i, the following generalized impulse–response function applies and the decay is said to be heterogeneous or multi–exponential.

$$I(t) = I_0 \sum_{i=1}^{n} \alpha_i e^{-t/\tau_i} \tag{5}$$

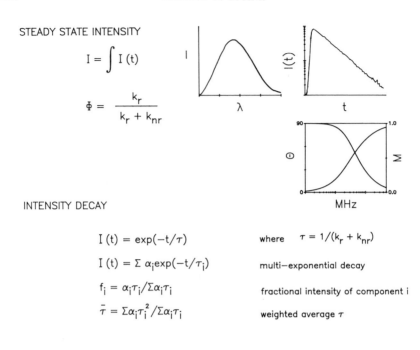

STEADY STATE INTENSITY

$$I = \int I(t)$$

$$\Phi = \frac{k_r}{k_r + k_{nr}}$$

INTENSITY DECAY

$$I(t) = \exp(-t/\tau) \qquad \text{where} \quad \tau = 1/(k_r + k_{nr})$$

$$I(t) = \Sigma\, \alpha_i \exp(-t/\tau_i) \qquad \text{multi-exponential decay}$$

$$f_i = \alpha_i \tau_i / \Sigma \alpha_i \tau_i \qquad \text{fractional intensity of component } i$$

$$\bar{\tau} = \Sigma \alpha_i \tau_i^2 / \Sigma \alpha_i \tau_i \qquad \text{weighted average } \tau$$

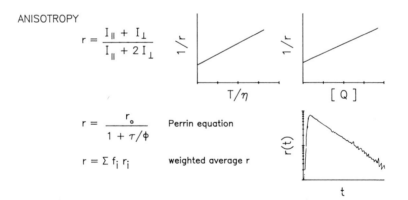

ANISOTROPY

$$r = \frac{I_\parallel + I_\perp}{I_\parallel + 2 I_\perp}$$

$$r = \frac{r_\circ}{1 + \tau/\phi} \qquad \text{Perrin equation}$$

$$r = \Sigma\, f_i\, r_i \qquad \text{weighted average } r$$

Fig. 2. Summary of important relationships and plots for fluorescence studies. For intensity decay, the typical appearance of both time-domain and phase-domain plots is shown. For solute quenching, a plot is shown for the progressive addition of quencher, and a Stern–Volmer plot is shown for the case in which upward curvature (static quenching) and downward curvature (multiple components) is observed. For energy transfer, spectra illustrate the overlap (hashed area) between the emission of the donor, D, and the absorbance of the acceptor, A. See text for further explanation of equations.

SOLUTE QUENCHING

$$I_o / I = (1 + K_{sv}[Q])(1 + K_A[Q])$$
$$\text{or} = (1 + K_{sv}[Q])\exp(V[Q])$$

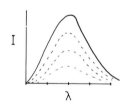

$$\tau_o / \tau_q = 1 + K_{sv}[Q]$$

$$K_{sv} = k_q\tau_o \quad \text{and} \quad k_q = \gamma 4\pi DR_o N'$$

$$I(t) = \exp(-t/\tau_q)$$

where $\tau_q = 1/(k_r + k_{nr} + k_q[Q])$

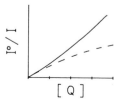

$$I / I_o = \sum \frac{f_i}{(1 + K_{svi}[Q])\exp(V_i[Q])} \qquad \text{multiple components}$$

ENERGY TRANSFER

$$E = 1 - I_{da}/I_{o,d} = 1 - \tau_{da}/\tau_{o,d}$$
$$k_{ret} = (1/\tau_{o,d})E/(1-E)$$
$$= (1/\tau_{o,d})(R_o/r)^6$$

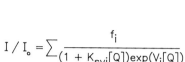

where $R_o = 9.79\times10^3 (\kappa^2\eta^{-4}\phi_d J)^{1/6}$ critical distance in Å

and $J = \displaystyle\int_0^\infty \frac{I_d(\nu)\epsilon_a d\nu}{\nu^4}$ overlap integral

and $\kappa^2 = (\cos\theta_{da} - 3\cos\theta_d \cos\theta_a)^2$ orientation factor

Here the α_i are amplitudes associated which each τ_i. Likewise, $I = I_0\Sigma\alpha_i\tau_i$ and the fractional contribution of component i to the steady-state emission is

$$f_i = \frac{\alpha_i\tau_i}{\Sigma\alpha_i\tau_i} \qquad (6)$$

and the weighted average fluorescence lifetime is

$$\overline{\tau} = \frac{\Sigma \alpha_i \tau_i^2}{\Sigma \alpha_i \tau_i} \tag{7}$$

The amplitudes, α_i, can be considered to be proportional to the excitation probabilities of the different ground-state species. These equations apply to the case in which all of the emission is collected and there are no excited-state reactions to interconvert the species. If only a narrow wavelength range is collected, the α_i (and f_i) may be wavelength dependent due to the different spectral contours of the emission spectra of the species. Equations for excited-state reactions are given below.

The $S_0 \rightarrow S_1$ absorptive transition will usually lead to the population of higher vibronic levels in S_1. Consequently, the absorption (or excitation) spectrum of a substance will sometimes show a series of peaks corresponding to the vibronic levels of S_1. A $0-0$ peak, corresponding to the transition between the lowest vibronic levels, can sometimes be observed at the red edge of the absorption spectrum. Often the absorption spectrum is too broad at room temperature to identify such features. From the lowest vibronic level of S_1, the radiative transition to S_0 will also lead to the population of higher vibronic levels of the ground state. As a consequence, (1) the vibronic features of the S_0 state can sometimes be observed as peaks (or, more often, shoulders) in the emission spectrum, (2) the excitation spectrum and emission spectrum often appear to be mirror images (unless there is overlap with additional $S_0 \rightarrow S_2$ electronic transitions), and (3) there will be a red shift, or so-called Stokes shift, in the emission spectrum. These features are all illustrated in Fig. 1. In solution at room temperature, the vibronic transitions are usually too broadened to be observed. Also, polar solvents will often interact with the enhanced dipole moment of the excited state to lead to a further red shift in an emission spectrum. In viscous solution this solvent relaxation can occur on the same time scale as $1/k_r$, resulting in emission spectra that red shift during the decay process (i.e., longer lived excited states having redder emission).

If the exciting light is made to be plane polarized, this will result in a photoselection of those fluorophores that have their excitation transition moment parallel to the plane of polarization. For polyaromatic hydrocarbons, the $S_0 \rightarrow S_1$ absorption transition moment is usually parallel to the longest axis of the molecule. Rotational diffusion of the excited molecule before emission will lead to a loss of the anisotropic distribution of states. By monitoring the emission that occurs parallel, $I_\parallel$, and perpendicular, $I_\perp$, to a vertically polarized excitation beam, the anisotropy, r, of the emission is

$$r = \frac{I_\parallel - I_\perp}{I_\parallel - 2I_\perp} \tag{8}$$

(A term used less often is polarization, P, which equals $(I_\parallel - I_\perp)/(I_\parallel + I_\perp)$.)

The steady-state anisotropy of the emission of a molecule depends on the average angle, Θ, in space between a vector representing the absorption transition and a vector representing the emission transition orientation for an ensemble of chromophores:

$$r = \frac{\overline{3 \cos^2\Theta} - 1}{5} \tag{9}$$

If the absorption and emission oscillators remain parallel (i.e., $\theta = 0$, no reorientation occurs during the excited state lifetime), then an anisotropy of 0.40 will be observed. This value is referred to as the limiting or fundamental anisotropy, r_0.

Several processes, however, can cause Θ to not be zero. Among these are (1) electronic effects, such as the existence of a difference in the orientation of the absorption $S_0 \to S_1$ and emission $S_1 \to S_0$ transition oscillators in the plane of the chromophore, or a difference in the $S_0 \to S_2$ and $S_1 \to S_0$ orientation (i.e., excitation into a higher energy level followed by internal conversion to S_0 and emission); (2) motional effects, which lead to a reorientation of the molecule during the excited-state lifetime; and (3) energy transfer effects, in which there is a difference in orientation between the donor and acceptor (see below). These depolarizing processes are illustrated in Fig. 3. Each of these processes will depolarize the emission by some depolarizing factor, d. If more than one process occurs, the observed r will be the product of r_0 and the various factors, that is,

$$r = r_0 \prod d_i \tag{10}$$

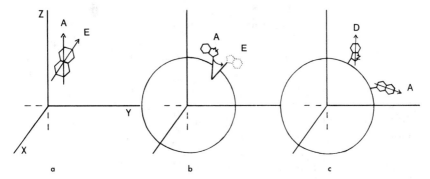

Fig. 3. Causes for depolarization of emission, following polarized excitation (in the Z direction). (a) The absorption moment, A, and emission moment, E, are not parallel in the chromophore. (b) Segmental rotation of the chromophore occurs (also global rotation of the sphere may occur). (c) Resonance energy transfer between the initially excited donor, D, and the acceptor, A, occurs.

The individual depolarizing factors, d_i, are related to the average angle, Θ_i, of reorientation by a given process through the following relationship

$$d_i = \frac{3 \cos^2\Theta_i - 1}{2} \tag{11}$$

For example, if neither motional nor energy transfer processes occur, and if the absorption and emission oscillators are not exactly parallel, but lie at a 21° angle with respect to one another, then $r = r_0 [3 \cos^2(21°) - 1]/2 = 0.32$. In fact, it is often observed that the apparent limiting anisotropy, $r_0(\text{app})$, in the absence of motion or energy transfer, is less than 0.4, suggesting that the two electronic oscillators are not parallel.

The relationship between r values and molecular motion is of particular imporatnce for the study of dynamics. For an isotopically rotating sphere, having a rigidly affixed chromophore, the steady-state anisotropy is related, by the modified Perrin equation, to the fluorescence lifetime, τ_0, and the rotational correlation time, ϕ, as follows:

$$r = \frac{r_0(\text{app})}{1 + \tau/\phi} \tag{12}$$

The value of ϕ is an inverse measure of the speed of rotation of the sphere and can be related to the rotational diffusion coefficient, D_R, and the effective hydrodynamic molar volume, V, of the sphere by

$$\phi = \frac{1}{6D_R} = \frac{V\eta}{RT} \tag{13}$$

where η is viscosity and R is the gas constant. By combining Eqs. 12 and 13 it can be seen that a plot of $1/r$ versus T/η should be linear, with intercept equal to $1/r_0(\text{app})$ and with a slope/intercept that is directly proportional to τ and indirectly proportional to V. If one of the latter two parameters is known, the other can be calculated from such data. The absence of a viscosity dependence of r indicates that some other depolarizing process dominates. A nonlinear plot of $1/r$ versus T/η may indicate the existence of more than one type of rotational mode (see below).

The anisotropy of a fluorescence signal may also decay as a function of time as follows, if motional depolarization occurs:

$$r(t) = r_0(\text{app})e^{-t/\phi} \tag{14}$$

The above equation applies for a spherical rotor with a monoexponential intensity decay. If the shape of the macromolecule is asymmetric, the anisotropy decay should appear to be a multiexponential process (see Section 4). If there are multiple emitting centers (i.e., ground-state heterogeneity), the anisotropy decay will likely be multiexponential, as follows:

$$r(t) = r_0(\text{app})\sum\beta_i\, e^{-t/\phi_i} \tag{15}$$

where β_i and ϕ_i are the amplitude and correlation times associated with each component. Analysis of such multiexponential anisotropy decays is complex. Steady-state r values will be the weighted sum of the anisotropy of the individual components,

$$r = \sum f_i\, r_i \tag{16}$$

Even for single-component systems, the anisotropy decay can be multi-exponential if different types of depolarizing motions exist. Often it appears that fluorophores attached to proteins depolarize both by (1) global rotation of the protein and by (2) localized side-chain rotational motion. The latter type of motion has been modeled as the wobbling of the fluorophore side chain within a cone (12–14); this model will be presented in Section 4.

Certain agents, known as quenchers, are able to strongly interact with excited states and lead to their deactivation with rate constant k_q (15). For quenchers that act by competing with the radiative process (so called dynamic or collisional quenchers), the ratio of the quantum yield in the absence, Φ_0, and presence, Φ, of quencher concentration, $[Q]$, is described by the Stern–Volmer equation, where the Stern–Volmer constant, K_{sv}, is equal to $k_q\,\tau_0$:

$$\frac{\Phi_0}{\Phi} = 1 + k_q\,\tau_0\,[Q] = 1 + K_{sv}[Q] \tag{17}$$

For a purely dynamic quencher, this Φ_0/Φ ratio will also be equal to τ_0/τ_q, the ratio of lifetimes in the absence and presence of quencher. Thus, from a plot of Φ_0/Φ (or the ratio of the fluorescence intensity, I_0/I) or τ_0/τ_q versus $[Q]$, the product $k_q\,\tau_0$ can be obtained. If τ_0 is known separately, the bimolecular quenching rate constant, k_q, can be determined. The magnitude of k_q is expected to be given by the product of the following factors:

$$k_q = \gamma 4\pi DRN' \tag{18}$$

where γ is the efficiency (ranging from 0 to 1.0) of the quenching reaction, D and R are the sum of the diffusion coefficients and molecular radii, respectively, for the quencher and fluorophore, and $N' = 6.02 \times 10^{20}$. (Equation 18 is the time-independent portion of the Smoluchowski equation for a diffusion limited reaction. There is also a transient term that is of some significance in fluorescence quenching reactions (16, 17). The diffusion coefficient for species i can be estimated from the Stokes–Einstein relationship to be

$$D_i = \frac{k_b T}{6\pi\eta R_i} \qquad (19)$$

where k_b is Boltzmann's constant and η is viscosity. For an efficient, collisional quenching reaction, the magnitude of k_q is predicted to be directly proportional to T/η.

Other quenchers act in a way that does not compete with the radiative process and does not involve diffusion. These are called static quenchers. An equation analogous to Eq. 18 applies, but for static quenchers no change in the fluorescence lifetime occurs. For most quenchers, both a dynamic and a static quenching component appears to exist. A relationship that can usually describe those cases in which both mechanisms operate is given, where V is the static quenching constant:

$$\frac{\Phi_0}{\Phi} = \frac{I_0}{I} = (1 + K_{sv}[Q])e^{V[Q]} \qquad (20)$$

The presence of both quenching modes results in upward curving plots of I_0/I against $[Q]$. The V value can be interpreted as an association constant for a "dark" quencher–fluorophore complex, or as a volume element surrounding the fluorophore (3, 15, 18). In the latter interpretation, if one or more Q molecules are located within this active volume, at the instant of excitation of the fluorophore, intantaneous quenching occurs. (An alternative to Eq. 20 replaces the exp($V[Q]$) term with its approximate, $(1 + K_A[Q])$, where K_A is an association constant for the 'dark' complex.)

The above equation applies for homogeneously emitting systems. If there is ground-state heterogeneity, then more than one $K_{sv,i}$ (and V_i) should be needed. The more general form of the Stern–Volmer equation is

$$\frac{I}{I_0} = \sum \frac{f_i}{(1 + K_{sv,i}[Q])\exp(V_i[Q])} \qquad (21)$$

and, since the fractional contribution of each component, f_i, may depend on wavelength, the ratio I/I_0 may also be wavelength dependent.

While the above types of quenchers usually appear to require contact with the excited state, resonance energy transfer (ret) is a process for quenching over larger distances. A ret acceptor, A, is a molecule that has an absorption transition that overlaps with the emission spectrum of the energy donor, D (see Fig. 2). This overlap integral, J, can be evaluated via

$$J = \int_0^\infty \frac{I_d(\nu)\ \varepsilon_a(\nu)\ d\nu}{\nu^4} \qquad (22)$$

$I_d(\nu)$ is the relative intensity of the fluorescence of the donor (with total intensity normalized, to unity) at wavenumber, v, and $\varepsilon_a(v)$ is the molar exctinction coefficient of the acceptor at that v. According to Forster theory, the critical distance, R_0, at which ret occurs with 50% efficiency is related to J and an orientation factor, κ^2, as follows (3):

$$R_0 \ (\text{Å}) = 9.79 \times 10^3 (\kappa^2 \eta^{-4} \Phi_d J)^{1/6} \tag{23}$$

where η is the refractive index of the medium separating D and A, and Φ_d is the fluorescence quantum yield of D (in the absence of A). The rate constant for energy transfer, k_{ret}, is predicted to depend on the $D \to A$ separation distance and the fluorescence lifetime of D (absence of A).

$$k_{ret} = \frac{1}{\tau_{0,d}} \left(\frac{R_0}{r} \right)^6 \tag{24}$$

The magnitude of k_{ret} can be determined from the efficiency of energy transfer, E, via

$$k_{ret} = \frac{1}{\tau_{0,d}} \frac{E}{(1 - E)} \tag{25}$$

and E, in turn, can be experimentally evaluated by measurement of the drop in fluorescence intensity, I_{da}, or lifetime, τ_{da}, of the donor, D, in the presence of the acceptor, A (19, 20):

$$E = 1 - \frac{I_{da}}{I_{0,d}} = 1 - \frac{\tau_{da}}{\tau_{0,d}} \tag{26}$$

Thus, by determining E and by knowing $\tau_{0,d}$, Φ_d, J, and κ^2, the $D \to A$ separation distance, r, can be calculated. Of key importance is the value of the orientation factor, κ^2, which depends on the relative orientation of the emission oscillator of D and absorption oscillator of A. These are depicted in Fig. 4, which defines the angles Θ_T, Θ_D and Θ_A (21). κ^2 is related to these angles by

$$\kappa^2 = (\cos\Theta_T - 3\cos\Theta_D \cos\Theta_A)^2 \tag{27}$$

and κ^2 can assume a value ranging from 0 to 4.0. As discussed in Section 6.2, evaluation (or setting a limit on the range) of κ^2 is needed in order to use ret to determine distances. Often a value of $\kappa^2 = \frac{2}{3}$ is assumed. This value corresponds to the condition in which there is rapid, isotropic rotation of the D and A moments, so that all mutual orientations are rapidly sampled before the ret process. When κ^2 is assumed to be $\frac{2}{3}$, the R_0 from Eq. 23 can be referred to as $R_0^{2/3}$.

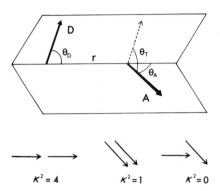

Fig. 4. Directions of the donor, D, and acceptor, A, transition moments, with definition of angles, Θ_D, Θ_A, and Θ_T. Lower part of diagram shows values of κ^2 for certain mutual orientations.

In addition to the above-mentioned collisional quenching and ret processes, other excited-state reactions include proton transfers, eximer or exiplex formation, dipolar relaxation with solvent or nearby polar groups, and isomerization processes (3). These processes differ from the previously described quenching reactions in that emission occurs from the product state. Here I briefly consider the fluorescence decay kinetics of a generalized excited-state reaction in which emission occurs from both the reactant and product states. (In some ret processes, emission from the acceptor is measured and so falls into this category.)

Consider an irreversible, two-state excited-state reaction, as given in Fig. 5. k_{ra} and k_{rb} are radiative rate constants for excited states A* and B*, respectively. Likewise, k_{nra} and k_{nrb} are the sum of nonradiative rate constants. k_{ba} is the rate constant for conversion of A* to B*. In general, the emission contours of A* and B* will not be the same. If the A* $\rightarrow$ B* reaction is irreversible, then the emission of B will be red-shifted with respect to that of A, as shown in Fig. 5. With slight modification, this two-state reaction could model a proton shift, eximer formation, or a ret reaction (3, 22, 23). Solvent relaxation reactions may be better described as multistate processes, but a two-state model at least presents the basic features of such a reaction (3).

The time and wavelength dependence of the intensity of A* and B* (following excitation into both states), for such a reacting sytem, is given by

$$I_A(\lambda, t) = \sigma_a C_a(\lambda) k_{ra} \exp[-t(k_{nra} + k_{ra} + k_{ba})] \tag{28a}$$

$$I_B(\lambda, t) = \frac{-\sigma_a C_b(\lambda) k_{rb} k_{ba} \exp[-t(k_{nra} + k_{ra} + k_{ba})]}{k_{nra} + k_{ra} + k_{ba} - k_{nrb} - k_{rb}} \tag{28b}$$

$$+ \left(\sigma_b C_b(\lambda) k_{rb} + \frac{\sigma_a C_b(\lambda) k_{rb} k_{ba}}{k_{nra} + k_{ra} + k_{ba} - k_{nrb} - k_{rb}} \right) \exp[-t(k_{nrb} + k_{rb})]$$

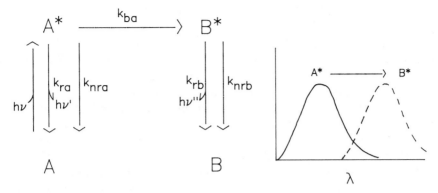

Fig. 5. Reaction scheme for a generalized excited-state reaction, with initial excitation into A and emission from both A* and B*, assuming the A* → B* reaction to be irreversible. The scheme can be expanded to include reversibility, excitation into B, and the excited state reaction can be written to be a dimerization, proton dissociation, etc.

Here, σ_a and σ_b are the normalized, relative absorption coefficients of A and B at the excitation wavelength (i.e., $\sigma_a + \sigma_b = 1.0$). $C_a(\lambda)$ and $C_b(\lambda)$ are the spectral shape and position functions for A* and B*.

Usually, direct measurement of the decay of A* and B* will not be feasible due to overlap. The emission at the red edge will be primarily that of B*. However, in most cases the total emission, $I(\lambda, t)$, will be observed and the impulse response function will appear to be a bi-exponential:

$$I(\lambda, t) = \alpha_1(\text{app})\, e^{-t/\tau_1} + \alpha_2(\text{app})\, e^{-t/\tau_2} \tag{29}$$

where $\tau_1 = 1/(k_{nra} + k_{ra} + k_{ba})$, $\tau_2 = 1/(k_{nrb} + k_{rb})$ and the α_i (app) are

$$\alpha_1(\text{app}) = \sigma_a \left(C_a(\lambda)k_{ra} - \frac{C_b(\lambda)k_{rb}k_{ba}\, \sigma_a/\sigma_b}{k_{nra} + k_{ra} + k_{ba} - k_{nrb} - k_{rb}} \right) \tag{30a}$$

$$\alpha_2(\text{app}) = \sigma_b \left(C_b(\lambda)k_{rb} + \frac{C_b(\lambda)k_{rb}k_{ba}\, \sigma_a/\sigma_b}{k_{nra} + k_{ra} + k_{ba} - k_{nrb} - k_{rb}} \right) \tag{30b}$$

It is important to note that the amplitudes for the two decay times are not simply related to the ground-state population of A and B (i.e., not related to the σ_a/σ_b ratio). Instead, the amplitudes are a function of rate constants. Notice that the amplitude for τ_1 contains a negative term. As a result it will sometimes be possible for $\alpha_1(\text{app})$ to be negative. Such a negative amplitude is a clear indication of an excited-state reaction.

For further discussion of the photophysics of excited-state reactions, see references (3, 7, 22, 23).

3. INTRINSIC AND EXTRINSIC PROBES

The amino acids phenylalanine, tyrosine, and tryptophan are capable of contributing to the fluorescence of proteins. When all three are present in a protein (so called class B protein), pure emission from tryptophan can be obtained by photoselective excitation at wavelengths above 295 nm. Below 290 and 270 nm there is an onset of absorption into tyrosine and phenylalanine, respectively, but resonance energy transfer can occur between these residues and tryptophans so that emission from the latter usually is dominate for class B proteins (25) (see (25) and (27) for a rare exception). Since there are relatively few tryptophan residues in a typical protein and since its fluorescence properties are responsive to its environment (see below), this residue is the most valuable intrinsic fluorescence probe in proteins. For proteins that have only phenylalanine and tyrosine (class A proteins), emission from the latter is usually the strongest. Shown in Fig. 6 are the emission spectra of tyrosine and tryptophan in neutral aqueous solution.

In addition to these three amino acids, there are a number of natural, fluorescent biomolecules that can bind covalently or noncovalently to proteins. For example, the coenzyme NADH (or NADPH) is an important ligand/ substrate for many enzymes and its fluorescence is often enhanced upon interaction with proteins. Other fluorescent coenzymes are the flavin mono- and dinucleotides, FMN and FAD. Table I, which updates a similar table from Cantor and Schimmel (28), presents spectral information for various intrinsic fluorescence probes that are pertinent to proteins. Table II includes several extrinsic probes that form covalent or noncovalent adducts with proteins and thus have been used in various biophysical studies of protein structure.

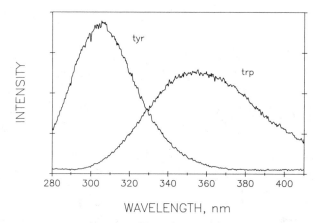

Fig. 6. Emission spectra of L-tyrosine (left) and L-tryptophan (right) in pH 7 aqueous buffer at 20 °C. Both excited at 280 nm; initial absorbance of 0.1 at this wavelength. Spectra collected with an intensified diode array detector.

TABLE I.

Fluorescence Properties of Intrinsic Fluorescence Probes and Natural Ligands for Proteins[a]

Substance	Absorption		Fluorescence			
	λ^{max} (nm)	ε ($\times 10^{-3} M^{-1} cm^{-1}$)	λ^{max} (nm)	Φ	τ_o (ns)	References
Tryptophan	280	5.6	355	0.13	<2.8>	9, 40, 41
Tyrosine	275	1.4	304	0.14	3.3	66, 67
Phenylalanine	258	0.2	282	0.02	6.8	28, 304
NMNH	337	6.5	473	0.0013	<0.36>	166, 168
NADH	339	6.2	470	0.02	<0.42>	166–169
NADH–LDH complex	339	5.4	435	0.32	6.2	200
Riboflavin	370,445	–	525	0.26	4.68	217
FMN	450	12.5	536	0.25	4.7	213
FAD	450	11	530	0.03	<2.11>	213
Pyridoxamine phosphate	325	8.3	390	0.14(0.55)	–	121–123
Lumazine	408	–	490	0.45	9.2	217
Guanosine[b]	275	8.1	329	3×10^{-4}	<0.02>	28
cis-Parinaric acid	320	80	420	0.001 (0.046)[c]	(4.6)[c]	124
Equilenin	335	22–27	360	–	5.8	125–127
Colchicine–tubulin complex	362	17–20	435	0.03	–	144, 145

[a] All data are for neutral aqueous solution at ~20°C unless stated otherwise. The angle brackets indicate a mean fluorescence lifetime, $\langle \tau \rangle = \Sigma \alpha_i \tau_i$, for a multiexponential decay.

[b] Other nucleosides have lower quantum yield.

[c] Values in hexane.

143

TABLE II.
Fluorescence Properties of Extrinsic Fluorescence Probes for Proteins[a]

Substance	Absorption λ_{max} (nm)	Absorption ε ($\times 10^{-3} M^{-1} cm^{-1}$)	Fluorescence λ_{max} (nm)	Fluorescence Φ	Fluorescence τ_o (ns)	References
Dansyl chloride	330	3.4	510	~0.1	~13	28, 98, 130
1, 5-IEDANS	360	6.8	480	~0.5	~15	28, 98
Prodan	364(342)[b]	14.5	531(401)[b]	—	2.2(1.7)[b]	128
Acrylodan	360	12.9	540	0.18	1.4	129
Fluorescein derivatives (i.e., FITC, IAF)	495	42–85	516–525	0.3–0.5	~4	28, 98, 130
Rhodamine derivatives	560	12	580	~0.7	~3	130
Pyrene derivatives	342	40	383	0.25	~100	28, 98
1, 8-ANS	355	~6	515	0.004	0.25	131, 132, 161
1, 8-ANS-apoMb	374	6.8	454	0.98	—	131
2, 6-TNS	317(366)	18.9(4.08)	500	0.0008	—	133, 134
NBD derivatives	470	25	530	0.1–0.5	—	98
Coumarin derivatives	~390	26–30	~460	~0.7	—	98
Salicylic acid derivatives	~320	—	~410	~0.44	~2	98
1, N^6-Ethenoadenosine[c]	294	3.1	410	0.56	20	135, 136
Formycin	295	9.6	340	0.06	<1	137
2-Aminopurine	303	7.1	370	0.68	<7	137
2, 6-Diaminopurine ribonucleoside	280	~10	350	0.01	<1	137
Protoporphyrin IX[d]	622	5.2	624	—	14.1	141, 142, 347
	(402 Soret)	143.7				

[a] All data are for neutral aqueous solution at ~20°C unless stated otherwise. The absorption λ_{max} is usually the wavelength of the lowest energy transition. Values for dansyl chloride, IEDANS, fluorescein, rhodamine, pyrene, coumarin, and salicylic acids are approximate values for protein adducts.

[b] For Prodan, the values in parentheses are for cyclohexane as solvent.

[c] A large variety of nucleotide derivatives of ethenoadenosine are available (136).

[d] Zinc and tin porphyrins also fluoresce and phosphoresce and show delayed fluorescence (142, 143).

144

We will not elaborate on most of these, but will discuss some of the properties of tryptophan, tyrosine, NADH, and flavin nucleotides, due to their importance.

3.1. Tryptophan

The fluorescence of tryptophan, and its parent indole, is complex and has been extensively studied (9, 29–51). Among the important characteristics of tryptophan fluorescence are (1) the strong solvent dependence of its Stokes shift, (2) the pH dependence of its quantum yield, (3) the double exponential kinetics of its fluorescence decay at neutral pH, and (4) the existence of two overlapping $S_0 \rightarrow S_1$ electronic transitions (^1L_a and ^1L_b). Our understanding of these properties for tryptophan is important for our interpretation of protein fluorescence.

The $S_0 \rightarrow {}^1L_a$ and $S_0 \rightarrow {}^1L_b$ transitions both occur in the 260- to 300-nm range. The ^1L_b absorption transition of indole has the lowest energy $0 \rightarrow 0$ onset in nonpolar solvent and in the gas phase (36, 47). This transition appears to be insensitive to the solvent environment and shows vibronic features (39). The ^1L_a absorption transition of indole, on the other hand, shows no vibronic features in condensed phases and its onset (or its maximum) depends greatly on the polarity of the solvent (33, 36, 47). In nonpolar solvents, ^1L_a has a higher energy level than ^1L_b. Polar solvents lower the energy level of ^1L_a, making this the lowest energy singlet state. This inversion of energy levels is believed to be due to the fact that the ^1L_a transition has the higher dipole moment of the two and is directed through the ring NH group, which enables this excited state moment to have dipole–dipole interactions with polar solvent molecules (see Fig. 7). The exact nature of this ^1L_a–solvent interaction has been a subject of much discussion (see below). Figure 8 shows the emission spectra of indole in nonpolar and polar solvents.

Whether the ^1L_a (polar solvent) or ^1L_b (nonpolar solvent) is the lowest S_1 state, equilibration between these two excited states is believed to be very rapid so that only emission from the lower S_1 state is observed. Fleming and coworkers (52) have provided evidence that this $^1L_a \rightarrow {}^1L_b$ interconversion occurs on the time scale of a few picoseconds. Dual emission from both ^1L_a

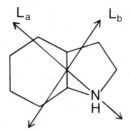

Fig. 7. Transition moment directions for the ^1L_a and ^1L_b transitions of indole. [Drawn following reference 38.]

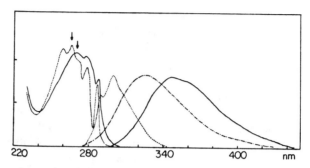

Fig. 8. Absorption and emission spectra of indole in various solvents: ---, hexane, .-., methanol (only emission shown); and — water. The ordinate refers to arbitrary units for absorption and fluorescence. [Taken from reference 32 with permission from the authors and Pergamon Press.]

and ^1L_b states (i.e., almost exact degeneracy of the two states) has been observed only for a few indole derivatives (37, 53). Since the ^1L_a and ^1L_b transition moments lie almost perpendicular with respect to one another in the plane of the indole ring (37, 38), excitation anisotropy spectra reveal the overlapping contributions of the two transitions (39).

As shown in Fig. 9 for indole derivatives in propylene glycol glass, absorption at the red edge photoselects the lowest energy S_1 (^1L_a in this case) and thus the anisotropy is high, since only depolarization due to small angular differences between the absorption and emission transition moments occurs. Excitation at lower wavelengths populates both ^1L_a and ^1L_b states. Equilibration between these two produces a depolarization due to the $\approx90°$ angular difference between the moments. Thus, near 289 nm there is sharp dip in the anisotropy due to the maximum in the of absorption into the ^1L_b state.

The large Stokes shift found for indole in polar solvents appears to also be due to the interaction between the ^1L_a excited dipole moment and the dipoles of solvent molecules, an interaction that has been described as an "exiplex" of defined stoichiometry by Walker et al. (31). A ground state indole—solvent complex can also be detected under some conditions (35). Meech et al. (47) have presented data and arguments that the Stokes shift of indole can best be described in terms of a partial charge-transfer complex between indole's ^1L_a state and polar solvent molecules. Time-resolved fluorescence measurements of indole in viscous solvents or in nonpolar:polar solvent mixtures have shown there to be a dynamic red shift in the emission of indole during its excited state, as expected for a dipolar relaxation between the excited indole and polar solvent molecules (54, 55).

For tryptophan residues in proteins, the range of Stokes shifts is extremely large. Fluorescence λ_{max} range from 308 nm for azurin (*Pseudomonas fluorescence*) to ~350 nm from glucagon and other small peptides, as shown in Fig. 10. This wide range of λ_{max} has provided a convenient means of gaining some information

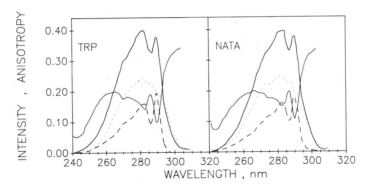

Fig. 9. Low-temperature (−50 °C) excitation and anisotropy spectra for L-tryptophan and *N*-acetyl-L-tryptophan amide (NATA) in propylene glycol. Also shown are the resolved spectra for the ^1L_a (...) and ^1L_b (---) transitions, as calculated from the data following the procedure of Valeur and Weber (39). [Reprinted from reference 53 with permission from the American Chemical Society.]

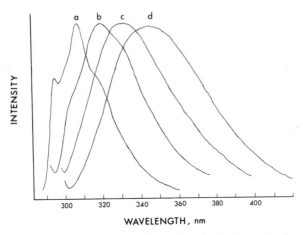

Fig. 10. Uncorrected emission spectra for selected single tryptophan proteins: a, apoazurin, Pfl; b, ribonuclease T_1; c, nuclease A; d, glucagon. Excitation at 280 nm, with 2-nm slits, for apoazurin. Excitation at 290 nm, with 5-nm slits for ribonuclease T_1. Excitation at 295 nm, with 5-nm slits, for nuclease and glucagon. Spectra were adjusted to be of similar height. All spectra obtained at 20 °C in aqueous buffer near pH 7.

about the environment of the tryptophan residues. Of course, for multitryptophan-containing proteins, the λ_{max} will be the sum of all contributions.

Since proteins provide an asymmetric, structured environment for the indole rings, the Stokes shift for a residue will depend on (1) the static polarity of the immediate microenvironment of the ring, and (2) the dynamic ability of polar

groups (solvent molecules) in the microenvironment to undergo dipolar relaxation with the excited indole dipole. The latter relaxation appears to
require accessibility to the imino NH of indole. The λ_{max} of protein fluorescence
has been shown to blue shift, from the room temperature value, as the solution
is frozen to -100 to $-200\,^{\circ}C$ (56, 57, 317), supporting the notion that thermal
motion is required for dipolar relaxation and that the λ_{max} is not simply a
measure of the hydrophobicity of the tryptophan environment. Further support
for the dynamic basis for the Stokes shift comes from the observation of time
resolved emission spectra (TRES) of single tryptophan proteins. These TRES
evolve from blue to red during the lifetime of the excited state (58). Also,
Demchenko (50) observed that the emission λ_{max} of proteins can red shift as
the excitation λ is increased. The interpretation offered for this finding is that
excitation at the red edge populates those tryptophan residues that have
already interacted with polar groups (or water molecules) in the ground state.

The lowest S_1 for tryptophan residues in proteins is considered to be a state
of ^1L_a origin. The identity of the lowest S_1 in very blue emitting proteins, such
as azurin, parvalbumin, asparaginase, and ribonuclease T_1, is not certain. The
fluorescence of these proteins shows a slight amount of fine structure, which is
suggestive of ^1L_b-type emission.

The fluorescence quantum yield and lifetime of tryptophan residues in
proteins can vary greatly and must reflect the different quenching processes
that can be experienced. Table III lists these and other fluorescence parameters
for several single-tryptophan-containing proteins. For example, the yield of
Cu(I)-azurin is only 0.052. Removal of the metal ion results in a quantum
yield of 0.31 for apoazurin (59). Likewise, the yield of tryptophan fluorescence
in heme proteins, such as tuna met-myoglobin, can be very low due to
resonance energy transfer to the heme group (60). Removal of the heme group
can produce a very large increase in tryptophan fluorescence. The fluorescence
intensity of Trp-113 of subtilisin Carlsberg is surprisingly low, in comparison
to the contribution from tyr residues in this protein (27). Trp-113 lies on the
surface of subtilisin and the crystal structure does not reveal any obvious
quenching mechanism. At the other extreme, one of the tryptophan residues of
3-phosphoglycerate mutase has a very long fluorescence lifetime (~ 16 ns);
again, there is not an obvious explanation for this long lifetime. There have
been attempts to draw correlations between the quantum yield (and lifetime)
of tryptophan residues and their solvent accessibility. In the opinion of this
reviewer, no reliable correlations exist. (In fact, plots of τ or Φ versus emission
λ_{max}, for the proteins in Table III, show no correlation.)

Whereas the fluorescence of indole and 3-methylindole has a monoexponential decay and no pH dependence between pH 3 and 11 (at lower and
higher pH, quenching by protons, hydroxide, or dissociation of the imino NH
occurs), the fluorescence of tryptophan shows biexponential decay kinetics and
a pH dependence attributable to dissociation of both its α-carboxylic acid and
α-amino group. Based on results from several laboratories, lifetimes of $\tau_1 =$
3.1 ns and $\tau_2 = 0.5$ ns are found for the zwitterion of tryptophan. The pH

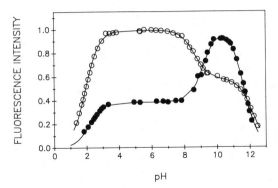

Fig. 11. pH dependence of the steady-state fluorescence intensity of aqueous solutions of L-tryptophan (●) and the rigid tryptophan analogue, 3-amino-3-carboxy-1,2,3,4-L-tetrahydro-carbazole (○). Conditions, 20 °C, 0.1 M NaCl, excitation at 295 nm. The solid lines are theoretical fits to a function that includes the excited state pK_a values for the α-carboxylic acid and α-amino groups. These values are $pK_1 = 2.6$ and $pK_2 = 9.21$ for L-trp.

dependence of the steady-state intensity of tryptophan is shown in Fig. 11.* These two effects will be considered together, since they involve interaction of the ground state or excited state of indole with the alanyl group.

The biexponential decay of tryptophan was early postulated to be due to dual emission from the ^1L_a and ^1L_b excited states (40). However, the observation of a monoexponential decay for indole, 3-methylindole, and N-acetyl-L-tryptophanamide (NATA) shifted attention to the interaction with the zwitterion alanyl side chain. The existence of different rotamers, about the $C^\alpha-C^\beta$ (and/or the $C^\beta-C^\gamma$) bond of tryptophan was also suggested by Szabo and Rayner (40) to provide the basis for explaining the biexponential decay. Rotamers having the α-amino group close to the indole ring were proposed to partially quench and produce the 0.5-ns decay component. This quenching process explains the biexponential decay and the increase in fluorescence intensity between pH 8 and 10, as the α-amino group undergoes proton dissociation.

There has been debate as to the actual mechanism of quenching by the side chain (40–45). This may involve proton transfer from the α-ammonium group to position 2 or 4 of the indole ring (48, 49). Alternatively, quenching may involve electron transfer from the excited indole to the ammonium or carboxylate groups (43). Since electron ejection into solvent water is known to occur for the indole ring alone, it is reasonable to expect the alanyl group to

---

* This figure also shows the pH dependence of the fluorescence of a "rigid" analog of tryptophan. This analog has been studied to test the rotamer model mentioned in the following paragraph. Tilstra et al (76) and our group find the fluorescence decay of rigid tryptophan to be more nearly a monoexponential than is the decay for tryptophan. The distinctly different pH dependence for rigid tryptophan can be attributed to the inability of a proton transfer, from the ammonium group to position 4 of the indole ring, to occur in this rigid analog.

TABLE III.

Fluorescence Parameters of Single-Tryptophan Containing Proteins[a]

Protein	Emission λ_{max} (nm)	Φ	τ (ns)	k_q ($\times 10^{-9}$ M⁻¹ s⁻¹) O₂	Acrylamide	I⁻	ϕ (ns)	Θ	References
Azurin, CuII, Pfl	308	0.051	<0.65>	—	<0.1	—	—	—	59, 311, 312
Apoazurin, Pfl	308	0.31	5.1	0.08	0.05	—	4.94	~0	59, 311, 312
Apoazurin, Afe	334	—	<2.05>	—	—	—	6.88	26°	311
Asparaginase, E. coli	319	0.17	<1.88>	1.0	0.01	—	>60	20°	b
Parvalbumin, cod (+Ca²⁺)	316	0.13	<3.83>	2.6	0.16	0.03	5.6	16°	364, 365
Parvalbumin, cod (−Ca²⁺)	340	—	<2.93>	—	1.7, 2.5	0.8	—	—	364, 408
Ribonuclease T₁	322	0.31	3.9	2.0	0.2, 0.17	0.008	6.5	0–12°	317–320
Myoglobin, tuna, apo	321	0.14	<2.04>	—	—	—	—	—	61
Myoglobin, tuna, met	—	—	<0.08>, 0.033	—	—	—	—	—	61, 409
Subtilisin, Carlsberg	322	—	<1.16>	—	—	—	—	—	27
Calpactin (−Ca²⁺)	325	0.064	<3.1>	—	0.79	0.11	—	—	401
Calpactin (+Ca²⁺)	314	0.076	<3.1>	—	0.54	~0	—	—	401
Phosphofructokinase, B. stearothermophilus	328	0.35	<4.06>	—	0.1	0.017	—	—	367
fd phage	328	—	<3.5>	2.5	0.64	—	>1000	10°	b
M13 coat protein									
In SDS micelles	—	—	—	—	—	—	9.8	31°	370
In DOC micelles	328	—	<3.58>	—	—	—	14.3	17°	369
Human serum albumin									
F form	334	—	<4.7>	5.8	1.0	—	—	—	396
N form	342	0.12–0.24	<3.6>, <5.1>	2.9	0.76	—	—	—	309, 396
Nuclease A, S. aureus	334	0.29	<4.69>	4.2	0.77	0.25	12.2	16°	340
Elongation factor Tu GDP complex	336	—	<0.72>	—	0.35	~0	63	21°	344
Human leutenizing hormone	336	—	<4.2>	—	1.6	—	(6.0)	20°	368
β subunit	344	—	<3.2>	—	3.2	—	(4.9)	23°	368
Lumazine binding protein (apo)	337	—	<2.52>	—	—	—	6.0	45°	217

150

Myelin coat protein	340	—	<3.47>	5.9	2.2	—	1.26	38°	314
α-Cobratoxin	340	0.10	<0.90>	—	2.8	1.7	3.8	~0	376, b
Phospholipase A₂	340	0.031	<1.76>	—	2.2, 4.0	3.2	6.35	23°	77, 338
Prophospholipase A₂	344	0.092	<3.56>	—	3.6	2.0	7.5	34°	77, 338
Monellin	342	—	<2.87>	7.7	1.65	2.1	6.15	24°	340
Melittin									
Monomer	346	0.115	<2.30>	11.0	3.2, 5.2	—	1.73	42°	331, 335, 372
Tetramer	334	0.116	<1.67>	8.2	1.3	—	3.40	31°	331, 335, 372
Calmodulin VU-9	348	0.19	<3.74>	—	2.0	—	—		305
Apolipoprotein C-I	347	—	<1.77>	—	—		1.4	33°	362, 373
Troponin I, skeletal, at 5 °C	350	—	<2.61>	—	1.1	3.1	8.89	36°	399
Apocytochrome C	350	—	<3.34>	—	—		5.3	47°	345
Bombesin	351	0.09	<2.4>	—	—		—	—	337
DMPS complex	339	0.09	<2.9>	—	—		—	—	337
Adrenocorticotrophin									
(1–24)	352	0.07	<2.17>	9.5	3.5	—	1.56	41°	327, 371
Glucagon	352	0.12	<2.44>	8.8	3.2	—	1.42	41°	327, 328

[a] All values are for studies near neutral pH and room temperature, unless indicated otherwise. The F form of serum albumin is found at low pH. See original references for exact conditions. The quantum yield values are based on a value of ca. 0.14 for tryptophan. Lifetime values indicated with angle brackets are the mean average, $\sum \alpha_i \tau_i$, for a multiexponential decay. Quenching rate constants for the quenchers oxygen, acrylamide, and iodide are obtained from either intensity or lifetime Stern–Volmer plots. References for most of the quenching rate constants can be found in (260). Anisotropy decay results are given as a rotational correlation time, ϕ, for the slowest (global) rotational mode, and as the semiangle, Θ, for a cone within which rapid restricted rotation occurs. Equation 32 is used to calculate Θ values from experimental β_2 and $r_{0,app}$ values. The values of Θ can be related to an order parameter, S, or rotation depolarization factor, $d = \beta_2/r_0$, by Eq. 32. Other single-Trp proteins and peptides, for which less extensive or unpublished fluorescence data exist, are scorpion neurotoxin variant 3 (170), entertoxins B and C1 (377), cholera toxin (378), erabutoxin sea krait (375), mastoparan X (374), gonadotrophin (326), subtilisin inhibitor (304), angiogenin (379), ribonuclease C2 (57), protease A (304), galactose binding protein (149), cardiac troponin I (382), S100 calcium binding protein (381), human superoxide dismutase (304), horse apoferritin (386), horse pancreatic colipase (407), iterleukin-1β (382), sterol carrier protein-2 (400), H-ras p21 (402), yeast thioredoxin (304), and Scenedesmus obliquus or chlorella fusca plastocyanin (383, 304). In addition, single-Trp site-directed mutants of nuclease A (346), ribonuclease T₁ (385), T4 lysozyme (308), phospholipase A2 (398), HIV-1 rev protein (403), insulin A chain (404). B. stearothermophilus lactate dehydrogenase (405), tet repressor (306), and lac repressor (384) have been prepared and studied.

[b] Unpublished data from our laboratory.

151

somehow affect this process. Also, photoinduced proton dissociation of the imino NH group of indole has been demonstrated, but it is difficult to see how the alanyl group would alter this process.

These studies of the biexponential decay and pH dependence of tryptophan fluorescence preview the decay kinetics of tryptophan residues in proteins, where various side chains, peptide bonds, and solvent molecules may interact (quench or undergo dipolar relaxation) with the excited indole ring. Indeed, the decay kinetics of tryptophan residues in proteins have been found to be complex in many cases and is the subject of a following section.

3.2. Tyrosine

Proteins and peptides that lack tryptophan show emission from tyrosine residues (25, 63). The emission λ_{max} of phenol or tyrosine occurs at ~305 nm and is very insensitive to environment, in contrast to indole. The emission band of phenol shows no fine structure and the spectral width also varies little with solvent.

The fluorescence quantum yield (and lifetime) of phenol derivatives, however, is very sensitive to environment and the type of side chains (62, 63, 65–67). Zwitterionic tyrosine decays as a monoexponential, but protonation of the α-carboxylic acid functional group quenches fluorescence and results in a biexponential decay (66, 67). Laws et al (66) and Gauduchon and Wahl (65) have attempted to relate such biexponential decays for tyrosine analogs and peptides to the population of various $C^{\alpha}-C^{\beta}$ rotamers.

Intramolecular quenching reactions in various tyrosine derivatives and peptides have been reviewed by Cowgill (63). In addition to the $-CO_2H$ group, $-NH_3^+$ groups appear to quench (i.e., in dipeptides, but not tyrosine), as do amide, disulfide, and sulfhydryl groups. Bimolecular quenching by these groups has also been demonstrated (62, 63). Quenching of phenol fluorescence by amide (peptide) groups has been shown to decrease as the polarity of the solvent decreases. In some cases, the quenching reactions are much less significant for anisole, indicating that hydrogen bonding between the phenolic hydroxyl and quencher is involved.

The lack of Stokes shift changes, the existence of several possible intramolecular quenching processes, the reduced fluorescence sensitivity, and the difficulty of working at very short wavelengths (and, of course, the masking by tryptophan in class B proteins) are factors contributing to there being fewer studies of tyrosine fluorescence in proteins, as compared to tryptophan. Cowgill (62) and Longworth (25) have reviewed this area. Libertini and Small (68) have presented a careful study of the fluorescence decay kinetics of the single tyrosine containing protein, histone H1. Time-resolved measurements of tyrosine in small peptides, such as oxytocin, have been performed (67, 73). Recently, Searcy et al. (397) have performed time-resolved fluorescence studies with an archaebacterial histone-like protein, which contains only phenylalanine and tyrosine residues. They presented evidence of resonance energy transfer from the former to the latter class of residues.

An important photophysical feature of phenol and tyrosine is that the proton dissociation pK_a of the phenolic hydroxyl group is much lower in the excited singlet state than in the ground state (i.e., p$K(S_0)$ = 10.3 and p$K(S_1)$ = 4.2 (69)). At pH 13, where the ground-state tyrosinate form prevails, very weak emission from this unprotonated species occurs with fluorescence λ_{max} ≈340 nm and a quantum yield of 0.006 (70). Since no tyrosinate emission is seen for tyrosine between pH 4 and 10 (where excited-state deprotonation is thermodynamically possible), the rate of proton dissociation must be much slower than the other radiative and nonradiative rates. In solutions having high concentrations of proton accepting buffer species, such as acetate and phosphate (69, 71), these proton acceptors kinetically facilitate the dissociation of the phenolic hydrogen and thus enhance tyrosinate emission at neutral pH. These findings have suggested that tyrosinate emission ($\lambda_{max} \approx 340$ nm) might be observed at neutral pH in proteins, if the hydroxyl group of a tyrosine residue happens to be near a basic side chain of another amino acid, so that intramolecular excited-state proton transfer can rapidly occur. Several examples of tyrosinate emission have, in fact, been reported for class A proteins. Among these are naja naja cobratoxin (72), bungarotoxin (74), and α- and β-purothions (75). (Each of these proteins lack tryptophan.) The implication of these observations is that tryptophan-like emission ($\lambda_{max} \sim 340$ nm) can be produced, in some instances, by tyrosine residues.

3.3 Dihydronicotinamide and Flavin Coenzymes

The reduced coenzyme, dihydronicotinamide adenine dinucleotide (β-NADH), and the related coenzyme, dihydronicotinamide mononucleotide (β-NMNH), are often encountered as substrate/ligands for the dehydrogenase class of enzymes. The fluorescence of these coenzymes has been used to characterize the thermodynamics and kinetics of formation of these protein−ligand complexes (196−198).

The $S_0 \rightarrow S_1$ absorption transition of NADH has λ_{max} = 340 nm, ε = 6.5 × 10^3 M^{-1} cm^{-1} in water. The fluorescence of NADH shows a large Stokes shift, with a maximum at 470 nm in water. This emission is relatively weak ($\Phi \approx 0.002$) and its fluorescence decay time is very short ($<\tau> \approx$ 0.4 ns) (199−203). The absorption and fluorescence λ_{max} both blue shift with decreasing solvent polarity (199, 203). Baumgarten and Hones (200) have demonstrated that both λ_{max} are linearly correlated with Reichardt's E_{TS} polarity scale. The molar extinction coefficient of NADH (or models for the dihydronicotinamide ring) is found to vary somewhat with solvent polarity (199). Also, these workers find a very small degree of fine structure in the absorption spectra of NADH, which can be correlated with the hydrogen bond accepting ability of the solvent (200).

The small Φ and τ values for NADH are caused by a dynamic intramolecular quenching process that involves the stacking of the adenine and dihydronicotinamide rings (203). Visser and van Hoek (202) have analyzed fluorescence decay data for NADH in aqueous solution in terms of a dynamic two-state (stacked and unstacked) model. The stacked form of NADH is thought to be the primary conformation in aqueous solution. When NADH binds to various dehydrogenases, the coenzyme unstacks. The fluorescence intensity and lifetime of NADH are usually found to increase substantially upon forming binary and/or ternary complexes with enzymes. In fact, measurement of the enhancement of NADH fluorescence is a convenient means of studying the kinetics and thermodynamics of the interaction of this coenzyme with enzymes (198). The unstacking of NADH upon binding may account for much of the enhancement, but specific protein−NADH interactions must also be important, since the degree of fluorescence enhancement varies from enzyme to enzyme.

The oxidized (NAD$^+$) form of the coenzyme does not absorb in the 340-nm region and does not fluoresce (although it does phosphoresce at low temperature (222)). The lack of fluorescence from NAD$^+$ provides a convenient way to monitor the kinetics of enzyme-catalyzed reduction reactions. Also, the 340-nm absorption transition of NADH provides for a large spectral overlap with tryptophan emission, so that resonance energy transfer occurs in protein−NADH complexes (204−206).

A second class of fluorescent coenzymes are the flavins. These include flavin adenine dinucleotide (FAD) and flavin mononucleotide (FMN) and are derived from the precursor, riboflavin (vitamin B$_2$). FAD and FMN are strongly bound cofactors for several oxidoreductases. The fluorophore of these coenzymes is an isoalloxazine ring and it fluoresces in its oxidized state.

The fluorescence properties of these coenzymes and their complexes with model compounds and proteins have been extensively studied (207−212). Reviews have been given by Weber (207−209), Yagi (210), and Visser (211). Riboflavin and FMN fluoresce with a quantum yield of 0.24−0.26 in neutral aqueous solution with an excitation λ_{max} at 450 nm and an emission λ_{max} at 530−540 nm (210, 211). Riboflavin and FMN both show a monoexponential fluorescence decay with $\tau \approx 4.7$ ns (213). Deprotonation of the 3-imino group of the isoalloxazine ring (p$K \approx 10$) causes quenching of riboflavin and FMN (210). A blue shift and increase in structure of the emission is seen upon decreasing solvent polarity (216).

The fluorescence yield and average lifetime of FAD is much lower than that for riboflavin or FMN (i.e., $\Phi = 0.025$ and $<\tau> = 2.1$ for FAD in water). As is the case with NADH, the reduced fluorescence of FAD has been attributed to intramolecular quenching by the adenine ring (211−215). Visser (213) has again analyzed the non-exponential fluorescence decay of FAD in terms of a stacked $\rightleftarrows$ unstacked dynamic model.

The flavins form complexes with several types of molecules, including phenols, indoles, thiols, purines, caffeine, phenobarbital, p-aminosalicylic acid, and chlorotetracycline (210). These complexes invariably quench the fluorescence of riboflavin.

When FMN or FAD are bound to proteins, their fluorescence is also usually lower than that for the free molecule. This is anticipated in light of the quenching of riboflavin by the side chains of tyrosine, tryptophan, and cysteine (see also (212)). Examples of flavoproteins in which quenching of the bound flavins occurs are lumazine apoprotein (216), D-amino acid oxidase (218), p-hydroxybenzoate hydroxylase (219), glutathione reductase (220), and flavodoxin (211). One of the few proteins for which enhanced fluorescence has been observed for the bound flavin is lipoamide dehydrogenase (220).

4. TIME-RESOLVED PROTEIN FLUORESCENCE: INTENSITY AND ANISOTROPY DECAYS

Time-resolved fluorescence measurements can reveal more detailed kinetic information about excited-state processes than steady-state fluorescence measurements, just as stopped flow enzyme kinetics studies can reveal more information than steady-state enzyme kinetics studies. The fluorescence lifetime, τ, of a compound is, as given by Eq. 2, equal to the reciprocal of the sum of the radiative and various nonradiative rate constants. Of practical importance, the fluorescence lifetime of an excited state establishes the time window during which other excited state processes (rotational and translational diffusion, energy transfer, dipolar relaxation, etc.), which alter the emission, can be detected. If this time window is very brief (i.e., short τ), then excited-state reactions that are much slower than τ will not have time to occur before emission. Likewise, if the time window is very long (i.e., long τ), very fast excited-state reactions may be completed before emission and thus be hard to detect; however, slower reactions may be monitored.

A fluorophore in a homogeneous environment is usually expected to show monoexponential fluorescence decay. As discussed above for the zwitterion of tryptophan, however, it is often found that a fluorophore has a biexponential (or higher order) decay pattern. When such nonexponential fluorescence decays are found for tryptophans in proteins, the molecular explanation can involve either ground-state heterogeneity or excited-state reactions. If there is more than one fluorophore in a protein (i.e., multiple tryptophan residues or contributions from both tyrosines and tryptophans), this provides a definite source of heterogeneity. If there is a single type of fluorophore in a protein, ground-state heterogeneity can be caused by the existence of multiple conformational states of the protein, which causes the fluorophore to experience a different environment and have a different decay time in each conformation. In this case, the individual τ_i values reflect the different quenching environments experienced in the conformations, and the α_i (see Eq. 5) can be approximately proportional to the relative fractional population of the conformations. In fact, the difference in τ_i values can provide a means of resolving the decay associated spectra (DAS, see Section 7 below) of each ground-state component.

A nonexponential fluorescence decay can also be caused by excited-state reactions, such as resonance energy transfer, dipolar relaxations, and conformational fluctuations which change the fluorophore's environment during the lifetime of the excited state. Thus, nonexponential decays can have either a static (i.e., equilibrium mixture of conformers) or dynamic (i.e., interconversion of conformers on nanosecond time scale) basis. In both cases, it seems logical than some types of ground-state and/or excited-state heterogeneity would not produce a bi- or triexponential fluorescence decay, but, instead, would produce a pseudo-continous distribution of decay times (77, 79–84). In fact, fluorescence decay data can usually be fitted almost equally well by a discrete number of decay terms (Eq. 5 for i values of τ_i and i - 1 values of α_i) or by a distribution model, provided that a comparable number of fitting parameters are used (78, 81). For distributions fits, the most common practice is to assume an arbitrary, symmetrical (i.e., Gaussian or Lorentzian on lifetime axis) distribution shape, for which the fitting parameters are the mean lifetime and the first moment (half-width) of the distribution. Depending on the system and the preferred interpretation (which may be influenced by auxiliary information), either the discrete or distributed model may be more appropriate, but, in general, the choice between the fitting models is not usually clear.

Grinvald and Steinberg (309) produced the first convincing reports of biexponential decays of tryptophan residues in proteins. Since their work, there have been vast improvements in the methodology. Below, I will review recent time-resolved fluorescence studies, with emphasis on the variety of interpretations of the multiexponential decay kinetics. Examining several cases (i.e., the work with azurins, heme proteins, subtilisin) shows that the methodology has reached a point where questions of possible impurities in protein samples can affect interpretations. Our biochemistry must be as good as our instruments.

Anisotropy decay studies will also be touched upon. If a fluororophore is rigidly attached to a spherical macromolecule (i.e., a globular protein), its fluorescence anisotropy is expected to decay as a monoexponential, with rotational correlation time, ϕ, equal to $1/(6D_R)$, where D_R is a rotational diffusion coefficient, which from the Stokes–Einstein relationship is equal to $k_b T/(6V\eta)$, where η is solvent viscosity and V is the effective molecular volume of the sphere. If a fluorophore is rigidly attached to an asymmetric structure, the anisotropy decay is expected to be much more complex, having five exponential terms (283–287). For ellipsoids of revolution, three exponential terms are expected (288). The three apparent rotational correlation times will be functions of the three principal diffusion coefficients of the ellipsoid; the preexponential terms will depend on the orientation of the fluorophore's absorption/emission moment and the major axis of the ellipsoid (283, 287, 288). While three to five terms are theoretically predicted, some of the exponential terms will be similar, and, in practice, only one or two rotational correlations times are resolvable (288). A more important factor is that fluorophores are usually not rigidly fixed to the macromolecule. Also, there may exist some flexing of domain regions in a protein (i.e., hinge–bending motions). Consequently, the anisotropy decay kinetics of protein fluorophores are usually found to be a double (or triple) exponential (Eq. 15), due to the contributions from internal motion and global rotation of the macromolecule:

$$r(t) = r_0(\text{app}) \sum_{j=1}^{2 \text{ or } 3} \beta_i \exp\left(-\frac{t}{\phi_i}\right) \tag{15}$$

If one models the internal fluorophore motion in terms of the wobbling of the fluorophore's transition moment within a cone, with the macromolecular substrate undergoing isotropic rotational diffusion (see Fig. 12), then the above equation can be rewritten as

$$r(t) = r_0(1 - A_\infty)\exp[-t(\phi_m^{-1} + \phi_i^{-1})] + r_0 A_\infty \exp(-t\phi_m^{-1}) \tag{31}$$

Here ϕ_m is the rotational correlation time for the isotropic, global rotation of the macromolecule and ϕ_i is the effective correlation time for internal motion with the cone (12–14, 284, 287). A_∞ is the fraction of the limiting anisotropy, r_0, which is associated with the internal, wobbling motion. Comparison of Eqs. 15 and 31 shows that $\beta_1 = r_0(1 - A_\infty)$, $\beta_2 = r_0 A_\infty$, $\phi_1 = (\phi_m^{-1} + \phi_i^{-1})^{-1}$ and $\phi_2 = \phi_m$. Also note that if $\phi_m \gg \phi_i$, Eq. 31 simplifies and $\phi_1 \approx \phi_i$. One can relate the value of ϕ_m to the rotational diffusion coefficient of the macromolecule, D_{Rm}, via $\phi_m = 1/(6D_{Rm})$. Likewise, ϕ_i can be related to the effective wobbling diffusion coefficient within the cone and the cone semiangle, Θ, can be calculated from A_∞:

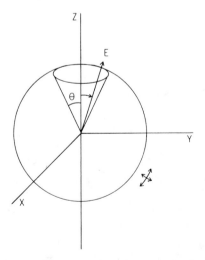

Fig. 12. Model for the rotational motion of a fluorophore that is attached to a spherical macromolecule. The electronic moment of the fluorophore is free to diffuse (wobble) within the volume of a cone of semiangle, Θ. This wobbling motion is superimposed upon the global rotation of the macromolecule.

$$A_\infty = \frac{\beta_2}{r_0} = \left[\frac{1}{2}(\cos \Theta)(1 + \cos \Theta)\right]^2 \tag{32}$$

The magnitude of this cone semiangle can give a description of the extensiveness of the wobbling motion. Alternatively, the extensiveness of the wobbling motion can be described by an order parameter, S, which is given by $S = (A_\infty)^{1/2} = (\beta_2/r_0)^{1/2}$. A value of S equal to unity describes no wobbling motion; values of S smaller than unity describe increasingly more extensive wobbling motion. (For a discussion of the assumptions and limitations of this model see (12).)

The above anisotropy decay model applies to cases where there is a single class of fluorophore in one type of environment (i.e., a single tryptophan containing protein with one major conformation). If there are multiple classes of fluorophores, analysis of anisotropy decay profiles becomes complicated, since each class of fluorophore may have different intensity decay parameters (different α_i and τ_i) and may also have different rotational diffusion kinetics (different β_i and ϕ_i). Even with proteins that have a single class of flourophores, there is the possible existence of different, slowly interconverting conformational states. If each of these putative conformational states has its own associated intensity decay (α_i and τ_i) and anisotropy decay (β_i and ϕ_i) parameters, then the observed anisotropy decay pattern can be very complex and difficult to relate back to the motional behavior of either state. Ludescher et al. (77) have illustrated these complications and have simulated anisotropy decay patterns that have both ascending and descending limbs. Macromolecules that behave as elongated anisotropic rotors can also show complex anisotropy decay patterns (289).

Below we review time-resolved intensity and anisotropy studies with various important proteins or classes of proteins. Attention is given primarily to proteins having single emitting centers. Various fluoroscence parameters for single tryptophan containing proteins are given in Table III.

4.1. Azurin

This protein, from *Pseudomonas aeruginosa* (Pae) or *Pseudomonas fluorescens* (Pfl), is the subject of several important fluorescence studies (59, 309–314). Azurin is a 14-kDa blue-copper electron-transfer protein with a single tryptophan residue (W48) that has very distinct properties. The fluorescence λ_{max} of azurin is 308 nm, which is the smallest Stokes' shift known for protein tryptophan. Structured emission is observed (see Fig. 10), similar to the emission of 3-methylindole in methylcyclohexane (59), indicating a very nonpolar environment for W48. This is borne out by x-ray crystallographic data, which show W48 to be in the core of a β-barrel structure, surrounded by alkyl side chains, and having no hydrogen bond with its imino NH (314). Apoazurin, produced by removal of the Cu(II), has a six-fold higher fluorescence quantum yield ($\Phi = 0.054$ for Cu(II) holoazurin and $\Phi = 0.31$ for apoazurin from Pfl (59)), but the apoprotein retains the same blue, structured emission spectrum.

Several time-resolved fluorescence studies of Pae, Pfl, and homologous azurins are available (59, 309, 311–314). Most researchers report apoazurin to have a monoexponential decay ($\tau \approx 5$ ns), making W48 of this protein a rare example of a homogeneously emitting tryptophan residue. The above mentioned structural features do not allow ground-state interactions of the indole ring with polar groups, nor do they allow any conformational dynamics which facilitate dipolar relaxation during the lifetime of the excited state.

The holoprotein shows a nonexponential decay (see Fig. 13 for the fluorescence decay of both holo and apoazurin). A long-lived component of 4–5 ns, along with one or two short-lived components of ~0.1 to 0.5 ns are observed. This long-lived component has an $\alpha \approx 0.02$ for Pae and $\alpha \approx 0.1$ for Pfl azurin, but it comprises >50% of the steady-state fluorescence of the proteins and thus the identification of this component is important. The facts that (1) the long lifetime is similar to that of apoazurin, (2) the emission spectrum of the long lifetime component is similar to that of both apoazurin and the short-lived components, and (3) the preexponential, α, for the long lifetime component varies between samples and laboratories (i.e., this α is only 0.03 or less for the most pure samples of Pae azurin (312)) has led to the argument that this long component is an "apo-like" impurity that cannot be reconstituted by simply adding Cu(II) ions to the solution (311). Hutnik and Szabo (312) carefully tested the purity of both Pae and Pfl azurin and concluded that the long-lived component is probably not due to a contaminant, but rather to the intrinsic

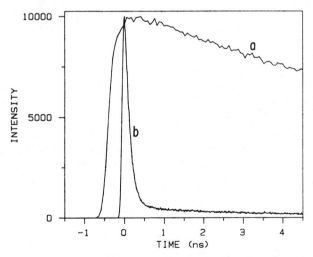

Fig. 13. Time-resolved fluorescence decay of apoazurin (a) and holoazurin (b) from Pae in pH 5.0, 50 mM acetate, at 20 °C. Apoazurin was fitted to a monoexponential decay with $\tau_1 = 5.16$ ns. Holoazurin was fitted with a biexponential decay with $\tau_1 = 0.102$ ns, $\tau_2 = 4.15$ ns, and $\alpha_1 = 0.97$. The instrument response function was ~0.1 ns FWHM. [This figure is reproduced from Petrich et al. (311) with permission from the authors and the American Chemical Society.]

conformational heterogeneity of azurin. That is, they attributed their triexponential decay fits to the existence of three conformational states of azurin.
Further, they showed that the equilibrium between these three states depends
on pH and temperature and differs between Pae and Pfl azurins. According to
this interpretation, the long-lived component corresponds to a very small
percentage of the protein in which W48 is not quenched by the presence of
Cu(II) in its binding site. The majority of the protein has a short lifetime and
is therefore quenched by the presence of Cu(II). Hutnik and Szabo suggest
that the coordination of Cu(II) may be different for the long-lived and short-
lived components (312). Regardless of the interpretation of the long−lived
component, the mechanism of the dynamic quenching by Cu(II) to produce
the short-lived components must be explained. Petrich et al. (311) propose
that this quenching occurs by electron transfer from W48 to the Cu(II) center
and argue that the observed rate of this process is consistent with Marcus
theory. However, similar short decay times are observed by Hutnik and Szabo
(313) for forms of azurin having Cu(I), Co(II), and Ni(II) at the metal site.
These researchers point out that these observations are inconsistent with an
electron-transfer mechanism, since an inverse correlation between the fluorescence lifetime and the reduction potential of the metal ions is predicted by
this model.

Early anisotropy decay studies by Munro et al. (314) indicated the existence of
rapid, internal motion by W48 of apoazurin. However, more recent measurement
by Petrich et al. (311) show only a single rotational correlation time, $\phi = 4.94$
ns, for apoazurin Pae. This observation demonstrates that W48 depolarizes
only by overall rotation of the protein, with no detectable internal motion of
the tryptophan side chain. Petrich et al. also studied two other apoazurins,
one with a single, surface tryptophan residue (W118), and the other with both
an interior (W48) and surface residue (these are apoazurins from *Alcaligenes
faecilis*, Afe, and *Alcaligenes denitrificans*, Ade, respectively). The surface residue
of apoazurin Afe has a redder fluorescence, with a nonexponential decay, and
enjoys a degree of rapid, internal rotational motion. The flourescence of
apoazurin Ade shows contributions from both the internal and surface trp
residues. In the latter protein, some resonance energy transfer appears to
occur between W48 and W118, as indicated by a negative amplitude in time-
resolved intensity measurements (311) and phase-resolved spectra (316). Figure
14 shows the emission spectra of the apoazurins from Pfl, Afe, and Ade.

4.2. Ribonuclease T₁

A second protein that has a single, buried tryptophan residue (W59) and
whose fluorescence properties have been extensively studied is ribonuclease T_1
from *Aspergillus oryzae*. Longworth (317) first observed that the emission of this
protein is unusually blue ($\lambda_{max} \approx 325$ nm) and shows vibronic structure.
Subsequent quenching studies (318) also indicate that W59 is relatively (but
not completely) inaccessible to probe molecules.

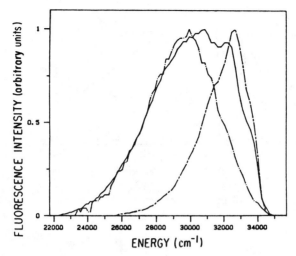

Fig. 14. Steady-state emission spectra of apoazurins from Pae ($-\cdot-$, the spectrum to the right), Afe ($---$, the spectrum to the left), and Ade (———). [This figure is reproduced from Petrich et al. (311) with permission from the authors and the American Chemical Society.]

The intensity decay of ribonuclease T_1, has been measured by several laboratories and can almost be considered a standard. There is general agreement that the fluorescence decay of this protein is a monoexponential at pH 5 to 5.5 and 20–25 °C with a lifetime of approximately 4 ns. This result is found in both time-domain and frequency-domain measurements (319–323), as shown in Fig. 15. At pH above 7, Chen et al. (320) report that the fluorescence decay becomes a double exponential, with the appearance of a 1.4-ns component in addition to the dominant long-lived component. The existence of this second component appears to be related to the proton dissociation of some group on the protein with a pK_a between 5 and 7. Chen et al. suggest that the loss of a proton from this group leads to two conformational states of the protein with two different microenvironments for the W59 indole ring. Protonating this group apparently produces a more homogeneous and nonpolar environment for W59.

A double exponential decay of ribonuclease T_1 at pH 5.3 is reported by MacKerell et al. (324), but their measurements were made at 40 °C, which is near the thermal denaturation temperature, T_m. Fluorescence decay measurements made as a function of temperature (and chemical denaturants) show that the decay kinetics become more complicated near T_m (or the denaturant concentration for 50% unfolding) (322). Above T_m (or at high denaturant concentrations) the decay is a double exponential, and the average lifetime is decreased.

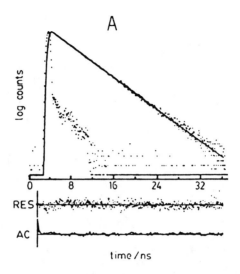

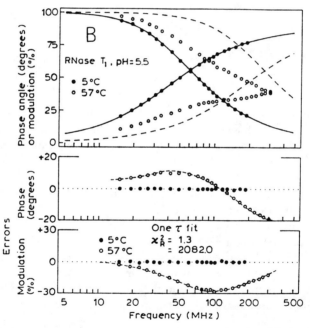

Fig. 15. Comparison of time- and frequency-domain intensity data for ribonuclease T$_1$. (*A*) Time-domain intensity decay data from ribonuclease T$_1$ (*Aspergillus oryzae*) at 25 °C. Data are fitted to a monoexponential decay law with $\tau = 3.87$ ns. RES and AC are the residual and autocorrelation functions for the fit. *Conditions*: excitation at 295 nm, emission at 350 nm, 0.05 M acetate buffer, pH 5.5, ionic strength 0.5 M. [This figure is reproduced from reference 319 with permission from the authors and the American Chemical Society.] (*B*) Frequency-domain intensity decay data for ribonuclease T$_1$. Data shown are for 5 and 57 °C. Similar data (not shown) at 20 °C were fitted to a monoexponential decay law with $\tau = 3.85$ ns. The bottom panels are the error profile in the modulation and phase angle for the fits. *Conditions*: excitation at 295 nm, emission through a WG 320 filter, 0.1 M acetate buffer, pH 5.5. [This figure is reproduced from reference 322 with permission from Elsevier Science Publishers.]

The binding of the specific ligand, 2'-GMP, causes a small degree of quenching and induces a double exponential decay at pH 5.5 (320).

Time-resolved (319, 320) (see Fig. 16) and quenching-resolved (325, 326) anisotropy decay studies also agree reasonably well. W59 appears to have little, if any, independent motion, as indicated by a monoexponential anisotropy decay fit and a relatively high r_0. Instead, W59 depolarizes by global rotation of the protein. At $20-25\,^\circ\text{C}$, values of ϕ_m of $5-6$ ns are found. James et al. (319) and Chen et al. (320) have both performed a temperature dependence study of ϕ_m for ribonuclease T_1 and find that ϕ_m does not follow the Stokes relationship; ϕ_m is approximately the expected value at $25\,^\circ\text{C}$ for a globular protein with a hydration layer, but ϕ_m becomes larger than expected at lower temperature. Since the ϕ_m values can be related to a molar volume, V, by Eq. 13, this result indicates that the hydrodynamic volume of RNAse T_1 increases at $0-5\,^\circ\text{C}$. Such a finding suggests increased hydration of the protein or some aggregation of the protein at the lower temperatures.

4.3. Flexible Polypeptides

At the other extreme, tryptophan residues may be part of a flexible polypeptide chain and fully exposed to the aqueous environment. The hormones adrenocorticotropin (ACTH), glucagon, bombesin, and the bee venom peptide, melittin, are believed to have little three-dimensional structure and to behave as flexible coils (although melittin exists as a tetramer under certain conditions, see below). These polypeptides have single tryptophan residues that thus serve as examples of fully solvent exposed residues. They each show the expected unstructured fluorescence with λ_{max} around 350 nm. Intensity and anisotropy decay measurements with these polypeptides are summarized below.

Ross et al. (327) performed time-resolved studies with the synthetic analog of adrenocorticotropin, ACTH-(1−24) tetracosapeptide, and obtained a biexponential intensity decay. Anisotropy decay measurements also gave a biexponential decay, with the major (defined as the exponential term with the larger βi) rotational correlation time having a subnanosecond value, which was ascribed to local rotation of the indole ring with respect to the polypeptide chain. A slower rotational correlation time was identified with more extensive motions of the polypeptide chain.

Similar nonexponential intensity and anisotropy decay data are reported for glucagon and the monomeric form of melittin (328−331). For each polypeptide, the subnanosecond rotational correlation time accounts for about 50% or more of the depolarizing process.

With globular proteins, a biexponential anisotropy decay can be interpreted via Eq. 31 and Fig. 12 in terms of the combination of global motion of the protein and internal, rapid wobbling of the trp side chain. For flexible polypeptides, however, this description is probably not adequate. Fleming and coworkers (361, 362) suggest the use of a model of Perico and Guenza (363) to describe the diffusional motion of a semiflexible linear chain and to interpret

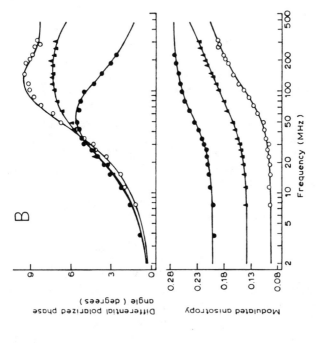

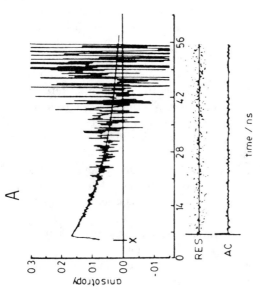

Fig. 16. Comparison of time- and frequency-domain anisotropy decay data for ribonuclease T₁. (*A*) Time-domain anisotropy decay data at −1.5°C, pH 5.5. A monoexponential anisotropy decay fit was obtained with φ = 20.9 ns. [This figure is reproduced from reference 319 with permission from the authors and the American Chemical Society.] (*B*) Frequency-domain anisotropy decay data at 5°C (●) and 52°C (○) at pH 5.5, and at 5°C with 6 M quanidine–HCl (▲). A monoexponential decay fit was obtained at 5°C with φ = 10.68 ns. [This figure is reproduced from reference 322 with permission from Elsevier Science Publishers.]

anisotropy decay data for flexible polypeptides. According to this model, the motion of tryptophan in these polypeptides can be described in terms of a persistence length of $7-10$ residues.

In fitting anisotropy decay data for such flexible tryptophan residues, an important point is whether the fitted r_0 is equal to the value expected for the complete absence of motion (i.e., in a vitrified solvent at $-50\,°C$). If the fitted r_0 is much lower than the "frozen" r_0, this indicates that very rapid depolarizing processes occur in a time scale shorter than can be measured by the instrument; as a result, fast depolarizing motion can go uncharacterized. The "frozen" r_0 for indole is found between 0.30 and 0.32 at an excitation wavelength of 300 nm or above (39, 53, 332). One can fit data by fixing this value for r_0, but in many experiments wavelengths other than 300 nm have been used for excitation. The "frozen" r_0 for indole varies sharply with wavelength between 290 and 300 nm and may depend on the polarity of the indole's environment. Advances in instrumentation enables one to observe shorter time windows, but the fluorescence lifetime of the fluorophore can also limit the observable time window. Lakowicz et al. (331) developed an elegant procedure in which the observable time window can be shifted by the addition of collision quenchers. By progressively shortening the fluorescence lifetime with quenchers, they were able to collect anisotropy decay data for shorter time windows and thus better determine the values of very short rotational correlation times. Using a 2-GHz multifrequency phase fluorometer and the quencher acrylamide, Lakowicz et al. (331) were able to resolve the anisotropy decay of the single tryptophan of melittin monomer into 0.16- and 1.73-ns components.

The monomer $\rightleftarrows$ tetramer equilibrium of melittin has also been studied by various fluorescence approaches (331, 334, 336). This equilibrium can be shifted by concentration, ionic strength, pH, temperature, and pressure. The intensity decay of the tetramer, like the monomer, is a multiexponential; the average lifetime of the tetramer is smaller than that of the monomer. As expected from its larger size and globular-like structure, the rotational correlation time (i.e., the longer ϕ_m for a biexponential fit) of the tryptophans of the tetramer is larger than that of the monomer (331).

The fluorescence decay kinetics of bombesin have also been carefully studied, both in solution and when bound to phospholipid vesicles (337). A complex (triexponential) decay is found in both cases for the single tryptophan of this poly-peptide. Using the DAS method (see below), the emission spectrum of each decay component is resolved. Other single-tryptophan, flexible polypeptides that have been carefully studied are apolipoprotein C-1 (362) and apocytochrome c (345).

4.4. Other Single-Tryptophan, Globular Proteins

We have reviewed the above examples of deeply buried and fully solvent exposed tryptophan residues in proteins/peptides. The environment and fluorescence properties of most tryptophan residues may lie between these extremes. That is, the indole ring may be partially exposed on the protein's surface, or, if it is beneath the surface, may interact with polar amino acid side chains or peptide bonds. A variety of proteins with such intermediate tryptophan residues have been studied, including porcine phospholipase A2 (77, 338), staphylococcus nuclease (339–342), subtilisin Carlsberg (27), lumazine binding protein (217), human serum albumin (35), and elongation factor T_u (344). The fluorescence λ_{max} of these proteins ranges from 330 to 340 nm and the intensity decays are nonexponential in each case. Here I comment on a few of these proteins.

In their studies with porcine phospholipase A2, Ludescher et al. (77) are probably the first to convincingly demonstrate a tryptophan intensity decay that is more complex than biexponential. They report that three or four decay times are needed to adequately describe this protein. They suggested that the decay kinetics are probably best represented as a pseudo-continuous distribution of decay times, an idea further promoted by Alcala et al. (79–80) for this protein. Ludescher et al. further characterize the intensity decay of the pro-protein form of phospholipase A2 and complexes of the mature enzyme with nonhydrolyzable phospholipid vesicles. Anisotropy decay studies show that Trp-3 of phospholipase A2 has subnanosecond motion in a cone of 20°. This rapid motion is attenuated upon complexation with vesicles (338).

The fluorescence of Trp-140 of nuclease A has been extensively studied (339–342). Multiexponential intensity decay kinetics are found and anisotropy decay data show a small degree of subnanosecond wobble. In recent years a set of site-directed mutants of nuclease have been prepared and studied with fluorescence techniques. These mutants include amino acids at various positions in the protein and reveal subtle influences on the time-resolved fluorescence of Trp-140. Several mutants have a replaced proline residue, allowing studies of the effect of proline isomerism on the fluorescence decay kinetics (342). Also, some mutants have reduced thermodynamic stability and are convenient for studies of the time-resolved fluorescence and anisotropy of their temperature, urea, and pressure unfolded states (346).

Subtilisin Carlsberg has been an important protein for fluorescence studies because it is rare among class B proteins in having tyrosine as its dominant emitter (27). This protein has 13 tyrosine residues and one tryptophan. The latter is located on the protein's surface; the crystal structure shows that this residue is not close to any obvious quenching residue. The emission (at 320–330 nm) from this residue, however, is very weak, resulting in a dominant contribution from tyrosine emission at 305 nm. Willis and Szabo (27) carefully removed any autohydrolytic impurities from this enzyme and characterized the DAS of the tyrosine and feeble tryptophan decay components (the latter being nonexponential). Even with excitation at 300 nm, a tyrosine-like emission component is seen in this protein.

4.5. Proteins with Other Intrinsic and Extrinsic Fluorophores

While tryptophan and tyrosine are usually available as intrinsic probes, other intrinsic and extrinsic fluorophores can sometimes provide important advantages in time-resolved studies by (1) extending the time window in cases of probes with longer fluorescence lifetime (i.e., with long-lived pyrene probes, one can measure longer rotational correlation times for larger proteins), (2) enabling the probe to be targeted at specific positions (i.e., chemical attachment to specific side chains), and (3) sometimes having higher quantum yields, larger Stokes shifts, and/or redder transitions, which can facilitate the fluorescence measurements (i.e., allowing use of a stronger laser line for excitation).

Some of the naturally available intrinsic fluorophores are mentioned in Section 3. Time-resolved measurements have been performed with protein-bound flavins (211, 217−219), NADH (204−206), and metal-depleted porphyrin (347, 348). One of the most extensively studied classes of intrinsic fluorophores is lumazine (and its analogues, 7-oxolumazine and riboflavin) bound to the protein referred to as lumazine binding protein from bioluminescent bacteria (217). It is noteworthy that the fluorescence of bound lumazine (or the analogues) has a monoexponential decay of about 14 ns at 20 °C. In addition, the anisotropy decay of bound lumazine indicates no independent motion; it rotates only with the entire protein ($\phi \sim 10$ ns at 20 °C). The bound fluorophore, therefore, appears to be in a rigid and relatively homogeneous environment.

Regarding extrinsic fluorophores, many examples are available of time-resolved intensity and anisotropy decay studies. In Section 6 a number of extrinsic probes are discussed in the context of their use in energy transfer measurements. Here we mention as examples the several studies of the use of time-resolved fluorescence to characterize the flexibility of myosin and immunoglobins (289−292).

4.6. Multitryptophan-Containing Proteins

For proteins with more than one tryptophan residue, the interpretation of time-domain fluorescence data obviously becomes more complicated, because each residue is usually expected to be a separate emitting center, with its own decay profile, Stokes shift, rotational correlation time, and so on. (Resonance energy transfer may be possible between tryptophan residues, further complicating matters. Even if this occurs, it appears to be eliminated by exciting at the red edge of the absorption spectrum (349)). It is incorrect to assume that each tryptophan will contribute equally to the fluorescence intensity of a native protein, since each residue will have a different fluorescence lifetime, and the contribution to the total intensity is the product $\alpha_i \tau_i$ for a given component. It is always tempting to assume that the α_i will be about the same for each residue, but this may not be true if (1) there is static quenching of components, (2) the absorption coefficient is not approximately the same for all components at the excitation wavelength, and (3) there is energy transfer between the fluorescing centers. Due to these uncertainties, it is usually only possible to draw qualitative interpretations of intensity and anisotropy decay measurements for multitryptophan containing proteins.

For proteins with two tryptophans, success can sometimes be achieved at resolving the fluorescence decays of the two centers, provided, of course, the two residues have different lifetimes. Good examples of such proteins are yeast 3-phosphoglycerate kinase (350, 351), horse liver alcohol dehydrogenase (352−358), and Ade apoazurin (311, 316). With alcohol dehydrogenase, a clear biexponential decay is found with $\tau_1 \approx 3.5$ ns and $\tau_2 \approx 7$ ns. The shorter-lived component has a DAS with $\lambda_{max} \sim 325$ nm, and the longer-lived component has a $\lambda_{max} \sim 340$ nm. The x-ray structure of this enzyme shows Trp-314 to be deeply buried at the intersubunit interface of this homodimeric protein; Trp-15 is on the surface of the protein. The 3.5-ns blue component has been assigned to Trp-314 and the 7-ns red component to Trp-15. This assignment is supported by the observation that the solute quenchers iodide and acrylamide (352−354, 357, 358) selectively lower the steady-state intensity and lifetime of the 7-ns red component. Also, the binding of NADH and NAD^+ to alcohol dehydrogenase selectively quenches the 3.5-ns blue component (356, 358).

Apoazurin from Ade, as mentioned above, is a two-tryptophan protein; one of its tryptophans is homologous with the extremely blue (~308 nm) Trp-48 of apoazurin from Pae or Pfl, and the other is homologous with Trp-118 from apoazurin from Afe. Figure 14 shows the steady-state spectrum of the two tryptophan Ade apoazurin, as well as the spectra of the single tryptophan Pae and Afe apoazurins. As can be seen, the two types of residues have distinctly different spectra, and the two-tryptophan Ade apoazurin appears to have a spectrum that is a composite of the two. Intensity decay measurements can separate the contributions of the two tryptophans in Ade apoazurin. Using phase-resolved spectroscopy, the subspectra (equivalent to DAS) of the two tryptophans of Ade apoazurin have been resolved (316). The intensity decay kinetics of this protein are interesting because they appear to reveal a negative α, which may be attributed to resonance energy transfer from Trp-48 to Trp-118 (311).

5. SOLUTE FLUORESCENCE QUENCHING

Quenching reactions are very useful for obtaining topographic and related types of information about proteins (11, 15, 17, 258−261). Dynamic quenching reactions involve a direct competition between the bimolecular quenching reaction and the unimolecular fluorescence decay. For aqueous solutions and for small, efficient quenchers, such as oxygen, acrylamide and iodide, the bimolecular quenching rate constant, k_q, will be on the order of $5-10 \times 10^9$ $M^{-1} s^{-1}$. Thus, for quencher concentrations, $[Q]$, from 0 to 1 M, the product of $k_q[Q]$ will be comparable to $1/\tau$ values. For efficient quenchers, the reaction is diffusion limited; consequently, steric factors are important, and protein-bound fluorophores can have a wide range of accessibility to quenchers, depending on how deeply a fluorophore is buried in a protein (262–264). Electrostatic interactions can also be important for charged quenchers (262, 265).

The apparent quenching rate constant for a collisional quenching reaction can be estimated by

$$k_q = \gamma 4\pi D R_0 N' \tag{18}$$

If $\gamma = 1.0$, then every encounter between quencher and the excited state results in quenching. For such a perfectly efficient quenching reaction, the experimental k_q value will be a direct measure of the frequency of collision between the reactants and k_q is expected to be directly proportional to T/η, since D in Eq. 18 is given by the Stokes–Einstein relationship as $D_i = k_b T/6\pi\eta R_i$. If $\gamma < 1.0$, this indicates that the quenching reaction is inefficient and that every collision between the reactants does not result in quenching. The k_q for an inefficient quenching reaction will not show Stokes–Einstein behavior and may also depend on solvent polarity (since γ may depend on solvent environment) (266, 267). For studies with proteins it is obviously desirable to use an efficient quenching reaction.

Three quenchers of indole fluorescence, which appear to have an efficiency near $\gamma = 1.0$, are molecular oxygen, acrylamide, and iodide (262–264). Taking the acrylamide–indole pair as an example, an experimental k_q value of 7.1×10^9 M^{-1} s^{-1} is found at $20\,^{\circ}C$. Taking D_Q and D_F to have values of 1.0×10^{-5} cm^2 s^{-1} and 0.5×10^{-5} cm^2 s^{-1} (i.e., $D = D_Q + D_F$) for the quencher and fluorophore, respectively, and R_Q and R_F to have values of 3.0 and 3.5 Å (i.e., $R = R_Q + R_F$), respectively, and $\gamma = 1.0$, then k_q is predicted to be 7.0×10^9 M^{-1} s^{-1}, almost identical to the observed value (18). When the fluorophore is attached to a large macromolecule, its diffusion coefficient will be greatly reduced. As a result, the maximum value of k_q, for a fully solvent exposed fluorophore attached to a large macromolecule, is expected to be approximately one-half the above value (268).

Oxygen, acrylamide, and iodide are certainly not the only quenchers that can be used in studies with proteins; in separate reviews we have discussed the characteristics of these and other solute quenchers (15, 269). These three quenchers are the most often used with proteins, especially in studies of tryptophan fluorescence, because they have a high efficiency and show minimal perturbation of proteins. With fluorophores other than tryptophan, it should be noted that acrylamide may be inefficient (267). For tryptophan in proteins, these three quenchers provide a useful, complementary set for studying the topography of the tryptophan residue. Oxygen is a small and apolar quencher and has the greatest ability to diffuse into the interior of globular proteins (264, 270). Acrylamide is a larger, polar quencher molecule. It has greater selectivity in that it preferentially quenches solvent exposed tryptophan residues (263, 266). Iodide, which is negatively charged and hydrated, has an even greater selectivity in being able to quench only exposed residues (262). Listed in Table III are some k_q values for these three quenchers for several single tryptophan proteins. It is generally found that tryptophan residues that are believed to be buried (i.e., blue λ_{max} or from inspection of an x-ray structure), such as the residues in apoazurin Pfl and ribonuclease T_1, have the lowest values of k_q for all three quenchers. Conversely, residues in flexible polypeptides show the largest k_q for all three quenchers.

To this point we have discussed only the dynamic quenching process, but studies with model systems clearly show that a static quenching component usually exists as well (18, 270, 271). By analyzing data using Eq. 20, one can resolve the dynamic and static components; using fluorescence lifetime data and Eq. 17, only the dynamic component is obtained. It is usually difficult to make much of the static constant, V, for a protein system, since the effects of heterogeneity in either k_q or τ_0 (due to multiple fluorophores, multiple protein conformations, or a nonexponential decay for a tryptophan residue) can mask the degree of static quenching. There may be certain quenchers that act primarily via a static mechanism, by binding to the protein. Usually the static component is less than the dynamic component. In fact, in some cases the apparent static component in intensity quenching data can be attributed to the transient term of the Smoluchowski equation (16, 272–274, 394).

The finding that internal tryptophan residues in proteins such as apoazurin, ribonuclease T_1, and parvalbumin can be dynamically quenched by oxygen and acrylamide has had important implications regarding the dynamics of globular proteins (264, 266, 270, 275–278). If it is accepted that the quenching process involves physical contact between the quencher and the excited fluorophore, which seems to be the case from model systems studies, then the question is how to explain the quenching of such interval residues. Two possible explanations are that (1) there is dynamic penetration of the quencher into the protein to reach the fluorophore, or (2) the protein transiently unfolds to expose the internal residue to the solvent (260, 264, 276, 277, 394). In either case, the quenching results indicate the existence of dynamic processes; the distinction between the penetration and unfolding models is essentially a question of the time scale and extensiveness of the fluctuations. Because of its small size, it is generally accepted that oxygen can penetrate into proteins, since only small-amplitude fluctuations, would be needed to facilitate the inward diffusion of oxygen (266, 270). Acrylamide is a larger quencher, so the amplitude of fluctuations would need to be larger to facilitate its diffusion. Yet, temperature, pressure, and bulk viscosity dependence studies are all more consistent with a penetration mechanism rather than an unfolding mechanism for acrylamide fluorescence quenching of internal tryptophan residues (277–279). Also, the greatly diminished quenching of internal residues by succinimide and iodide (diminished as compared to acrylamide) are inconsistent with an unfolding mechanism (260, 266). We have therefore concluded that the penetration mechanism is probably the more realistic model for describing the quenching of internal residues (260). It has been suggested that the quenching of internal residues by acrylamide and other quenchers may be due to the fact that the quenching reaction does not require collision but can occur over a distance (320). Because acrylamide appears to quench indole fluorescence by an electron-transfer reaction mechanism (267), the suggestion is that electron transfer from an internal tryptophan, through the protein fabric, to a surface acrylamide molecule may occur (320). This proposal merits consideration. We have prepared covalent adducts in which an acrylamide moiety is fixed at short distances from an indole ring. The rate constant for intramolecular quenching drops from 3×10^{10} s^{-1} to 3×10^9 s^{-1} as the separation between the groups is increased from two to four C$-$C sigma bonds (280). Because of the conduction of electrons should be better in these covalent adducts that in proteins, where a covalent path would be absent, we still consider it unlikely that electron transfer over a distance can completely explain the quenching of the fluorescence of internal residues. Additional tests of this mechanism are needed.

In addition to steric effects, the action of charged quenchers can be influenced by the electrostatic interaction between the quencher and charges on the protein in the vicinity of the fluorophore. Ando et al. (265, 281) illustrate this effect in their studies of the iodide, I$^-$, and thallium, Tl$^+$, quenching of an extrinsic fluorophore covalently attached to reactive sulfhydryls of heavy meromyosin. By comparing anionic, neutral, and cationic quenchers, or by studying the effect of ionic strength or pH on the quenching by an ionic quencher, the local charge density around the fluorophore can be revealed.

Due to the large range of k_q for acrylamide and iodide quenching of tryptophans in proteins, these quenchers can be used with multitryptophan proteins (1) to determine if there is heterogeneity in the accessibility of the tryptophan residues to the quencher, and, if there is heterogeneity, (2) to separate the fluorescence contributions of accessible residues from inaccessible residues. Heterogeneity is conveniently detected by a downward curving Stern$-$Volmer plot of I_0/I versus $[Q]$. Figure 17 shows data for the iodide and acrylamide quenching of several two tryptophan-containing proteins, including horse liver alcohol dehydrogenase (LADH). As described in Section 4, Trp-314 of this protein is deeply buried in the intersubunit interface of LADH, while Trp-15 lies on the surface of the protein. The Stern$-$Volmer plots are consistent with the facile quenching of Trp-15 and with the very limited quenching of Trp-314 ($K_{sv} = 0$ for iodide and $K_{sv} = 0.04$ M^{-1} for acrylamide) (282, 352$-$354). There is a blue shift in the protein's fluorescence as these quenchers are added, as expected with the loss of Trp-15 emission ($\lambda_{max} \approx 340$ nm) and the persistence of Trp-314 emission ($\lambda_{max} \approx 325$ nm) (282, 352, 353). The wide difference in the accessibilities of Trp-15 and Trp-314 in LADH are particularly advantageous, enabling these tryptophan residues to be used as intrinsic probes to study induced conformational changes in this protein (i.e., upon ligand binding (356)).

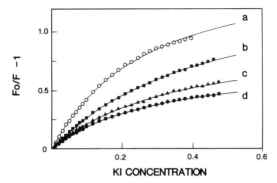

Fig. 17. Stern—Volmer plots for the iodide quenching of the fluorescence of selected two-tryptophan containing proteins at 20 °C. (*a*) LADH NAD$^+$ TFE ternary complex; (*b*) apomyoglobin; (*c*) metalloprotease; and (*d*) LADH. Solid lines are fits of Eq. 21 for two components (and no static quenching). See reference 316 for details. [This figure is reproduced from this reference with permission from the American Chemical Society.]

In general, the Stern—Volmer equation for a heterogeneously emitting system is given by

$$\frac{I}{I_0} = \sum_{i=1}^{n} \frac{f_i}{(1 + K_{sv,i}[Q])\exp(V_i[Q])} \tag{21}$$

where f_i is the fractional contribution of component i to the steady-state intensity, and $K_{sv,i}$ and V_i are the dynamic and static quenching constants for component i. Nonlinear regression analysis can be used to fit this equation to data; usually the fitting must be limited to two components ($i = 2$), unless one assumes that the $V_i = 0$ or that the V_i are some fraction (i.e., 0.1 $K_{sv,i}$) of the $K_{sv,i}$. It should be noted that since even single-tryptophan-containing proteins may have nonexponential fluorescence lifetimes and since $K_{sv,i} = k_{q,i}\tau_{0,i}$, such single-tryptophan proteins can, in principle, show downward curving Stern—Volmer plots, although static quenching will usually mask this trend (260).

While the values of k_q for quenching are of intrinsic interest, fluorescence quenching reactions can also be very useful for other purposes. As mentioned, they may be used to facilitate the resolution of the emission spectra of accessible and inaccessible components. Similarly, dynamic quenching reactions can be used in combination with steady-state anisotropy measurements to (1) resolve the anisotropy of accessible and inaccessible components, and (2) determine the rotational correlation time of a fluorophore via the Perrin equation (Eq. 12) (325, 326). Dynamic quenching reactions can be used in conjunction with time-resolved intensity and anisotropy decay studies to help resolve components and/or to reduce the average lifetime to enable shorter rotational correlation times to be measured (331). Probably the most frequent and valuable use of solute quenching is to indicate conformational changes in proteins (i.e., induced by ligand binding) in terms of changes in fluorophore accessibility (260).

6. RESONANCE ENERGY TRANSFER

Whereas other fluorescence measurements respond to interactions between an excited state and its immediate environment, singlet−singlet resonance energy transfer (RET) differs in that it involves interactions over a distance of 20−70 Å. RET measurements have been widely used in mapping distances between sites in macromolecular assemblies and in monitoring the thermodynamics and kinetics of protein association reactions and unfolding reactions (3, 19−21). RET measurements have been considered to provide a "molecular ruler" for determining distances within proteins. However, the limitations of the method, particularly with regard to the importance of orientation effects, as discussed by Dale and Eisenger (21, 85−87), must be considered. As we review below, fluorescence anisotropy data can be used to define the range of possible κ^2 values and thus set limits on the distance determinations. In addition to questions regarding the relative orientation of the donor and acceptor, analyses have also been presented with a model in which there is a distribution of donor−acceptor distances, rather than a discrete distance (88−93). Examples of these various applications and interpretations of RET measurements will be presented below.

The efficiency, E, of energy transfer between donor, D, and acceptor, A, (or the rate constant, k_{ret}, for energy transfer) can be experimentally determined by performing steady-state or time-resolved measurement of the emission of either the donor or acceptor (if the acceptor fluoresces). Steady-state measurements of the emission of D, in the absence and presence of A, is the most common procedure for such experiments, with analysis for the separation distance, r, using Eqs. 22−26. In such cases it is important to correct for nonstoichiometric attachment of the acceptor (19, 94). Alternatively, one can measure the steady-state, sensitized emission of the acceptor, in the absence and presence of donor, with excitation at some wavelength where both absorb. Determining the relative absorbance for the donor and acceptor can be difficult in such cases (19). In time-resolved measurements, the decay of the donor will be shortened by the presence of the acceptor, since the energy-transfer process competes directly with the emission process (19). (See Eq. 28 for the impulse−response function of the donor and acceptor.) In the simplest case of a monoexponentially decaying donor (in the absence of acceptor) and a fixed D−A distance and dynamically random orientation factor ($\kappa^2 = \frac{2}{3}$), the energy-transfer process will result in a shortened, monoexponential decay for D. If there is more than one D→A orientation and/or distance (i.e., different, slowly interconverting conformational states of the protein or multiple D→A pairs), then the fluorescence decay of the donor may become multiexponential in the presence of the acceptor.

If the time-resolved sensitized emission of the acceptor can be measured, then a biexponential (or higher order) decay will be observed. The model for a two-state excited-state reaction applies (see Fig. 5 and Eq. 29 in Section 2), and a negative preexponential (a buildup) for one of the decay times should be obtained (provided there is little overlap between the emission of the donor and acceptor). Further, the apparent "decay" time associated with this negative preexponential will be equal to $1/(k_{ret} + \tau_{0,d}^{-1})$. If $\tau_{0,d}$ is also known, then r can be calculated, since $k_{ret} = (R_0/r)^6/\tau_{0,d}$. Such negative preexponentials for sensitized acceptor emission have been used to characterize energy-transfer kinetics between the single tryptophan and lumazine chromophore of lumazine binding protein (217), between the buried and solvent exposed tryptophans of Ade apoazurin (311), and between the two individual tryptophans of alcohol dehydrogenase and bound NADH (206).

6.1. Selection of Donor and Acceptor Chromophores

Among the intrinsic D−A pairs in proteins, transfer from tyrosine to tryptophan is very facile, as has been amply demonstrated by the weak tyrosine emission in most class B proteins (25). Since there are almost always more tyrosine residues (donors) than tryptophans (acceptors), and since the spectra of these fluorophores is overlapped, analysis of data in terms of k_{ret} and intramolecular distances is problematic.

Energy-transfer studies with tryptophan as donor are much more common. Tryptophan to tryptophan homotransfer has been often speculated upon in the interpretation of the fluorescence of multitryptophan proteins (118), but demonstration of the existence of such homotransfer is rare. Ghiron and Longworth (332) showed that the low-temperature anisotropy spectrum of trypsin is indicative of homotransfer between this protein's four tryptophans. Observing negative preexponentials in time-resolved studies with apoazurin Ade, as mentioned above, is another type of evidence for homotransfer, as is a difference between decay-associated (or phase-resolved) spectra and quenching-resolved spectra for a multitryptophan protein (316). For each type of evidence, dimunition of energy transfer at the red edge of the excitation spectrum is expected, as demonstrated by Weber and Shinitzky (349).

Other intrinsic acceptors for tryptophan are NADH (204−206), riboflavin (217), lumazine (217), pyridoxal phosphate (121, 123), and hemes (60, 95−97). Tryptophan→heme energy transfer occurs with high efficiency in myoglobin and hemoglobin and has been carefully studied by time-resolved fluorescence methods (60, 95−97). Hochstrasser and coworkers (96) studied the fluorescence decay of tryptophan(s) in three homologous myoglobins, which contain one (tuna) or two (sperm whale and Aplysia limacina) tryptophan residues. The dominant ($\alpha \approx 0.95$) component has a decay time of ~20 ps; minor components have decay times of ~800 ps and 3 ns. Using x-ray structural data for the tryptophan→heme distances and allowing for some uncertainty in the orientation factor for energy transfer, these workers calculate fluorescence lifetimes of 20–100 ps for the various tryptophan positions (60). The 800-ps and 3-ns lifetimes thus are not predicted, suggesting that they may be due to a conformational state(s) dissimilar to the crystal state and/or to an impurity. The 20-ps decay time is sensitive to the oxidation/ligation state of the heme iron and to temperature. As temperature is increased, the value of the short lifetime increases progressively and its amplitude decreases. This decrease in energy-transfer quenching with increasing temperature was interpreted as being due to an increased flexibility of the tryptophan residues. Above the thermal transition temperature, the energy-transfer process appears to be further decreased (i.e., reduced amplitude for the short component), as expected if the tryptophan→heme average distance increases upon unfolding of the protein structure.

A variety of extrinsic probes can be selectively or nonselectively attached to proteins (98). These probes can serve as acceptors for energy transfer from tryptophan. Alternatively, one can attach a complementary pair of probes, or one can introduce a new D→A pair. Extrinsic probes can be either noncovalently or covalently bound to a protein; the latter is, of course, preferred, since it enables stoichiometries and attachment sites to be more directly determined (i.e., peptide mapping). Covalent attachment can be done by random labeling or by a site-specific reaction. There are some advantages to random attachment in studies of distance mapping of multisubunit aggregates (19, 20), but most studies attempt to employ specific attachment. The sulfhydryl group of cysteine residues is the most useful site for covalent attachment. Often only one or a few reduced cysteine residues are present in a protein and it is also often found that one of these cysteine residues is very reactive. Table IV lists a number of sulfhydryl-reactive fluorescence probes.

Lysine residues are another target for covalent attachment. Because there are usually several lysine residues in a protein, they will usually be a site for random labeling (19, 98). The neutral ε-amino (or N-terminal amino) group will react with fluorescence probes that possess isocyanates, sulfonyl chlorides, anhydrides, or hydroxysuccinate esters (19, 98). Also, the amino group can be reacted with an aldehyde, followed by borohydride reduction to form a secondary amine.

Other extrinsic probes can be attached by taking advantage of the reactivity of an enzyme's active site (99) or by using a separate enzyme to catalyze the specific linkage of an chromophore to a protein. An interesting example of the latter type is the use of a bacterial toxin to catalyze the ADP-ribosylation (with an etheno-NAD^+) of the T_α subunit of transducin (a GTP-binding protein) (102). Substituting lanthanides (i.e., Tb^{3+}, which is an acceptor for tryptophan) in place of Ca^{2+} or other metal ions in a protein's structure is another way to introduce an extrinsic probe (100, 101). For all extrinsic probes, it is essential to determine the stoichiometry of attachment (i.e., number of probes per protein) in order to allow corrections to be made, if necessary (see (94) for an example of this correction).

The various extrinsic probes have different spectral positions, extinction

TABLE IV

Common Covalent Fluorescent Probes

Abbreviation	Meaning	References
	Sulfhydryl-Modifying Fluorescent Probes	
IAF[b,d,e]	5-(Iodoacetamido)fluorescein	151, 94, 109
IAEDANS[a,d]	5-[[[(Iodoacetyl)amino]-ethyl]amino]naphthalene-1-sulfonic acid	153, 156, 148, 94, 120
ISAL[b]	5-(Iodoacetamido)salicylic acid	151
IANBD[a,d]	4-[N-[(Iodoacetoxy)ethyl]-N-methylamino]-7-nitrobenz-2-oxa-1,3-diazole	148, 154
IATR	Tetramethylethylrhodamine-5 (and 6-) iodoacetamide	148
CPM	7-Diethylamino-3-[4'-maleimidophenyl]-4-methylcoumarin	152, 158
NPM[a,c]	N-[1-Pyrenyl]maleimide	152
Acrylodan	6-Acryloyl-2-(dimethylamino)naphthalene	129
IAE[b]	5-(Iodoacetamido)eosin	159
IAANS[b]	2-((4'-Iodoacetamido)anilino)naphthalene-6-sulfonic acid	154
ANM	1-Anilinonaphthyl-4-maleimide	155
	Amino-Modifying Fluorescent Probes	
FITC	Fluorescein isothiocyanate (also rhodamine isothiocyanate)	130, 156
	Dansyl chloride, Mamsyl chloride, Dabsyl chloride, TNS chloride	130, 137
Fluorescamine		160
Coumarin isothiocyanate		98
NBD chloride (also reacts with SH groups)		157
	Methionine-Modifying Fluorescent Probes	
DANZ	Dansylaziridine (also reacts with SH groups)	150

Note. See reference 98 for listings and properties and other covalent fluorescent probes and practical suggestions on their use.
[a] Also, acid or sulfonyl chlorides available, which react with both −SH and −NH$_2$ groups.
[b] Also, maleimide derivatives available.
[c] Also, iodoacetate or iodoacetamide derivatives available.
[d] Also, disulfide derivatives available.
[e] Can modify met groups (149).

coefficients, and lifetimes and thus there is some selection of $R_0^{2/3}$ values for different D→A pairs. The combination of IAEDANS as donor and IAF as acceptor is a useful pair with a very long $R_0^{2/3}$.

6.2. Orientation Effects

In RET experiments the focus is usually on determining of the D→A separation distance, r. It is important to realize, however, that r is only one of the two molecular parameters involved in RET; the other molecular parameter is the orientation factor, κ^2. A dynamic average value of $<\kappa^2> = \frac{2}{3}$ is often assumed to enable an r value to be calculated from E or k_{ret} data. Dale and Eisinger (21, 85−87) warn, however, that uncertainty about the appropriate value of κ^2 (which can range from 0 to 4) can greatly affect our calculation of r. Dale and

Eisinger demonstrate that at best we can limit the range of κ^2, and hence the range of calculated r values.

To understand the nature of the problem, first consider an essemble of molecules (having D→A pairs separated by a fixed distance, r) in the isotropic static limit (85). In the static limit, the D and A moments do not reorient during the excited state lifetime. Further, assume there is an isotropic distribution of relative D→A orientations, as illustrated in Fig. 18 (right panel). There will thus be a range of κ_i^2, for individual molecules, from 0 to 4. the average efficiency of energy transfer, over the ensemble of molecules, would be

$$<E>_s = \left\langle \frac{\kappa_i^2}{C^{-1}r^6 + \kappa_i^2} \right\rangle \tag{33}$$

(where angle brackets indicate average). Steady-state fluorescence measurements would reveal energy transfer, but, as pointed out by Dale and Eisinger (85), it is generally not possible to determine the true value of r from the observed E. The fluorescence decay of the donor would appear to be multiexponential. This isotropic static limit should apply to a frozen (i.e., low-temperature and/or high-viscosity) solution of randomly coiled polypeptide having D and A groups attached on either end. If there is a random distribution of both orientations and separation distances for D-A pairs in the ensemble, then a static average $\kappa^2 = 0.476$ may be used (89), but a multiexponential decay of donor would still be expected.

For D−A pairs in globular proteins, a more likely static model is the anisotropic static limit depicted in Fig. 18 (center panel). Here the D and A transition dipoles are all rigidly held at the same relative orientation, defined by Θ_A, Θ_D, and Θ_T (see Fig. 4), giving a unique value of κ^2 for the system. This value of κ^2 may range from 0 to 4, and, because κ^2 will be unknown, there will be no way to relate observed E values to separation distances.

The other extreme case is the dynamic averaging limit (Fig. 18, left panel), in which the D and/or A dipoles can sample all orientations (isotropic rotation) during the excited state D lifetime. The transfer efficiency of D−A pairs in the ensemble is the dynamically averaged value, $<E>_d$, given by Eq. 34, where $<\kappa^2> = \frac{2}{3}$ is the average κ_i^2 over all orientations:

$$<E>_d = \frac{<\kappa_i^2>}{C^{-1}r^6 + <\kappa_i^2>} \tag{34}$$

This, of course, is the ideal case for relating energy-transfer data to r. In reality, however, it is more likely that D and A chromophores attached to proteins will undergo rapid rotations that are not isotropic, but that are sterically restricted (i.e., rotation limited to the surface or the volume of a cone, see Fig. 12). If this is the case, then Eq. 34 applies, but the appropriate value of $<\kappa_i^2>$ may not be the dynamic, isotropic value of $\frac{2}{3}$. Instead, the

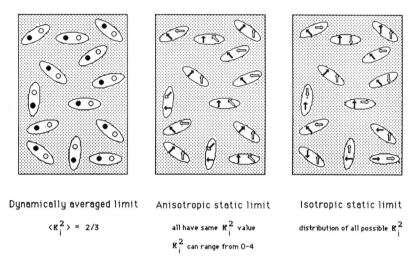

Dynamically averaged limit   Anisotropic static limit   Isotropic static limit

$$\langle \kappa_i^2 \rangle = 2/3$$

all have same κ_i^2 value   distribution of all possible κ_i^2

κ_i^2 can range from 0–4

Fig. 18. Representations of ensembles of macromolecules (large ovals) having resonance energy transfer donors (dark small circles or arrows) and acceptors (open circles or arrows). For each macromolecule the donor and acceptor groups are separated by the same distance r. *Left*: Dynamically averaged limit. Donor and acceptor moments can assume all possible mutual orientations during the lifetime of the excited state. This can result from rapid, independent rotation of the donor and acceptor groups or to the lack of an intrinsic polarization of the absorption and emission moments of the donor and acceptor. In this dynamic limit, $\langle \kappa \rangle^2$ will be equal to 2/3. *Center*: Anisotropic static limit. Donor and acceptor moments are fixed at one relative orientation in all macromolecules, as defined by the angles Θ_A, Θ_D and Θ_T in Fig. 4. The orientation factor κ^2 will have a unique value and can range from 0 to 0.40. *Right*: Isotropic static limit. The mutual orientation of donor acceptor pairs can assume all possible values (of Θ_A, Θ_D, Θ_T) for different macromolecules; however, within a given macromolecule this orientation does not change during the excited lifetime of the donor. κ_i^2 values will differ for each macromolecule. [This figure is redrawn after that in reference 85.]

average orientation factor may be greater than or lesser than $\frac{2}{3}$ (likewise, the true R_0 may differ from $R_0^{2/3}$).

For such cases of limited, dynamic averaging, Dale and Eisenger (21, 86, 87) present a strategy for delimiting the range of $\langle \kappa_i^2 \rangle$ and R_0. This strategy draws upon fluorescence anisotropy measurements for the attached D and A groups (provided that A fluoresces) to provide independent information about the degree of rapid rotational freedom of these centers. If the degree of rotational freedom of both D and A can be determined, then, by assuming certain models for the motional freedom, minimum and maximum limits can be set for $\langle \kappa^2 \rangle$ and R_0. As discussed in Section 4, time-resolved anisotropy measurements of D and A (directly excited) may reveal the existence of rapid, restricted motion, in terms of a depolarization factor, d, equal to β_2/r_0. A depolarization factor of unity indicates the absence of restricted motion; a factor of zero indicates complete orientational freedom. One can interpret this motion in terms of rapid rotation of the emission dipole within a cone of half-angle, Θ (or on the surface of a cone) using equation 32. With these d_D and d_A values (or Θ_A and Θ_D values), one can calculate $<\kappa^2>_{min}$ and $<\kappa^2>_{max}$ from the following equations. For example, if $d_D = d_A = 1.0$, the case for completely

$$<\kappa^2>_{min} = \tfrac{2}{3} [1 - \tfrac{1}{2} (<d_D> + <d_A>)] \tag{35}$$

$$<\kappa^2>_{max} = \tfrac{2}{3} [1 + (<d_D> + 3<d_D><d_A>)] \tag{36}$$

immobilized D and A dipoles, then $<\kappa^2>_{min} = 0$ and $<\kappa^2>_{max} = 4.0$, as mentioned above for the anisotropic, static limit, and it will be impossible to relate RET data to r. Conversely, if $d_D = d_A = 0.0$, then both $<\kappa^2>_{min}$ and $<\kappa^2>_{max}$ will be $\tfrac{2}{3}$. If one d factor is 1.0 and the other is 0.0, then the range of $<\kappa^2>$ values is 0.167 to 1.333. From the anisotropy data for single-tryptophan residues in proteins (Table III), we find dynamic depolarization factors of 0.35 to ~1.0 for the rapid motion of various tryptophan residues. If RET acceptors have a similar range of d_A values, then it must be realized that there will be a wide range of $<\kappa^2>_{min}$ and $<\kappa^2>_{max}$ values in studies with D−A pairs attached to native proteins.

Dale and Eisenger also present many contour maps of d_A versus d_D (or Θ_A versus Θ_D) for several assumed orientations of the cone axis of the D and A pair (21). From these contours one can estimate $<\kappa^2>_{min}$ and $<\kappa^2>_{max}$.

Further refinement of $<\kappa^2>$ is possible if the anisotropy of the sensitized emission of A can be measured, so that a transfer depolarization factor, d_T, can be obtained. From contour maps of d_A versus d_D for various values of d_T, one can further delimit the range of $<\kappa^2>_{min}$ and $<\kappa^2>_{max}$ (87).

To summarize the approach suggested by Dale and Eisenger (21, 97−102) to analyze RET data, anisotropy measurements of D and A emission must first be performed to determine whether the static isotropic, dynamic, or restricted-dynamic averaging regimes apply. In the latter case, such measurements also allow d_D and d_A to be determined (and possibly d_T). These depolarization factors can then be used to set limits on the range of $<\kappa^2>$, using either Eq. 35 and 36 or the published contour maps. Once $<\kappa^2>_{min}$ and $<\kappa^2>_{max}$ are established, minimum and maximum R_0 values can be calculated (i.e., $R_{0,min} = R_0^{2/3} [3<\kappa^2>_{min}/2]^{1/6}$ and $R_{0,max} = R_0^{2/3} [3<\kappa^2>_{max}/2]^{1/6}$), and the minimum and maximum r can then be evaluated from E or k_T values. Because of the cumbersome problem of evaluating $<\kappa^2>$, it has been pointed out that its range of values can also be reduced by using fluorophores that have intrinsically low limiting anisotropies, r_0, and/or by exciting into higher electronic transitions, which are often polarized orthogonally to that of the $S_0 \rightarrow S_1$ transition (20). Of course, the 1/6th power dependence of r on $<\kappa^2>$ minimizes the error of assuming a $\tfrac{2}{3}$ value, unless the true value is near zero. When using extrinsic probes, one can also compare r values obtained with different D−A pairs; if there is agreement, this supports the use of $<\kappa^2> = \tfrac{2}{3}$ (19, 20).

6.3. Distribution of Separation Distances

The above discussion considered variations in the orientation factor with a single distance, r. Haas, Steinberg, and coworkers (88−91) raised the possibility that, in some cases, a distribution of D→A distances may exist and that RET data (time-resolved decay of donor) may reveal these distributions. If there is a set of r_i values, then (in the dynamically averaged isotropic limit, $\langle \kappa_i^2 \rangle = \frac{2}{3}$) the $\langle E \rangle_{d,i}$ and $k_{\text{ret},i}$ values will be

$$\langle E \rangle_{d,i} = \frac{\langle \kappa_i^2 \rangle}{C^{-1} r_i^6 + \langle \kappa_i^2 \rangle} \tag{37}$$

$$k_{\text{ret},i} = \frac{C \langle \kappa_i^2 \rangle}{(r_i^6 \tau_D)} \tag{38a}$$

$$= \frac{(R_0^{2/3})^6}{(r_i^6 \tau_D)} \tag{38b}$$

where C is a product of constants in Eq. 23. For an ensemble of molecules with D-A pairs, the impulse−response function of the donor will be (assuming a monoexponential decay in the absence of the acceptor)

$$I_D(t) = \sum P_i \exp\left[-\frac{t}{\tau_D} - \frac{t}{\tau_D} \left(\frac{R_0^{2/3}}{r_i} \right)^6 \right] \tag{39}$$

where P_i is the normalized population of species with separation distance r_i. The practice has been to model the set of species as a continuous distribution, $P(r)$, of distances; this distribution has been assumed to be a Gaussian function:

$$P(r) = \frac{1}{\sigma(2\pi)^{1/2}} \exp\left[-\frac{1}{2} \left(\frac{r - \bar{r}}{\sigma} \right)^2 \right] \tag{40}$$

where $\bar{r}$ is the average separation and σ is the standard deviation of the distribution. Using these assumptions, the above impulse−response function becomes

$$I_D(t) = \int_0^\infty P(r) \exp\left[-\frac{t}{\tau_D} - \frac{t}{\tau_D} \left(\frac{R_0^{2/3}}{r} \right)^6 \right] dr \tag{41}$$

Equations 37−41 predict a nonexponential decay of D in the presence of RET to A. Haas et al. (88) demonstrate this for a flexible polypeptide chain having D and A groups attached at opposite ends and analyzed intensity decay data to obtain an average separation distance, $\bar{r}$, and a standard deviation of the distribution, σ. Amir and Haas also studied the RET between D−A pairs covalently linked to the protein bovine pancreatic trypsin inhibitor and analyzed the nonexponential $I_D(t)$ decay in terms of a distribution of intramolecular distances (91). Lakowicz and coworkers (92) recently used multifrequency phase fluorometry to collect similar data for flexible D—A model systems. Also, this group studied the RET between a single tryptophan residue of troponin I and covalently attached IAEDANS acceptor and analyzed their data in terms of a distribution of r values (average $\bar{r} = 23$ Å, width of 12 Å) (93). Upon thermal denaturation, the average $\bar{r}$ (27 Å) and width (47 Å) increased, as expected for an unfolded protein.

Haas et al. (90) also argued that the end-to-end diffusion coefficient can be obtained from time-resolved RET studies. The existence of such motion leads to a skewing of a distribution to shorter values (i.e., shorter than that for a frozen distribution). To demonstate this effect, they studied a D—A-labeled flexible chain as a function of solvent viscosity. As predicted, the distribution center increased at increased viscosity. This analysis assumes that the dynamic averaging limit is maintained at all viscosities. Dale and Eisinger (85) criticized that assumption by pointing out that, in the static limit, distribution of κ_i^2 as well as r_i, may contribute to the observed nonexponential decays.

While one can debate the relative significance of dispersion in distance versus orientation factors, it is clear that these molecular features can be determined only with time-resolved data, not with steady-state data (unless quenching-resolved measurements are used (103)). If only steady-state fluorescence data is obtained, the above complex features will not be indicated. Instead, standard analysis of steady-state data will yield an apparent separation distance that may differ significantly from the average of a distribution of distances (104). Also, if the average fluorescence lifetime of the donor is used to determine a distance, this value will usually overestimate the distribution average as well.

6.4. Examples of the Use of RET in Studies of Distance Mapping and Association Reactions

Stryer (20) and Fairclough and Cantor (19) have provided very thorough reviews of the use of RET to study the structure of biomacromolecular assemblies; we will mention here only a few recent applications to protein systems.

RET has been used advantageously in studies of the muscle proteins, myosin, actin, and troponin (94, 105—108, 119, 120, 395). There are reactive sulfhydryl groups on the subfragment 1 (S1) portion of myosin (two reactive cysteines, referred to as SH1 and SH2), the alkali light chains, actin, and both the Ca^{2+}-binding and inhibitory subunits of troponin. In addition, etheno analogues of ATP and ADP can be bound specifically to sites on both S1 and actin, and a labeled "antipeptide" can be covalently attached to region 633—642 of S1 (395). This has enabled the attachment of probes for RET and anisotropy decay studies. For example, Trayer and Trayer (109) labeled the SH1 of S1 and Cys-373 of actin with dansyl and fluorescein groups and studied the effect of binding ADP and nonhydrolyzable ATP analogues on the D→A distance in the F-actin—S1 complex. During steady-state hydrolysis of ATP the extent of RET was diminished, consistent with the dissociation of these two proteins during the hydrolysis cycle. Oriented muscle myofibrils have also been studied by RET (119).

Topographical studies of membrane-associated proteins have employed RET between tryptophan residues of the proteins and acceptors attached to the fatty acid chains of phospholipids (110–113). Kleinfeld (110) has developed a "tryptophan imagining" procedure, using a series of n-(9-anthroyloxy)fatty acid acceptors, to determine information about the depth of penetration of protein tryptophans into membrane bilayers. Other studies with membranes have used RET to characterize the degree of aggregation of peptides or proteins upon their interaction with bilayers (111, 112).

In studies with monomeric proteins, inserting Tb^{3+} and Co^{2+}, in place of Ca^{2+} and Zn^{2+}, respectively, in metalloproteins provides useful RET acceptors of tryptophan excitation (100, 101). These metal ions are particularly valuable because they have electronic transitions along three perpendicular axes. This limits the range of $<\kappa^2>$ when these are used as acceptors, thus removing uncertainty in distance determinations. RET has also been used to study the kinetics of protein unfolding reactions (114, 115), protein–protein association reactions (116), and ligand-induced conformational changes in proteins (117).

7. SEPARATION OF FLUORESCENCE COMPONENTS

The assignment of fluorescence properties to individual fluorophores or conformational states is of prime importance is using this method to obtain any molecular insights. Nature has dealt us an interesting set of single tryptophan proteins (see Table III), but most proteins, of course, contain multiple-tryptophan residues. A major task in fluorescence studies with proteins is to separate different components, and, if possible, to assign the signals to individual fluorescence centers.

The multidimensional nature of fluorescence signals facilitates this resolution. Fluorescence signals can be measured as a function of *wavelength* (excitation or emission), *time* (following an excitation pulse), or *modulation frequency* (with sinusoidally modulated excitation), and solute *quencher* concentration. Also, with polarized excitation, the parallel and perpendicular fluorescence signals (i.e., *anisotropy*) can be obtained as a function of wavelength, time, quencher concentration and so on. Other useful variables include temperature, pressure, pH, and solvent composition. Thus, there are many dimensional axes that can be used to resolve components. That is, two components can potentially be resolved if they have different fluorescence λ_{max}, decay times, rotational correlation times, or quenching constants.

Two-dimensional steady-state measurements (i.e., emission spectra vs. excitation wavelength) can, in principle, be analyzed by matrix techniques (singular value decomposition) to resolve components, provided there is sufficient difference in the excitation and emission spectra of the components (293, 294). Most successful attempts to resolve components of protein fluorescence have exploited differences in decay times or quencher accessibility of fluorophore (i.e., tryptophan) components, since the latter parameters vary over a wider range than do the excitation and emission spectra and are expected to obey established functions (i.e., Eqs. 5 and 21).

In the time or frequency domain, the modest range of fluorescence decay times (see Table III) enables, in optimum situations, two or more components to be resolved. In Section 4 we mention studies with the protein LADH, where the biexponential decay is interpreted as lifetimes for the protein's two tryptophan residues, Trp-15 (τ = 7 ns) and Trp-314 (τ = 3.5 ns). Time-resolved measurements for LADH made as a function of emission wavelength provide a two-dimensional (λ and t) family of data sets for the following impulse–decay function (295, 355):

$$I(\lambda,\ t)\ =\ \sum_\lambda \sum_i \alpha_{\lambda,i}\ \exp\!\left(\frac{-t}{\tau_i}\right) \tag{42}$$

Fitting this function yields the amplitudes, $\alpha_{\lambda,i}$, for the two components (i = 2) as a function of λ. The product of $\alpha_{\lambda,i}$ times the steady-state spectrum, $I(\lambda)$, represents the spectrum of each decaying component. These are known as the decay-associated spectra (DAS) (295). An analogous method using phase fluorometry yields phase-resolved spectra (PRS) (296, 297, 316). Figure 19

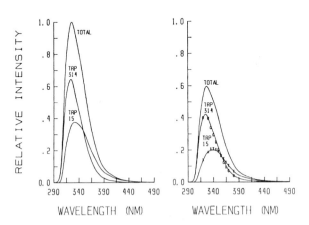

Fig. 19. Steady-state (total) and decay-associated emission spectra (DAS) for LADH. *Left*: LADH alone at pH 7.4, 10 °C. The DAS assigned to Trp-15 and Trp-314 are shown. *Right*: Ternary complex between LADH NAD$^+$ pyrazole. Notice the quenching of total emission caused by formation of this complex. See reference 295 for details. [This figure is reproduced from this reference with permission from the authors and the American Chemical Society.]

shows the DAS for the short and long lifetime components of LADH. The short-lived component is the bluer of the two, as expected for the buried Trp-314. Changes in the DAS of each tryptophan are found upon forming a ternary complex with the coenzyme, NAD^+, and pyrazole. In analyzing these multiple λ data sets, the practice is to fit Eq. 42 to all data sets simultaneously and to require the τ_i to be the same at all λ. This procedure is referred to as a global analysis, with the τ_i considered to be global fitting parameters and the $\alpha_{\lambda,i}$ considered to be local (i.e., λ-dependent) parameters. One can debate the assumption that the τ_i for a given fluorescent center is independent of λ. However, the DAS are, as the name implies, the spectra associated with each τ_i used to fit the data and not necessarily the spectra of individual (directly excited) fluorescent centers. Recovered DAS will clearly depend on the number of components assumed in the analysis.

In an analogous manner, it is possible to resolve time-resolved, wavelength-dependent anisotropy data to obtain the spectra associated with different rotational correlation times, ϕ_i. This method is referred to as ADAS, anisotropy decay-associated spectra (298). An example of the use of ADAS to separate the components of a mixture of a TNS—LADH complex and a TNS—cyclodextrin complex is shown in Fig. 20. The intensity decay is similar in these two complexes. However, the great difference in molecular size (LADH, M_r ~84 000; β-cyclodextrin, M_r ~1135) and concomitant difference in rotational correlation times (71 ns; 0.4 ns) enable a resolution of the more mobile component (TNS—cyclodextrin) from the less mobile component (TNS—LADH).

The combined use of steady-state anisotropy measurements and solute quenching can be used to resolve the anisotropy values of individual fluorescing

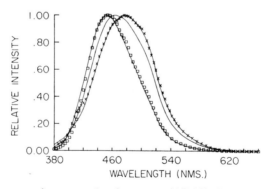

Fig. 20. Anisotropy decay-associated spectra (ADAS) for a mixture of the TNS/LADH complex and the TNS/cyclodextrin complex at 10 °C. □, ADAS for the slowly rotating TNS/LADH complex. x, ADAS for the rapidly rotating TNS/cyclodextrin complex. The middle (———) spectrum is a mixture. See reference 298 for details. [This figure is reproduced from reference 298 with permission from the authors and the American Chemical Society.]

centers (325, 326, 356). By selectively quenching the fluorescence of a more accessible fluorophore, the residual anisotropy data will be that of the inaccessible fluorophores.

Solute quenching is the simplest experimental method to use in attempting to resolve fluorescing components. As demonstrated by Lehrer's classic study of the iodide quenching of lysozyme (262), if there are classes of tryptophans in a protein with widely different degrees of solvent accessibility, measuring the emission spectrum as a function of $[Q]$ will reveal the residual spectrum of the inaccessible class and, by difference, the spectrum of the accessible class. Selective quenching can also be used to resolve fluorescence lifetime data (320, 351, 353, 354). Several important quenching studies have been performed to resolve the steady-state and time-resolved fluorescence properties of the two tryptophan residues of LADH (352−356).

The resolution of steady-state spectra can be improved by performing simultaneous "global", nonlinear, least-squares fitting of Eq. 21 for data obtained at multiple-emission wavelengths (300−302). This practice is illustrated in Fig. 21 for the iodide quenching of the fluorescence of a pyrene labeled sulfhydryl group of enzyme I of the bacterial PTS. Figure 21A shows a series of Stern−Volmer plots, and Fig. 21B shows the resolution of the spectra associated with the accessible and inaccessible components.

A number of other methods can be used to separate fluorescence components and aid in their assignment to specific tryptophan (or other fluorophore) residues. Among these are selective chemical modification (i.e., N-bromosuccinimide oxidation of tryptophan residues), comparison of homologous

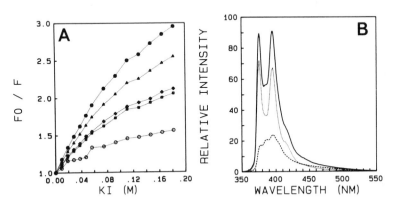

Fig. 21. (A) Stern−Volmer plots of the iodide quenching of pyrene labeled PM-Enzyme I at different emission wavelengths: 376 nm (○); 386 nm (◆); 396 nm (■); 406 nm (▲); and 460 nm (●). Pyrene maleimide was used to preferentially label the 'fast-reacting'-SH group of the enzyme. Global analysis via equation 21 yielded K_{sv} values of 0.4 and 23 M^{-1}. (B) Quenching decay-associated spectra of the above pyrene labeled PM-Enzyme I. ——, total emission spectrum; ..., accessible component with $K_{sv} = 23$ M^{-1}; ---, inaccessible component with $K_{sv} = 0.4$ M^{-1}. [Figures reproduced from reference 303 with permission of the authors.]

proteins which differ in tryptophan content, and the solvent isotope effect on tryptophan fluorescence. Longworth (304) has provided a general summary of these and other approaches. With the development of site-directed mutagenesis techniques, it is now possible to delete or insert tryptophan residues at specific sites in proteins to aid in the resolution of components and to introduce an "intrinsic" probe (305—307).

8. DEVELOPMENTS IN INSTRUMENTATION AND DATA ANALYSIS

Improvements in fluorescence instrumentation, particularly that for time-resolved measurements, have been significant over the past 25 years. There are two basic instrumental designs, the first being pulse-decay or time-correlated single-photon counting (TCPC) and the second being harmonic response or phase/modulation fluorometry. (See (3, 7, 162—166) for descriptions of these techniques.)

Early TCPC fluorometers employed flash lamps, with a pulse width of typically 1—2 ns, low power and repetition rate, and erratic pulse shape. Synchronously pumped picosecond dye lasers are now available and these provide very narrow, uniform pulses with high power and repetition rate.

Likewise, the performance of phase/modulation fluorometers has improved, from the early instruments that employed Debye—Sears modulation tanks and operated at only 1—3 frequencies, to the present day, multifrequency instruments that employ Pockels cells or the harmonic content of mode-locked lasers to provide the modulation of the excitation light (166). Modulation frequencies up to 6—10 GHz are achieved using the harmonic content (166, 300, 301). A train of synchrotron pulses has been used as the light source for both phase/modulation and TCPC methods (185—187).

For both techniques the time response of the photomultipliers is becoming a critical factor. Microchannel plate (MCP) detectors are now used with both types of instruments (166, 300) to increase fast time resolution.

A number of strategies are being developed to collect multidimensional fluorescence data. For example, arrays of fibers (188, 189) or multianode MCPs (190) are used to simultaneously collect time-domain data at multiple-emission wavelengths. Gratton et al. (168) are investigating the possibility of modulating intensified diode array detectors for use with phase/modulation fluorometry. This would enable multiwavelength (i.e., 512—1024 channels) phase-resolved measurements to be made. Streak cameras continue to provide advantages for very rapid time-resolved measurements, and the technology is available to collect decay patterns versus emission wavelength (169). Efforts are being made to enable both time-domain and frequency-domain instruments to acquire fluorescence lifetime data sets on a milliseconds to seconds time scale, thus allowing transient (stopped flow) measurements of fluorescence lifetimes. This method, which is referred to as KINDK (kinetic decay), has been employed by Brand, Knutson, and coworkers (183, 184) to study the kinetics of protein unfolding and modification of proteins.

A hybrid TCPC-phase/modulation instrument, recently assembled by Hedstrom et al. (170), is diagrammed in Fig. 22. Similar fluorescence decay parameters are found for a variety of systems for the two types of measurements.

Analysis of TCPC data is no simple task; it involves 'deconvolution' of the instrument response from the true impulse—response function, selecting a model to describe the data (i.e., a sum of n exponentials, or a distribution of decay times, see below) and determining the decay parameters for this selected model. A number of analysis procedures for such decay data are available. An iterative nonlinear least-squares method is most frequently used (162, 171, 180). A second, closely related group of procedures involves calculation of integral transforms. The method of moments (172—174) and various Laplace

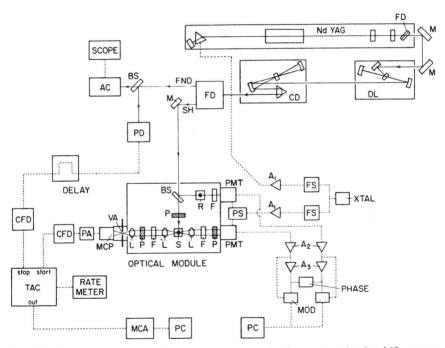

Fig. 22. Schematic for a hybrid time-correlated single-photon counting/multifrequency phase-modulation fluorometer. Components include Nd:YAG pump laser; frequency doubler, FD; mirror, M; dye laser, DL; cavity dumper, CD; beam splitter, BS; real-time autocorrelator, AC; photodiode, PD; constant fraction discriminator, CFD; polarizer, P; sample, S; lens, L; filter, F; variable aperature, VA; preamplifier, PA; microchannel plate, MCP; time-to-amplitude converter, TAC; reference, R; photomultiplier tube, PMT; power splitter, PS; rf amplifier, A_1; DC amplifier, A_2; AC tuned amplifier, A_3; frequency synthesizer, FS; 10-MHz quartz oscillator, XTAL; computer, PC; multichannel analyzer, MCA. [This figure is reproduced from reference 170 with permission from the authors and the American Chemical Society.]

transform methods (175, 181) fall into this category. Recently, a method has been developed that uses Pade approximants with the Laplace transform (176); a purported advantage of this method is the ability to determine the number of components in a multiexponential decay law. Another procedure first uses of a Fourier transform to deconvolute the impulse—response function and then employs one of the above methods to obtain the decay parameters (177). The maximum entropy method is yet another recently developed method for analyzing fluorescence decay data (178). Some of these methods have been used for both TCPC and phase/modulation data (176, 179).

In recent years the question has arisen as to whether to describe non-exponential fluorescence decay data in terms of a sum of exponential components or a continuous distribution of decay times (77−84). The sum of discrete exponentials model is given by Eq. 5. In a continuous distribution model, the impulse—decay response is represented by

$$I(t) = \int_{\tau=0}^{\infty} \alpha(\tau)e^{-t/\tau}d\tau \tag{43}$$

where the integration is from $\tau = 0$ to some high value of nanoseconds, and $\alpha(\tau)$ is a function that describes the variation of the amplitudes with τ (such that $\int \alpha(\tau)d\tau = 1.0$). The distribution shape for $\alpha(\tau)$ is usually assumed to be a Gaussian (G) or Lorentzian (L) distribution, as given by either of the following equations:

$$\alpha_G(\tau) = \frac{1}{\sigma(2\pi)^{1/2}} \exp\left[-\frac{1}{2}\left(\frac{\tau - \bar{\tau}}{\sigma}\right)^2\right] \tag{44a}$$

$$\alpha_L(\tau) = \frac{1}{\pi}\frac{\Gamma_L/2}{(\tau - \bar{\tau})^2 + (\Gamma_L/2)^2} \tag{44b}$$

where $\bar{\tau}$ is the central lifetime value for the distribution, σ is the standard deviation of the Gaussian distribution, and Γ_L is the full-width−half-maximum of the Lorentzian distribution. (For comparison, the full-width−half-maximum of a Gaussian distribution, Γ_G, is 2.354σ.) The above equations describe a unimodal symmetrical distribution in lifetime space. A bi- or trimodal distribution may be considered, for which equations are given in Lakowicz et al. (81). It will usually be difficult or impossible to distinguish, based on comparison of goodness of fit, between a discrete exponential model and certain continuous distribution models. In particular, Lakowicz et al. (81) show that a double exponential decay fit (3 parameters, τ_1, τ_2, α_1) and a unimodal distribution fit (2 parameters, $\bar{\tau}$ and Γ) should usually give similar reduced chi squares values. Likewise, a triple exponential decay fit (5 parameters) and a bimodal distribution (5 parameters) are virtually indistinguishable. In these cases, auxiliary information may help to distinguish between the discrete and distribution models. More often, the interpretation will be the prerogative of the researcher, but the possibility of the two interpretations should be considered. The discrete exponentials model may appeal to the reductionist attitude, whereas the distribution models may appeal to our sense that some proteins are conformationally flexible.

Preferably, the two models should be distinguishable based on fluorescence data and fitting alone. To this end, James, Ware, and coworkers (82−84) introduced the exponential τ series method of analyzing impulse−response data. By fixing a series of τ_i (i.e., every 0.5 ns or so from 0 to 10 ns) and fitting data with the α_i for each fixed τ_i, a pseudo-distribution can be revealed. A similar, "filter-tau" method by Small, Libertini, and coworkers (173) for recovering distribution features is available. The maximum entropy method also allows for analysis of data in terms of a distribution of amplitudes in lifetime space. Livesey and Brochon (178) have used this method to obtain distribution fits (in log τ space) for tryptophan and tryptophan residues in proteins. These latter procedures do not assume a particular shape to the function, as do the above Gaussian or Lorentzian least-squares analyses.

9. OTHER USES OF PROTEIN FLUORESCENCE

To this point we have focused on how fluorescence studies can provide information about the dynamics, structure, and transitions in proteins. In this section we briefly mention some of the other practical uses of protein fluorescence in biochemistry.

The binding of ligands to proteins will often result in a change in the fluorescence intensity of either the protein or the ligand (223, 196−198), providing a convenient means of determining association constants. The thermal, urea, or quanidine HCl denaturation of proteins can be monitored by changes in tryptophan fluorescence as the protein unfolds (224, 225). An example of thermal transitions of proteins determined by fluorescence measurements is shown in Fig. 23. Effects of pressure on protein interactions can also be studied, using a specialized high pressure chamber (226−228, 333, 336).

A variety of kinetics studies can utilize protein fluorescence changes. Dandliker et al. (223) expound upon the application of fluorescence techniques in transient kinetics studies. Enzyme kinetics studies can involve changes in the fluorescence of the substrate molecule as it is converted to product (229).

Fluorescence photobleaching recovery (FPR) can provide information on the lateral diffusion of protein molecules embedded in membranes (230).

Due to its intrinsic sensitivity, selectivity, ready readout of results, and ability to be adapted to study samples in a variety of physical states, fluorescence techniques are widely used to quantitate analytes ranging from small molecules to cells. Of particular importance is the use of fluorescence to quantitate antibody−hapten complexes by observing quenching of the intrinsic antibody fluorescence (i.e., with 2,4-dinitrophenyl-labeled haptens) or the enhancement or quenching of labeled haptens (i.e., labeled with dansyl, ANS, or fluorescein)

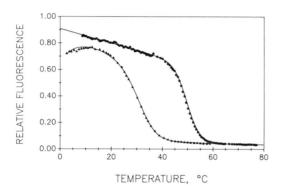

Fig. 23. Fluorescence thermal transition from staphylococcal nuclease A ($\bullet$) and its mutant nuclease-conA 528G ($\blacktriangle$). Data are fitted, as described in references 224 and 346, to obtain the thermal transition temperature, T_m, and the enthalpy change, $\Delta H°$, for the unfolding transition. For the nuclease-conA 528G mutant, the data were also fitted to an expression that includes the heat capacity change, ΔC_p, for the transition. See reference 346 for details.

(231). In recent years much effort has gone into developing the technique of fluorescence microscopy, which combines the sensitivity and selectivity of fluorescence with the spatial resolution of the microscope (232−234). These topics are outside of our present focus, but are mentioned to illustrate the continued expansion of fluorescence techniques into related biochemical and biomedical areas.

Table V lists a number of acronyms for various fluorescence techniques.

10. PHOSPHORESCENCE OF PROTEINS

At low temperature and in vitrified solvents, phosphorescence can easily be observed from Trp (maxima at ~410 and ~435 nm) and Tyr (broad, with maximum at ~385 nm) residues in proteins. Longworth (25) has thoroughly reviewed this area. Recently, studies have focused on measuring room temperature phosphorescence of trp in proteins. Papp and Vanderkooi (235) and Geacintov and Brenner (255) have reviewed this topic so the treatment here will be brief.

The first, clear demonstration of room temperature phosphorescence of Trp in proteins was made by Saviotti and Galley for LADH (236). Because molecular oxygen is a potent quencher of Trp phosphorescence, much effort must be made to remove it from solutions (237, 238) and/or determine the concentration of residual/added oxygen (239). When oxygen is removed, Trp phosphorescence decay times range from ~0.5 to 2000 ms for proteins in aqueous solution at room temperature (235, 237, 240, 248). Phosphorescence from protein crystals has also been studied (241). There appears to be an inverse relationship between the Trp phosphorescence decay time and the

TABLE V

Some Luminescence Acronyms (Alphabet Soup)

Abbreviation	Meaning	References
ADAS	Anisotropy decay-associated spectra	298
DAS	Decay-associated spectra	295
DERET	Diffusion-enhanced resonance energy transfer	191
DF	Delayed fluorescence	
EA	Emission anisotropy	
EXDAS	Excitation decay-associated spectra	190
FCS	Fluorescence correlation spectroscopy	195
FDCD	Fluorescence detected circular dichroism	192
FPR (or FRAP)	Fluorescence photobleaching recovery	230
FRET (or RET)	Fluorescence resonance energy transfer	
FRS	Fluorescence recovery spectroscopy	252
KINDK	Kinetic measurements of fluorescence decays	183, 184
MPF	Multifrequency phase/modulation fluorometry	165, 166
ODMR	Optically detected magnetic resonance	193, 256, 257
PRS	Phase-resolved spectra	296, 297, 316
QDAS	Quenching-resolved decay-associated spectra	303
QREA	Quenching-resolved emission anisotropy	325, 326
TCPC (or TCSPC)	Time-correlated single-photon counting	162
TIRF	Total internal reflection fluorescence	194
TRES	Time-resolved emission spectra	295

Stokes shift of the fluorescence of single Trp proteins (235). The inference is that a long phosphorescence lifetime indicates that a Trp residue is shielded from solvent and external quenchers (i.e., residual oxygen) and/or is immobilized, so that intramolecular collisions with quenching amino acid side chains do not occur.

The longer time scale of Trp phosphorescence, relative to fluorescence, offers the opportunity to use phosphorescence to monitor competitive excited state processes in the millisecond time range. Measurements of phosphorescence anisotropy decay have been made with proteins in viscous solvents and in protein crystals. These measurements yield rotational correlation times and are useful for studying very large biomacromolecular structures. The rotational correlation times for large oligomeric proteins (glutamate dehydrogenase) (243), lipoprotein complexes (244), and a virus (245) are obtained from phosphorescence studies. Measurements on crystalline LADH demonstrate that the internal Trp-314 is immobilized on the seconds time scale (241).

Phosphorescence quenching studies, using oxygen and other solute quenchers, can also take advantage of the longer time window to enable buried Trp residues to be studied, and lower quencher concentrations to be used. Calhoun et al. (239, 270, 276) studied the phosphorescence quenching of a number of proteins by oxygen and a set of other quenchers, including NO,

CO, NO_2^-, ethanethiol, and nicotinamide. Ghiron et al. (246, 247) compared the O_2 and acrylamide quenching of the phosphorescence and fluorescence of several single-Trp proteins. Strambini and coworkers have used O_2 to quench the phosphorescence of deeply buried Trp residues in proteins such as LADH (in solution and in crystals) (241), glutamate dehydrogenase (243), and alkaline phosphatase (237). The consensus is that O_2 (and probably NO) quench internal Trp residues by a penetration mechanism. The mechanism of quenching by larger, polar quenchers is questionable. Calhoun et al., noting the similarity of phosphorescence quenching by neutral polar and charged quenchers, argue against a penetration mechanism for these and suggest that unfolding (276) and long-range quenching mechanisms (239) are responsible. Ghiron et al. (246, 247) point out that both oxygen and acrylamide are not unitary efficient quenchers of Trp phosphorescence (i.e., $\gamma \approx 0.5$ and 0.3 for oxygen and acrylamide quenching of indole triplet states in water) and that O_2 phosphorescence quenching rate constants do not show Stokes–Einstein T/η behavior for model systems. These workers caution that any Trp phosphorescence quenching rate constants determined using these quenchers may be sensitive to the microenvironment of the Trp residue (i.e., particularly for acrylamide as quenchers, since its γ is reduced >10-fold in going from water to acetonitrile) and will underestimate the true collision frequency between quencher and the Trp triplet states.

Extrinsic triplet state probes, such as eosin and erythrosin, can also be used to obtain information about translational and rotational motion of proteins on the microseconds to milliseconds time scale (249–254). Motion of large protein complexes, particularly in membrane systems, occurs on this time scale, eliciting much interest in the development of technology and methods in this area. The decay of triplet states can be observed directly, as phosphorescence emission (or as delayed fluorescence following intersystem crossing back to the singlet manifold (253)), or by measuring the transient absorbance of the triplet state (249, 251). Additional, elegant methods measure the transient depletion (or the recovery of the depletion) of the ground state following an intense laser pulse (which populates the triplet state of probes having a high quantum yield for intersystem crossing to the triplet state) (250, 251). The time course of ground-state recovery, which is a measure of the triplet-state lifetime, can be continuously monitored by the absorbance of the ground state or by the steady-state fluorescence signal of the probe, using a second, analyzing beam. This last use of fluorescence detection, known as fluorescence recovery spectroscopy (FRS), combines the high sensitivity of steady-state fluorescence measurements with the long time scale for triplet decay. Corin et al. (252) demonstrate the use of FRS in studying the rotational motion of eosin, fluorescein, and tetramethylrhodamine labeled proteins, both as free proteins in solution and when attached to cells. Yoshida and Barisas (254) assembled a microscope-based system, which can be used to perform both polarized FRS or FPR (fluorescence photobleaching recovery) measurements. This enables the determination of both rotational and translational diffusion for proteins labeled with appropriate triplet probes.

11. CLOSING COMMENTS

Research in the area of protein fluorescence has seen significant advances in instrumentation, rapid and multidimensional data acquisition, and data analysis procedures. The various methods allow dynamic luminescence measurements to be made from the second to the subpicosecond time scale. The sensitivity of fluorescence has always been a major advantage to biochemists, along with the adaptability of the methodology to many kinds of measurements. The resolution of the method has sometimes been a problem. Now, however, scientists have been able to resolve multiple fluorescence components in a protein so that the problem becomes one of assigning and correlating fluorescence parameters to structural/dynamic features of the protein. Quoting Axelson and Prendergast (391), "our basic understanding of protein fluorescence has advanced slowly in comparison to our appreciation of its complexity (as anticipated by Longworth, 1971)".

To improve our understanding, several groups have attempted to look closely at the x-ray structure of proteins for clues and perform molecular dynamics calculations to try to simulate and discover the molecular bases for such fluorescence observations as the Stokes shift (polarity of fluorophore microenvironment), fluorescence decay (i.e., presence or absence of intramolecular quenching processes), anisotropy decay (rotational motion of fluorophore), accessibility to external quenchers (transient exposure of fluorophore side chains), and energy transfer efficiency (relative orientation and separation distance of D−A pairs) (308, 387−393). The recent availability of site-directed mutant proteins, furthermore, makes it possible for researchers to investigate the molecular fluorescence properties by enabling the selective alteration of amino acid residues in the fluorophore's environment and inserting/deleting fluorophores (305−308, 342, 346).

Understanding fluorescence properties is often difficult, but the techniques and interesting systems are at hand. Due to the important time window and adaptability of luminescence methods, improvements in our understanding of the molecular basis of fluorescence properties should have significant impact on our fundamental knowledge of protein structure/dynamics/function interrelations and in our confident use of fluorescence techniques in more applied areas.

ACKNOWLEDGMENTS

Some of the unpublished work described here was supported by National

Science Foundation Grant DMB 88–06113. I thank C.A. Ghiron, University of Missouri, for his many helpful discussions and Rita Brown for her capable and untiring assistance in preparing this manuscript.

References

1. R.F. Chen and H. Edelhoch, Eds., *Biochemical Fluorescence: Concepts*, Vol. 1 and 2, Dekker, New York, 1975.

2. R.F. Steiner, Ed., *Excited States of Biopolymers*, Plenum, New York, 1983.

3. J.R. Lakowicz, *Principles of Fluorescence Spectroscopy*, Plenum, New York, 1983.

4. D.M. Jameson and G.D. Reinhart, Eds., *Fluorescent Biomolecules*, Plenum, New York, 1989.

5. G.G. Hammes, in *Protein–Protein Interactions*, C. Frieden and L.W. Nichol, Eds., Wiley, New York, 1981, pp. 257–287.

6. J.R. Lakowicz, Eds., *Time-Resolved Laser Spectroscopy in Biochemistry*, SPIE Proceedings, Vol., 909 (1988).

7. M.G. Badea and L. Brand, *Methods Enzymol.* **61**, 378–425 (1979).

8. B.S Hudson, D.L. Harris, R.D. Ludescher, A. Ruggiero, A. Cooney-Freed, and S.A. Cavalier, in *Fluorescence in the Biological Sciences*, D.L. Taylor, A.S. Waggoner, F. Lami, R.F. Murphy, and R. Birge, Eds., Liss, New York, 1986, pp. 159–202.

9. J. Beechem and L. Brand, *Annu. Rev. Biochem.* **9**, 4723–4729 (1985).

10. G. Weber, in *Spectroscopic Approaches to Biomolecular Conformation*, D.W. Urry, Ed., American Med. Assn., Chicago, 1970, pp. 23–31.

11. A.P. Denchenko, *Ultraviolet Spectroscopy of Proteins*, Springer-Verlag, New York, 1987.

12. G. Lipari and A. Szabo, *Biophys. J.* **30**, 489–506 (1980).

13. Y. Gottlieb and P. Wahl, *J. Chim. Phys.* **60**, 849–856 (1963).

14. R.F. Steiner, in *Excited States of Biopolymers*, R.F. Steiner, Ed., Plenum, New York, 1983. pp. 117–162.

15. M.R. Eftink and C.A. Ghiron, *Anal. Biochem.* **114**, 199–227 (1981).

16. J.R. Lakowicz, M.L. Johnson, I. Gryczynski, N. Joshi, and G. Laczko, *J. Phys. Chem.* **91**, 3277–3284 (1987).

17. M.R. Eftink, in *Fluorescence Spectroscopy: Principles and Techniques*, Vol. 1, J.R. Lakowicz, Ed., Plenum, New York, in press.

18. M.R. Eftink and C.A. Ghiron, *J. Phys. Chem.* **80**, 486–493 (1976).

19. R.H. Fairclough and C.R. Cantor, *Methods Enzymol.* **48**, 347–379 (1978).

20. L. Stryer, *Annu. Rev. Biochem.* **47**, 819–846 (1978).

21. R.E. Dale and J. Eisinger, in *Biochemical Fluorescence: Concepts*, Vol. 1, R.F. Chen and H. Edelhoch, Eds., Dekker, New York, 1975, pp. 115–284.

22. W.R. Laws and L. Brand, *J. Phys. Chem.* **83**, 795–802 (1979).

23. J.R. Lakowicz and A. Balter, *Biophys. Chem.* **16**, 99–115, 117–132 (1982).

24. J.B. Birks, in *Organic Molecular Photophysics*, Vol. 2, J.B. Birks, Ed., Wiley, New York, 1973, pp. 409–613.

25. J.W. Longworth, in *Excited States of Proteins and Nucleic Acids*, R.F. Steiner and I. Weinryb, Eds., Plenum, New York, 1971, 319–484.

26. F.W.J. Teale, *Biochem. J.* **76**, 381–388 (1960).

27. K.J. Willis and A.G. Szabo, *Biochemistry* **28**, 4902–4908 (1989).

28. C.R. Cantor and P.R. Schimmel, *Biophysical Chemistry. Part II, Techniques for the Study of Biological Structure and Function*, Freeman, San Francisco, 1980.

29. I. Weinryb and R.F. Steiner, in *Excited States of Proteins and Nucleic Acids*, R.F. Steiner and I. Weinryb, Eds., Plenum, New York, 1971 pp. 277−318.

30. B.L. Van Duuren, *J. Org. Chem.* **26**, 2954−2960 (1961).

31. M.S. Walker, T.W. Bednar, and R. Lumry, *J. Chem. Phys.* **47**, 1020−1028 (1967).

32. M. Sun and P.-S. Song, *Photochem. Photobiol.* **25**, 3−9 (1977).

33. I. Tatischeff, R. Klein, T. Zemb, and M. Duquesne, *Chem. Phys. Lett.* **54**, 394−397 (1978).

34. J.-P. Privat, P. Wahl, and J.-C. Auchet, *Biophys. Chem.* **9**, 223−233 (1979).

35. B. Skalski, D.M. Rayner, and A.G. Szabo, *Chem. Phys. Lett.* **70**, 587−590 (1980).

36. H. Lami and N. Glasser, *J. Chem. Phys.* **84**, 59−66 (1986).

37. P.-S. Song and W.E. Kurtin, *J. Am. Chem. Soc.* **91**, 4892−4906 (1969).

38. Y. Yamamoto and J. Tanaka, *Bull. Chem. Soc. Japan* **45**, 1362−1366 (1972).

39. B. Valeur and G. Weber, *Photochem. Photobiol.* **25**, 441−444 (1977).

40. A.G. Szabo and D.M. Rayner, *J. Am. Chem. Soc.* **102**, 554−563 (1980).

41. R.J. Robbins, G.R. Fleming, G.S. Beddard, G.W. Robinson, P.J. Thistlethwaite, and G.J. Woolfe, *J. Am. Chem. Soc.* **102**, 6271−6279 (1980).

42. M.C. Chang, J.E. Petrich, D.B. McDonald, and G.R. Fleming, *J. Am. Chem. Soc.* **105**, 3819−3824 (1983).

43. J.W. Petrich, M.C. Chang, D.B. McDonald, and G.R. Fleming, *J. Am. Chem. Soc.* **105**, 3824−3832 (1983).

44. J.W. Petrich, M.C. Chang, D.B. McDonald, and G.R. Fleming, *J. Am. Chem. Soc.* **105**, 3832−3836 (1983).

45. A. Engh, L.X.-Q. Chen, and G.R. Fleming, *J. Am. Chem. Soc.* **126**, 365−372 (1986).

46. E. Gudgin, R. Lopez-Delgado, and W.R. Ware, *J. Phys. Chem.* **9**, 1559−1565 (1983).

47. S.R. Meech, D. Phillips, and A.G. Lee, *Chem. Phys.* **80**, 317−328 (1983).

48. I. Saito, H. Suqiyama, A. Yamamoto, S. Muramatsu, and T. Matsuura, *J. Am. Chem. Soc.* **106**, 4286−4287 (1984).

49. H. Shizuka, M. Serizawa, T. Shimo, I. Saito, and T. Matsuura, *J. Am. Chem. Soc.* **110**, 1930−1934 (1988).

50. A.P. Demchenko, *Eur. J. Biochem.* **16**, 121−129 (1988).

51. D. Creed, *Photochem. Photobiol.* **39**, 537−562 (1984).

52. A.J. Ruggiero, D.C. Todd, and G.R. Fleming, *J. Am. Chem. Soc.* **112**, 1003−1014 (1990).

53. M.R. Eftink, L.A. Selvidge, P.R. Callis, and A.A. Rehms, *J. Phys. Chem* **94**, 3469−3479 (1990).

54. I. Johnson and B. Hudson, *Biophys. J.* **55**, 189a (1989).

55. J.R. Lakowicz, H. Szmacinski, and I. Gryczynski, *Photochem. Photobiol.* **47**, 31−41 (1988).

56. J.W. Longworth and M.C. Battista, *Photochem. Photobiol.* **12**, 29−35 (1970).

57. T.L. Bushueva, E.P. Busel, and E.A. Burstein, *Biochim. Biophys. Acta* **534**, 141−152 (1978).

58. H. Szmacinski, J.R. Lakowicz, and M.L. Johnson, *SPIE* **909**, 293−298 (1988).

59. A.G. Szabo, T.M. Stepanik, D.M. Wagner, and N.M. Young, *Biophys. J.* **41**, 233−244 (1983).

60. J.A. Schauerte and A. Gafni, *Biochemistry* **28**, 3948−3954 (1989).

61. R.M. Hochstrasser and D.K. Negus, *Proc. Natl. Acad. Sci. USA* **81**, 4399−4403 (1984).

62. R.W. Cowgill, *Arch. Biochem. Biophys.* **104**, 84−92 (1964).

63. R.W. Cowgill, in *Biochemical Fluorescence: Concepts*, Vol. 2, R.F. Chen and H. Edelhoch Eds., Dekker, New York, 1976. pp. 441−486.

64. J.W. Longworth, in *Time Resolved Fluorescence Spectroscopy in Biochemistry and Biology*, R.B. Cundall and R.E. Dale, Eds., Plenum, New York, 1983, pp. 651−725.

65. P. Gauduchon and P. Wahl, *Biophys. Chem.* **8**, 87−104 (1978).

66. W.R. Laws, J.B.A. Ross, H.R. Wyssbrod, J.M. Beechem, L. Brand, and J.C. Sutherland, *Biochemistry* **25**, 599−607 (1986).

67. J.R. Lakowicz, G. Laczko, and I. Gryczynski, *Biochemistry* **26**, 82−90 (1987).

68. L.J. Libertini and E.W. Small, *Biophys. J.* **47**, 765−772 (1985).

69. D.M. Rayner, D.T. Krajcarski, and A.G. Szabo, *Can. J. Chem.* **56**, 1238−1245 (1978).

70. J.L. Cornog and W.R. Adams, *Biochim. Biophys. Acta* **66**, 356−365 (1963).

71. O. Shimizu, J. Watanabe, and K. Imakubo, *Photochem. Photobiol.* **29**, 915−919 (1979).

72. A.G. Szabo, K.R. Lynn, D.T. Krajcarski, and D.M. Rayner, *Can. J. Chem.* **56**, 1238−1245 (1978).

73. J.B.A. Ross, W.R. Laws, A. Buku, J.C. Sutherland, and H.R. Wyssbrod, *Biochemistry* **25**, 607−612 (1986).

74. A.G. Szabo, K.R. Lynn, D.T. Kracjarski, and D.M. Rayner, *FEBS Lett.* **94**, 249−252 (1978).

75. F.G. Prendergast, P.D. Hampton, and B. Jones, *Biochemistry* **23**, 6690−6697 (1984).

76. L. Tilstra, M.C. Sattler, W.R. Cherry, and M.R. Barkley, submitted for publication.

77. R.D. Ludescher, J.J. Volwerk, G.H. deHaas, and B.S. Hudson, *Biochemistry* **24**, 7240−7249 (1985).

78. J.R. Alcala, E. Gratton, and F.G. Prendergast, *Biophys. J.* **51**, 587−596 (1987).

79. J.R. Alcala, E. Gratton, and F.G. Prendergast, *Biophys. J.* **51**, 597−604 (1987).

80. J.R. Alcala, E. Gratton, and F.G. Prendergast, *Biophys. J.* **51**, 925−936 (1987).

81. J.R. Lakowicz, H. Cherek, I. Gryczynski, N. Joshi, and M.L. Johnson, *Biophys. Chem.* **28**, 35−50 (1987).

82. D.R. James, Y.-S. Liu, A. Siemiarczuk, B.D. Wagner, and W.R. Ware, *SPIE Proc.* **909**, 90−96 (1988).

83. D.R. James and W.R. Ware, *Chem. Phys. Lett.* **120**, 455−459 (1985).

84. D.R. James and W.R. Ware, *Chem. Phys. Lett.* **126**, 7−11 (1986).

85. H.E. Dale and J. Eisinger, *Proc. Natl. Acad. Sci. USA* **73**, 271−273 (1976).

86. R.E. Dale and J. Eisinger, *Biopolymers* **13**, 1573−1605 (1974).

87. J. Eisinger, W.E. Blumberg, and R.E. Dale, *Ann. New York Acad. Sci.* **366**, 155−175 (1981).

88. E. Haas, H. Wilchek, E. Katchalski-Katzir, and I.Z. Steinberg, *Proc. Natl. Acad. Sci. USA* **72**, 1807−1811 (1975).

89. E. Haas, E. Katchalski-Katzir, and I.Z. Steinberg, *Biochemistry* **17**, 5064−5070 (1978).

90. E. Haas, E. Katchalski-Katzir, and I.Z. Steinberg, *Biopolymers* **17**, 11−31 (1978).

91. D. Amir and E. Haas, *Biopolymers* **25**, 235−240 (1986); D. Amir and E. Haas, *Biochemistry* **26**, 2162−2175 (1987).

92. J.R. Lakowicz, M.L. Johnson, W. Wiczk, A. Bhat, and R.F. Steiner, *Chem. Phys. Lett.* **138**, 587−593 (1987).

93. J.R. Lakowicz, I. Gryczynski, H.C. Cheung, C.-K. Wang, and M.L. Johnson, *Biopolymers* **27**, 821−830 (1988).

94. D.J. Marsh and S. Lowey, *Biochemistry* **19**, 774−784 (1980).

95. A.G. Szabo, D. Krajcarski, M. Zuker, and B. Alpert, *Chem. Phys. Lett.* **108**, 145−149 (1984).

96. S.M. Janes, G. Hottom, P. Ascenzi, M. Brunori, and R.M. Hochstrasser, *Biophys. J.* **51**, 653−660 (1987).

97. E. Bucci, H. Malak, C. Fronticelli, I. Gryczynski, and J.R. Lakowicz, *J. Biol. Chem.* 6972−6977 (1988).

98. R.P. Haughland in *Excited States of Biopolymers*, R.F. Steiner, Ed., Plenum, New York, 1983, pp. 29–58.

99. J.M. Connellan, S.I. Chung, N.K. Whetzel, L.M. Bradley, and J.E. Folk, *J. Biol. Chem.* **246**, 1093–1098 (1971).

100. M.-J. Rhee, D.R. Sudnick, V.K. Arkle, and W.D. Horrocks, Jr., *Biochemistry* **20**, 3328–3334 (1981).

101. G. Desie, D. Van Deynse, and F.C. DeSchryver, *Photochem. Photobiol.* **45**, 67–77 (1987).

102. V.N. Hingorani and Y.-K. Ho, *J. Biol. Chem.* **263**, 19804–19808 (1988).

103. I. Gryczynski, W. Wiczk, M.L. Johnson, and J.R. Lakowicz, *Chem. Phy. Lett.* **145**, 439–446 (1988).

104. S. Albaugh, J. Lan, and R.F. Steiner, *Biophys. Chem.* **33**, 71–76 (1989).

105. R.E. Dalbey, J. Weiel, and R.G. Yount, *Biochemistry* **22**, 4696–4706 (1983).

106. C.-K. Wang and H.C. Cheung, *J. Mol. Biol.* **190**, 509–521 (1984).

107. H.C. Cheung, F. Gonsoulin, and F. Garland, *Biochim. Biophys. Acta* **832**, 52–62 (1985).

108. T. Tao, E. Gowell, G.M. Strasburg, J. Gergely, and P.C. Leavis, *Biochemistry* **28**, 5902–5908 (1989).

109. H.R. Trayer and I.P. Trayer, *Eur. J. Biochem.* **135**, 47–59 (1983).

110. A.M. Kleinfeld, *Biochemistry* **24**, 1874–1882 (1985).

111. A. Hermetter and J.R. Lakowicz, *J. Biol. Chem.* **261**, 8243–8248 (1986).

112. J.C. Talbot, J.F. Faucon, and J. DuFourcq, *Eur. J. Biochem.* **15**, 147–157 (1987).

113. E.A. Haigh, K.R. Thulborn, and W.H. Sawyer, *Biochemistry* **18**, 3525–3532 (1979).

114. C.A. McWherter, E. Haas, A.R. Leed, and H.A. Scheraga, *Biochemistry* **25**, 1951–1963 (1986).

115. J.M. Beechem, L. James, and L. Brand, *SPIE*, **1204**, 686–698 (1990).

116. J.C. Talbot, J.F. Faucon, and J. Dufoureq, *Eur. Biophys. J.* **15**, 147–157 (1987).

117. F. Garland, F. Gonsoulin, and H.C. Cheung, *J. Biol. Chem.* **263**, 11621–11623 (1988).

118. G. Desie, N. Boens, and F.C. DeSchryver, *Biochemistry* **25**, 8301–8308 (1986).

119. T.P. Burghardt and N.L. Thompson, *Biochemistry* **24**, 3731–3735 (1985).

120. R.A. Mendelson, M.F. Morales, and J. Botts, *Biochemistry* **12**, 2250–2255 (1973).

121. J. Churchich, *Biochemistry* **4**, 1405–1410 (1965).

122. R.F. Chen, *Science* **150**, 1593–1595 (1965).

123. S. Matsumoto and G.G. Hammes, *Biochemistry* **14**, 214–224 (1975).

124. L.A. Sklar, B. Hudson, M. Peterson, and J. Diamond, *Biochemistry* **16**, 813–819 (1977); L.A. Sklar, B. Hudson, and R.D. Simoni, *Biochemistry* **16**, 5100–5108 (1977).

125. L. Davenport, J.R. Knutson, and L. Brand, *Biochemistry* **25**, 1186–1195 (1986).

126. A. Orstan, M.F. Lulka, B. Eide, P.H. Petra, and J.B.A. Ross, *Biochemistry* **25**, 2686–2692 (1986).

127. T.C.M. Eames, R.M. Pollack, and R.F. Steiner, *Biochemistry* **28**, 6269–6275 (1989).

128. G. Weber and F. Farris, *Biochemistry* **18**, 3075–3078 (1979).

129. F.G. Prendergast, M. Meyer, G.L. Carlson, S. Iida, and J.D. Potter, *J. Biol. Chem.* **258**, 7541–7544 (1983).

130. R.F. Chen, *Arch. Biochem. Biophys.* **133**, 263–276 (1969).

131. L. Stryer, *J. Mol. Biol.* **13**, 482–495 (1965).

132. L. Brand and J.R. Gohlke, *Annu. Rev. Biochem.* **41**, 843–868 (1972).

133. W.O. McClure and G.M. Edelman, *Biochemistry* **5**, 1908–1918 (1966).

134. A. Gafni, R.P. DeToma, E. Manrow, and L. Brand, *Biophys. J.* **17**, 155–168 (1977).

135. J.A. Secrist III, J.R. Barrio, N.J. Leonard, and G. Weber, *Biochemistry* **11**, 3499–3506 (1972).

136. N.J. Leonard, *CRC Crit. Rev. Biochem.* **15**, 125–199 (1983).

137. D.C. Ward, E. Reich, and L. Stryer, *J. Biol. Chem.* **244**, 1228–1237 (1969); D.C. Ward, T. Horn, and E. Reich, *J. Biol. Chem.* **247**, 4014–4020 (1972).

138. W.R. Dawson, J.L. Kropp, and M.W. Windsor, *J. Chem. Phys.* **45**, 2410–2418 (1966).

139. D.W. Horrocks, B. Holmquist, and B.L. Valee, *Proc. Natl. Acad. Sci. USA* **72**, 4764–4768 (1975).

140. L. Stryer, D.D. Thomas, and C.F. Meares, *Annu. Rev. Biophys. Bioeng.* **11**, 203–222 (1982).

141. P. Seban, M. Coppey, B. Alpert, L. Lindquist, and D.M. Jameson, *Photochem. Photobiol.* **32**, 727–731 (1980).

142. J.J. Leonard, T. Yonetani, and J.B. Callis, *Biochemistry* **13**, 1460–1464 (1974).

143. J.M. Vanderkooi, F. Adar, and M. Erecinska, *Eur. J. Biochem.* **64**, 381–387 (1976).

144. B. Bhattacharyya and J. Wolff, *Proc. Natl. Acad. Sci. USA* **71**, 2627–2631 (1974).

145. B. Bhattacharyya and J. Wolfe, *J. Biol. Chem.* **259**, 11836–11843 (1984).

146. B.R. Dean and R.B. Homer, *Biochim. Biophys. Acta* **322**, 141–144 (1973).

147. N.W. Woodbury and W.W. Parsons, *Biochim. Biophys. Acta* **850**, 197–210 (1986).

148. K. Ajtai and T.P. Burghardt, *Biochemistry* **28**, 2204–2210 (1989).

149. R.S. Zukin, P.R. Hartig, and D.E. Koshland Jr, *Proc. Natl. Acad. Sci. USA* **74**, 1932–1936 (1977).

150. J.D. Potter and J. Gergely, *J. Biol. Chem.* **220**, 4628–4633 (1975).

151. R. Aquirre, F. Gonsoulin, and H.C. Cheung, *Biochemistry* **25**, 6827–6835 (1986).

152. P.J. Fay and T.M. Smudzin, *J. Biol. Chem.* **264**, 14005–14010 (1989).

153. E.N. Hudson and G. Weber, *Biochemistry* **12**, 4154–4161 (1973).

154. R.P. Haugland, *J. Supramol. Struct.* **3**, 338–347 (1975).

155. T. Iio and H. Kondo, *J. Biochem. (Tokyo)* **86**, 1883–1886 (1979).

156. W. Birmachu, F.L. Nisswandt, and D.D. Thomas, *Biochemistry* **128**, 3940–3947 (1989).

157. P.B. Ghosh and M.W. Whitehouse, *Biochem J.* **108**, 155–156 (1968).

158. K. Yamamoto, Y. Okamoto, and T. Sekine, *Anal. Biochem.* **84**, 313–318 (1978).

159. R.J. Cherry, *Biochim. Biophys. Acta* **559**, 289–340 (1979).

160. S. Udenfriend, S. Stein, P. Bohlen, W. Dairman, W. Leimgruber, and M. Weigele, *Science* **178**, 871–872 (1972).

161. G.W. Robinson, R.J. Robbins, G.R. Fleming, J.M. Morris, A.E.W. Knight, and R.J.S. Morrison, *J. Am. Chem. Soc.* **100**, 7145–7150 (1978).

162. D.V. O'Conner and D. Phillips, *Time-Correlated Single Photon Counting*, Academic, New York, 1984.

163. A.J.W.G. Visser, *Anal. Instrum.* **14**, 193–546 (1985).

164. M.C. Chang, S.H. Courtney, A.J. Cross, R.J. Gulotty, J.W. Petrich, and G.R. Fleming, *Anal. Instrum.* **14**, 433–464 (1985).

165. D.M. Jameson, E. Gratton, and R.D. Hall, *Appl. Spectrosc. Rev.* **20**, 55–106 (1984).

166. E. Gratton and M. Limkeman, *Biophys. J.* **44**, 315 (1983); J.R. Lakowicz, and B.P. Maliwal, *Biophys. Chem.* **21**, 61 (1985).

167. I. Yamazaki, H. Tamai, H. Kume, H. Tsuchiya, and K. Oba, *Rev. Sci. Instrum.* **56**, 1187–1194 (1985).

168. B. Fedderson, M. vandeVen, and E. Gratton, *Biophys. J.* **55**, 190a (1989).

169. T.M. Nordlund, *SPIE Proc.* **909**, 35–50 (1988).

170. J. Hedstrom, S. Sedarous, and F.G. Prendergast, *Biochemistry* **27**, 6203−6208 (1988).

171. A. Grinvald and I.Z. Steinberg, *Anal. Biochem.* **59**, 583−598 (1974).

172. I. Isenberg, *Biophys. J.* **43**, 141−148 (1983).

173. E.W. Small, L.J. Libertini, D.W. Brown, and J.R. Small, *SPIE Proc.* **1054**, 36−53 (1989).

174. E.W. Small and I. Isenberg, in *Time-Resolved Fluorescence Spectroscopy in Biochemistry and Biology*, R.B. Cundall and R.E. Dale, Eds., Plenum, New York, 1983, pp. 199−222.

175. A. Gafni, R.L. Modlin, and L. Brand, *Biophys. J.* **15**, 263−280 (1975).

176. Z. Bajzer, A.C. Myers, S.S. Sedarous, and F.G. Prendergast, *Biophys. J.* **56**, 79−93 (1989).

177. J.C. Andre L.M. Vincent, D. O'Conner, and W.R. Ware, *J. Phys. Chem.* **83**, 2285−2294 (1979).

178. A.K. Livesey and J.C. Bronchon, *Biophys. J.* **52**, 693−706 (1987).

179. J.R. Lakowicz, G. Laczko, H. Cherek, E. Gratton, and M. Linkeman, *Biophys. J.* **46**, 463−477 (1983).

180. A.E. McKinnon, A.G. Szabo, and D.R. Miller, *J. Phys. Chem.* **81**, 1564−1570 (1977).

181. M. Ameloot, J.M. Beechem, and L. Brand, *Biophys. Chem.* **23**, 155−171 (1986).

182. G. Laczko and J.R. Lakowicz, *Biophys. J.* **55**, 190a (1989).

183. D.W. Walbridge, J.R. Knutson, and L. Brand, *Anal. Biochem.* **161**, 467−478 (1987).

184. M.K. Han, D.W. Walbridge, J.R. Knutson, L. Brand, and S. Roseman, *Anal. Biochem.* **161**, 479−486 (1987).

185. E. Gratton, D.M. Jameson, N. Rosato, and G. Weber, *Rev. Sci. Instrum.* **55**, 486−494 (1984).

186. I.H. Munro and N. Schwentrer, *Nucl. Instrum. Methods* **208**, 819−834 (1983).

187. W.R. Laws and J.C. Sutherland, *Photochem. Photobiol.* **44**, 343−348 (1986).

188. D.J.S. Birch, A.S. Holmes, R.E. Imhof, and B.Z. Nadolski, *SPIE Proc.* **909**, 8−14 (1988).

189. D.J.S. Birch, A.S. Holmes, R.E. Imhof, and K. Suhling, *J. Phys. E: Sci. Instrum* **21**, 415−417 (1988).

190. J.R. Knutson, *SPIE Proc.* **909** 51−60 (1988).

191. L. Stryer, D.D. Thomas, and C.F. Meares, *Annu. Rev. Biophys. Bioeng.* **11**, 203−222 (1982).

192. E.W. Lobenstine, W.C. Schaefer, and D.H. Turner, *J. Am. Chem. Soc.* **103**, 4936−4940 (1981).

193. A.H. Maki, in *Biological Magnetic Resonance*, L.J. Berliner and J. Reuben, Eds., Plenum, New York, 1984, pp. 187−294.

194. D. Axelrod, T.P. Burghardt, and N.L. Thompson, *Annu. Rev. Biophys. Bioeng.* **13**, 247−268 (1984).

195. E.L. Elson and D. Magde, *Biopolymers* **13**, 1−27 (1974); D. Magde, E.L. Elson, and W.W. Weble, *Biopolymers* **13**, 29−61 (1974).

196. H. Thoerell and A.D. Winer, *Arch. Biochem. Biophys.* **83**, 291−308 (1959).

197. H. Thoerell and K. Tatemoto, *Arch. Biochem. Biophys.* **142**, 69−82 (1971).

198. J.J. Holbrook and R.G. Wolfe, *Biochemistry* **11**, 2499−2502 (1972).

199. P. Fischer, J. Fleckenstein, and J. Hones, *Photochem. Photobiol.* **47**, 193−199 (1988).

200. B. Baumgarten and J. Hones, *Photochem. Photobiol.* **47**, 201−205 (1988).

201. A. Gafni and L. Brand, *Biochemistry* **15**, 3165−3171 (1976).

202. A.J.W.G. Visser and A. van Hoek, *Photochem. Photobiol.* **33**, 35−40 (1981).

203. T.G. Scott, R.D. Spencer, N.J. Leonard, and G. Weber, *J. Am. Chem. Soc.* **92**, 687−695 (1970).

204. J.C. Brochon, Ph. Wahl, J.M. Jallon, and M. Iwatsubo, *Biochemistry* **15**, 3259−3265 (1976).

205. J.C. Brochon, Ph. Wahl, J.-M. Jallon, and M. Iwatsubo, *Biochim. Biophys. Acta* **462**, 759−769 (1977).

206. M.R. Eftink, *Biophys. J.* **51**, 278a (1987).

207. G. Weber, in *Flavins and Flavoproteins*, E.C. Slater, Ed., Elsevier, Amsterdam, 1966, pp. 15–21.

208. R.D. Spencer and G. Weber, in *Structure and Function of Oxidation-Reduction Enzymes*, A. Akeson and A. Ehrenberg, Eds., Pergamon, Oxford, 1972, pp. 393–399.

209. G. Weber, *Biochem. J.* **47**, 114–121 (1950).

210. K. Yagi, in *Biochemical Fluorescence: Concepts*, Vol. 2, R.F. Chen and H. Edelhoch, Eds., Dekker, New York, 1976, pp. 639–658.

211. A. Visser, in *Fluorescent Biomolecules: Methodologies and Applications*, D.M. Jameson and G.D. Reinhart, Eds., Plenum, New York, 1989, pp. 319–341.

212. R.E. MacKenzie, W. Fory, and D.B. McCormick, *Biochemistry* **8**, 1839–1844 (1969).

213. A.J.W.G. Visser, *Photochem. Photobiol.* **40**, 703–706 (1984).

214. A.J.W.G. Visser, T.M. Li, H.G. Drickamer, and G. Weber, *Biochemistry* **16**, 4883–4886 (1977).

215. A.J.W.G. Visser, T.M. Li, H.G. Drickamer, and G. Weber, *Biochemistry* **16**, 4879–4882 (1977).

216. A. Kotaki and K. Yagi, (1970) *J. Biochem.* **68**, 509–516 (1970).

217. T. Kulinski, A.J.W.G. Visser, D.J. O'Kane, and J. Lee, *Biochemistry* **26**, 540–549 (1987).

218. K. Yagi, F. Tanaka, N. Nakashima, and K. Yoshihara, *J. Biol. Chem.* **258**, 3799–3802 (1983); F. Tanaka, N. Tamia I. Yamasaki, N. Nakashima, and K. Yoshihara, *Biophysc. J.* **56**, 901–909 (1989).

219. A.J.W.G. Visser, N.H.G. Penners, W.J.H. vanBerkel, and F. Miller, *Eur. J. Biochem.* **143**, 189–197 (1984).

220. A. deKok and A.J.W.G. Visser, *FEBS Lett.* **218**, 135–138 (1987).

221. D.M. Jameson, V. Thomas, and D. Zhou, *Biochim. Biophys. Acta* **994**, 187–190 (1989).

222. K. Rousslang, L. Allen, and J.B.A. Ross, *Photochem. Photobiol.* **49**, 137–143 (1989).

223. W.B. Dandliker, J. Dandliker, S.A. Levison, R.J. Kelly, A.N. Hicks, and J.U. White, *Methods Enzymol.* **48**, 380–415 (1978).

224. D. Shortle, A.K. Meeker, and E. Freire, *Biochemistry* **27**, 4761–4768 (1987).

225. C.N. Pace, *Meth. Enzymol.* **131**, 266–280 (1986).

226. A.A. Paladini and G. Weber, *Rev. Sci. Instrum.* **52**, 419–427 (1981).

227. G. Weber and H.G. Drickamer, *Q. Rev. Biophys.* **16**, 89–112 (1983).

228. C.A. Royer, G. Weber, T.J. Daly, and K.S. Matthews, *Biochemistry* **25**, 8308–8315 (1986).

229. W.-Y. Lin and H.E. VanWart, *Biochemistry* **27**, 5054–5061 (1988).

230. D. Axelrod, D.E. Koppel, J. Schlessinger, E. Elson, and W.W. Webb. *Biophys. J.* **16**, 1055–1059 (1976).

231. W.T. Shearer and C.W. Parker, in *Biochemical Fluorescence: Concepts*, Vol. 2, R.F. Chen and H. Edelhoch, Eds., Dekker, New York, 1976, pp. 811–843.

232. D.L. Taylor, A.S. Waggoner, R.F. Murphy, F. Lanni and R.R. Bridge, Eds., *Applications of Fluorescence in the Biomedical Sciences*, Liss, New York, 1986.

233. D.L. Taylor, P.A. Amato, K. Luby-Helps, and P. McNeil, *Trends Biochem. Sci.* **9**, 88–91 (1984).

234. J.E. Wampler, J. Chen, L.G. DeMendoza, R.H. Furukawa, D. McCurdy, L. Pruett, R.A. White, and M. Fechheimer, in *Time-Resolved Laser Spectroscopy in Biochemistry*, J.R. Lakowicz, Eds., *SPIE Proc.* **909**, 319–327.

235. S. Papp and J.M. Vanderkooi, *Photochem. Photobiol.* **49**, 775–784 (1989).

236. M.L. Saviotti and W.C. Galley, *Proc. Natl. Acad. Sci. USA* **71**, 4154–4158 (1974).

237. G.B. Strambini, *Biophys. J.* **52**, 23–28 (1987).

238. S.W. Englander, D.B. Calhoun, and J.J. Englander, *Anal. Biochem.* **161**, 300–306 (1987).

239. D.B. Calhoun, S.W. Englander, W.W. Wright, and J.M. Vanderkooi, *Biochemistry* **27**, 8466–8474 (1988).

240. J.M. Vanderkooi, D.B. Calhoun, S.W. Englander, *Science* **236**, 568–569 (1987).

241. G.B. Strambini and E. Gabellieri, *Biochemistry* **26**, 6527–6530 (1987).

242. G.B. Strambini and W.C. Galley, *Biopolymers* **19**, 383–394 (1980).

243. G.B. Strambini, P. Cioni, and R.A. Felicioli, *Biochemistry* **26**, 4968–4975 (1987).

244. H. Kim and W.C. Galley, *Can. J. Biochem.* **61**, 46–53 (1983).

245. J.W. Berger and J.M. Vanderkooi, *Photochem. Photobiol.* **47**, 10S (1988).

246. C.A. Ghiron, M. Basin, and R. Santus, *Biochim. Biophys. Acta* **957**, 207–216 (1988).

247. C.A. Ghiron, M. Basin, and R. Santus, *Photochem. Photobiol.* **48**, 539–543 (1988).

248. Y. Kai and K. Imakubo, *Photochem. Photobiol.* **29**, 261–265 (1979).

249. R.J. Cherry, *Biochim. Biophys. Acta* **559** 289–327 (1979).

250. A.F. Corin, E.D. Matayoshi, and T.M. Jovin, in *Spectroscopy and the Dynamics of Molecular Biological Systems*, P.M. Bayley and R.E. Dale, Eds., Academic, New York, 1985, pp. 53–93.

251. J.J. Birmingham and P.B. Garland, *SPIE Proc.* **909**, 370–376 (1988).

252. A.F. Corin, E. Blatt, and T.M. Jovin, *Biochemistry* **26**, 2207–2217 (1987).

253. P.B. Garland and C.H. Moore, *Biochem. J.* **183**, 561–572 (1979).

254. T.M. Yoshida and B.G. Barisas, *Biophys. J.* **50**, 41–53 (1986).

255. N.E. Geacintov and H.C. Brenner, *Photochem. Photobiol.*, **50**, 841–858 (1989).

256. S. Ghosh, L.-H. Zhang, and A.H. Maki, *J. Chem. Phys.* **88**, 2769–2775 (1988).

257. S. Ghosh, L.-H. Zang, and A.H. Maki, *Biochemistry* **27**, 7816–7820 (1988); L.-H. Zang, S. Ghosh, and A.H. Maki, *Biochemistry* **27**, 7820–7825 (1988).

258. S.S. Lehrer and P.C. Leavis, *Methods Enzymol.* **49**, 222–236 (1978).

259. S.S. Lehrer, in *Biochemical Fluorescence: Concepts* Vol. 2, R.F. Chen and H. Edelhoch, Eds., Dekker, New York, 1976 pp. 515–544.

260. M.R. Eftink, in *Fluorescence Spectroscopy: Principles and Techniques*, Vol. 1, J.R. Lakowicz, Ed., Plenum, New York, in press.

261. E.A. Burstein, N.S. Vendenkina, and M.N. Ivkova, *Photochem. Photobiol.* **18**, 263–279 (1973).

262. S.S. Lehrer, *Biochemistry* **10**, 3254–3263. (1971).

263. M.R. Eftink and C.A. Ghiron, *Biochemistry* **15**, 672–680 (1976).

264. J.R. Lakowicz and G. Weber, *Biochemistry* **12**, 4171–4179 (1973).

265. T. Ando and H. Asai, *J. Biochem. (Tokyo)* **88**, 255–264 (1980).

266. M.R. Eftink and C.A. Ghiron, *Biochemistry* **23**, 3891–3899 (1984).

267. M.R. Eftink, T. Selva, and Z. Wasylewski, *Photochem. Photobiol.* **46**, 23–30 (1987).

268. D.A. Johnson and J. Yguerabide, *Biophys. J.* **48**, 949–955 (1985).

269. M.R. Eftink, in *Fluorescence in Biochemistry and Cell Biology*, G. Dewey, Ed., Plenum, New York, in press.

270. D.B. Calhoun, J.M. Vanderkooi, G.V. Woodrow III., and S.W. Englander, *Biochemistry* **22**, 1526–1532 (1983).

271. D. Peak, T.C. Werner, R.M. Dennin Jr., and J.R. Baird, *J. Chem. Phys.* **79**, 3328–3335 (1983).

272. T.L. Nemzek and W.R. Ware, *J. Chem. Phys.* **62**, 477–489 (1975).

273. J.R. Lakowicz, N.B. Joshi, M.L. Johnson, H. Szmacinski, and I. Gryczynski, *J. Biol. Chem.* **262**, 10907–10910 (1987).

274. M.R. Eftink, *SPIE Proc.* **1204**, 406–414 (1990).

275. D.B. Calhoun, S.W. Englander, W.W. Wright, and J.M. Vanderkooi, *Biochemistry* **27**, 8466–8474 (1988).

276. D.B. Calhoun, J.M. Vanderkooi, and S.W. Englander, *Biochemistry* **22**, 1533–1539 (1983).

277. M.R. Eftink, and C.A. Ghiron, *Biochemistry* **16**, 5546–5551 (1977).

278. M.R. Eftink, and K. Hagaman, *Biophys. Chem.* **22**, 173–180 (1985).

279. M.R. Eftink, and Z. Wasylewski, *Biophys. Chem.* **32**, 121–130 (1988).

280. M.R. Eftink, and Y.-W. Jia, D.E. Graves, W. Wiczk, I. Gryczynski, and J.R. Lakowicz, *Photochem. Photobiol.* **49**, 725–729 (1989) and unpublished results.

281. T. Ando, H. Fujisak, and H. Asai, *J. Biochem. (Tokyo)* **88**, 265–276 (1980).

282. M.R. Eftink and L.A. Selvidge, *Biochemistry* **21**, 117–125 (1982).

283. G.C. Belford, R.L. Belford, and G. Weber, *Proc. Natl. Acad. Sci. USA* **69**, 1392–1393 (1972).

284. P. Wahl, *Biochemical Fluorescence*, Vol. 1, R.F. Chen and H. Edelhoch, Eds., Plenum, New York, 1975, pp. 1–41.

285. J.Y. Yguerabide, *Methods Enzymol.* **26**, 498–578 (1972).

286. L. Brand, J.R. Knutson, L. Davenport, J.M. Beechem, R.E. Dale, D.G. Walbridge, and A.A. Kowalczyk, in *Spectroscopy and the Dynamics of Molecular Biological Systems*, P.M. Bayley and R.E. Dale, Eds., Academic, London, 1985, pp. 259–305.

287. E. Bucci and R.F. Steiner, *Biophys. Chem.* **30**, 199–224 (1988).

288. E.W. Small, and I. Eisenberg, *Biopolymers* **16**, 1907–1928 (1977).

289. S.C. Harvey and H.C. Cheung, *Biochemistry* **16**, 5181–5187 (1977).

290. B.M. Liu, H.C. Cheung, and J. Mestecky, *Biochemistry* **20**, 1997–2003 (1981).

291. J. Yguerabide, H.F. Epstein, and L. Stryer, *J. Mol. Biol.* **51**, 573–590 (1970).

292. C. Hanson, J. Yguerabide, and V.N. Schumaker, *Biochemistry* **20**, 6842–6852 (1981).

293. G. Weber, *Nature* **190**, 27–29 (1961).

294. R.I. Schrager and R.W. Hendler, *Anal. Chem.* **54**, 1147–1152 (1982).

295. J.R. Knutson, D.G. Walbridge, and L. Brand, *Biochemistry* 4671–4679 (1982).

296. E. Gratton and D.M. Jameson, *Anal. Chem.* **57**, 1694–1697 (1985).

297. S.M. Keating-Nakamoto, H. Cherek, and J.R. Lakowicz, *Biophys. Chem.* **24**, 79–85 (1986).

298. J.R. Knutson, L. Davenport, and L. Brand, *Biochemistry* **25**, 1805–1810 (1986).

299. R.D. Ludescher, L. Peting, S. Hudson, and B. Hudson, *Biophys. Chem.* **28**, 59–75 (1987).

300. J.R. Lakowicz, G. Laczko, and I. Gryczynski, *Rev. Sci. Instrum.* **57**, 2499–2506 (1986).

301. G. Laczko, J.R. Lakowicz, *Biophys. J.* **55**, 190a (1989).

302. Z. Wasylewski, H. Koloczek, and A. Wasniowska, *Eur. J. Biochem.* **172**, 719–724 (1988).

303. J.R. Knutson, M.K. Han, D.G. Walbridge, A. Cappuccino, S. Baker, A. Folgueras, and L. Brand, personal communication.

304. J.W. Longworth, in *Time-Resolved Fluorescence Spectroscopy in Biochemistry and Biology*, R.B. Cundall and R.E. Dale, Eds., Plenum, New York, 1983, pp. 651–725.

305. M. Chabbert, M.-C. Kilhoffer, D.M. Watterson, J. Haiech, and H. Lami, *Biochemistry* **28**, 6093–6098 (1989).

306. D. Hansen, L. Altschmied, and W. Hillen, *J. Biol. Chem.* **262**, 14030–14035 (1987).

307. A.D.B. Waldman, A.R. Clarke, D.B. Wigley, K.W. Hart, W.N. Chia, D. Barstow, T. Atkinson, I. Munro, and J.J. Holbrook *Biochim. Biophys. Acta* **913**, 66–71 (1987).

308. B. Hudson, A. Ruggiero, D. Harris, I. Johnson, X.-M. Dou, T. Novet, L. McIntosh, C. Phillips, and T. Nester, *SPIE Proc.* **909**, 113–120 (1988); B. Hudson, and D. Harris, *SPIE Proc.* **1204**, 80–91 (1990).

309. A. Grinvald and I.Z. Steinberg, *Biochim. Biophys. Acta* **427**, 663–678 (1976).

310. A. Finazzi-Agro, G. Rotilio, L. Avigliano, P. Guerrieri, V. Boffi, and B. Mondovi, *Biochemistry* **9**, 2009−2014 (1970).

311. J.W. Petrich, J.W. Longworth, and G.R. Fleming, *Biochemistry* **26**, 2711−2722 (1987).

312. C.M. Hutnik and A.G. Szabo, *Biochemistry* **28**, 3923−3934 (1989).

313. C.M. Hutnik and A.G. Szabo, *Biochemistry* **28**, 3935−3939 (1989).

314. I. Munro, I. Pecht, and L. Stryer, *Proc. Natl. Acad. Sci. USA* **76**, 56−60 (1979).

315. E.T. Adman, R.E. Stenkamp, L.C. Seiker, and L.H. Jensen, *J. Mol. Biol.* **123**, 35−47 (1978).

316. M.R. Eftink, Z. Wasylewski, and C.A. Ghiron, *Biochemistry* **26**, 8338−8346 (1987).

317. J.W. Longworth, *Photochem. Photobiol.* **7**, 587−592 (1968).

318. M.R. Eftink and C.A. Ghiron, *Proc. Natl. Acad. Sci. USA* **72**, 3290−3294 (1975).

319. D.R. James, D.R. Demmer, R.P. Steer, and R.E. Verrall, *Biochemistry* **24**, 5517−5526 (1985).

320. L.X.-Q. Chen, J.W. Longworth, and G.R. Fleming, *Biophys. J.* **51**, 865−873 (1987).

321. M.R. Eftink and C.A. Ghiron, *Biophys. J.* **52**, 467−473 (1987).

322. I. Gryczynski, M.R. Eftink, and J.R. Lakowicz, *Biochim. Biophys. Acta* **954**, 244−252 (1988).

323. J. Hedstrom, S. Sedarous, and F.G. Prendergast, *Biochemistry* **27**, 6203−6208 (1988).

324. A.D. MacKerell Jr., R. Rigler, L. Nilsson, U. Hahn, and W. Saenger, *Biophys. Chem.* **26**, 247−261 (1987).

325. M.R. Eftink, *Biophys. J.* **43**, 323−333 (1983).

326. J.R. Lakowicz, B. Maliwal, H. Cherek, and A. Balter, *Biochemistry* **22**, 1741−1752.

327. J.B.A. Ross, K. Rousslang, and L. Brand, *Biochemistry* **20**, 4361−4369 (1981).

328. S.A. Cockle and A.G. Szabo, *Photochem. Photobiol.* **34**, 23−27 (1981).

329. C.D. Tran, G.S. Beddard, and A.D. Osborne, *Biochim. Biophys. Acta* **491**, 155−159 (1982).

330. C.D. Tran, and G.S. Beddard, *Eur. J. Biochem.* **13**, 59−64 (1985).

331. J.R. Lakowicz, H. Cherek, I. Gryczynski, N. Joshi, and M.L. Johnson, *Biophys. J.* **51**, 755−768 (1987).

332. C.A. Ghiron and J.W. Longworth, *Biochemistry* **18**, 3828−3832 (1979).

333. R.B. Thompson and J.R. Lakowicz, *Biochemistry* **23**, 3411−3417 (1984).

334. J.C. Talbot, J. Dufourcq, J. deBony, J.F. Faucon, and C. Lussan, *FEBS Lett.* **102**, 191−193 (1979).

335. S.C. Quay and C.C. Condie, *Biochemistry* **22**, 695−700 (1983).

336. M.R. Eftink and Z. Wasylewski, *Biophys. Chem.* **32**, 121−130 (1988).

337. L. Masotti, P. Cavatorta, A.G. Szabo, G. Farruggia, and G. Sartor, in *Fluorescent Biomolecules: Methodologies and Applications*, D.M. Jameson and G.D. Reinhart, Eds., Plenum, New York, 1989, pp. 173−193.

338. R.D. Ludescher, I.D. Johnson, J.J. Volwerk, G.H. deHaas, P.C. Jost, and B.S. Hudson, *Biochemistry* **27**, 6618−6628 (1988).

339. J.-C. Bronhon, Ph. Wahl, and J.-C. Auchet, *Eur. J. Biochem.* **41**, 577−583 (1974).

340. J.R. Lakowicz, G. Laczko, I. Gryczynski, and H. Cherek, *J. Biol. Chem.* **261** 2240−2245 (1986).

341. J. Rudzki, J. Beechem, A. Kimball, D. Implicito, A. Chun, and L. Brand, *Biophys. J.* **49**, 33a (1986).

342. M.R. Eftink, C.A. Ghiron, R.A. Kautz, and R.O. Fox, *Biophys. J.* **55**, 575−579 (1989).

343. J.M. Brewer, P. Bastiaens, and J. Lee, *Biophys. Chem.* **28**, 77−88 (1987).

344. D.M. Jameson, E. Gratton, and J.F. Eccleston, *Biochemistry* **26**, 3894−3901 (1987).

345. M. Vincent, J.-C. Brochon, F. Merola, W. Jordi, and J. Gallay, *Biochemistry* **27**, 8752−8761 (1988).

346. M.R. Eftink, C.A. Ghiron, R.A. Kautz, and R.O. Fox, in preparation; M.R. Eftink, I. Gryczynski, W. Wiczk G. Laczko, and J.R. Lakowicz, in preparation.

347. P. Sebban, M. Coppey, B. Alpert, L. Lindgrist, and D.M. Jameson, *Photochem. Photobiol.* **32**, 727–731 (1980).

348. D.M.E. Jameson, G. Gratton, G. Weber, and B. Alpert, *Biophys. J.* **45**, 795–803 (1984).

349. G. Weber and M. Shinitzky, *Proc. Natl. Acad. Sci. USA* **65**, 823–830 (1970).

350. J.-P. Privat, P. Wahl, and J.-C. Auchet, *Biophys. Chem.* **11**, 239–248 (1980).

351. Z. Wasylewski and M.R. Eftink, *Eur. J. Biochem.* **167**, 513–518 (1987).

352. M.A. Abdallah, J.-F. Biellmann, P. Wiget, R. Joppich-Kuhn, and P.L., Luisi, *Eur. J. Biochem.* **89**, 397–405 (1978).

353. J.B.A. Ross, C. Schmidt, and L. Brand, *Biochemistry* **20**, 4369–4377 (1981).

354. D.R. Demmer, D.R. James, R.P. Steer, and R.E. Verrall, *Photochem. Photobiol.* **45**, 39–48 (1987).

355. J.R. Knutson, D.G. Walbridge, and L. Brand, *Biochemistry* **21**, 4671–4679 (1982).

356. M.R. Eftink and K.A. Hagaman, *Biochemistry* **25**, 6631–6637 (1986).

357. M.R. Eftink and D.M. Jameson, *Biochemistry* **21**, 4443–4449 (1982).

358. W.R. Laws and J.D. Shore, *J. Biol. Chem.* **253**, 8593–8597 (1978).

359. C.-I. Branden, H. Jornwall, H. Eklund, and B. Furugren, *Enzymes (3rd ed.)* **11**, 103–190 (1975).

360. L.X.-Q. Chen, J.W. Petrich, A. Perico, and G.R. Fleming, *SPIE Proc.* **909**, 216–222 (1988).

361. L.X.-Q. Chen, R.A. Engh, and G.R. Fleming, *SPIE Proc.* **909**, 223–230 (1988).

362. M.C. Chang, G.R. Fleming, A.M. Scanu, and N.-C.C. Yang, *Photochem. Photobiol.* **47**, 345–355 (1988).

363. A. Perico and M. Guenza, *J. Chem. Phys.* **84**, 510–516 (1986).

364. M.R. Eftink and Z. Wasylewski, *Biochemistry* **28**, 382–391 (1989); M.R. Eftink and K. Hagaman, *Biophys. Chem.* **22**, 173–180 (1985).

365. F. Castelli, H. White, and L. Forster, *Biochemistry* **27**, 3366–3372 (1988).

366. E.A. Permyakov, A.V. Ostrovsky, E.A. Burstein, P.G. Pleshanov, and Ch. Gerday *Arch. Biochem. Biophys.* **240**, 781–792 (1985).

367. S.J. Kim, B.C. Valdez, S.H. Chang, E.S. Younathan, and M.D. Barkley, *Biophys. J.* **55**, 515a (1989).

368. G. Sanyal, C. Charlesworth, R.J. Ryan, and F.G. Prendergast, *Biochemistry* **26**, 1860–1866 (1987).

369. I.E. Johnson and B.S. Hudson, *Biochemistry* **28**, 6392–6400 (1989).

370. K.P. Datema, A.J.W.G. Visser, A. vanHoek, C.J.A.M. Wolfs, R.B. Spruijt, and M.A. Hemminga, *Biochemistry* **26**, 6145–6152 (1987).

371. J. Gallay, M. Vincent, C. Nicot, and M. Waks, *Biochemistry* **26**, 5738–5747 (1987).

372. E. John and F. Jahnig, *Biophys. J.* **54**, 829–844 (1989).

373. A. Jonas, J.-P. Privat, P. Wahl, and J.C. Osborne Jr, *Biochemistry* **21**, 6205–6211 (1982).

374. L. McDowell, G. Sanyal, and F.G. Prendergast, *Biochemistry* **24**, 2979–2984 (1985).

375. M. Nakanishi, M. Kobayashi, M. Tsuboi, C. Takasaki, and N. Tamiya, *Biochemistry* **19**, 3204–3208 (1980).

376. A. Grinvald and I.Z. Steinberg, *Biochim. Biophys. Acta* **427**, 663–678 (1976).

377. B.R. Singh, M.L. Evenson, and M.S. Bergdoll, *Biochemistry* **27**, 8735–8741 (1988).

378. M.J.S. DeWolf, M. Fridkin, and L.D. Kohn, *J. Biol. Chem.* **256**, 5489–5496 (1981).

379. F.S. Lee, D.S. Auld, and B.L. Vallee, *Biochemistry* **28**, 219–224 (1989).

380. C.-K. Wang, R.S. Mami, C.M. Kay, and H.C. Cheung, *SPIE Proc.* **1204**, in press.

381. R.D. Moreno and G. Weber, *Biochim. Biophys. Acta* **703**, 231−240 (1982).

382. R. Liao, C.-K. Wang, and H.C. Cheung, *SPIE Proc.* **1204**, 699−705 (1990).

383. D.P. Millar and G.G. Dupuy, *Biophys. J.* **55**, 347a (1989).

384. C. Royer, *SPIE Proc.* **1204**, 112−123 (1990).

385. B.A. Shirley, P. Stanssens, J. Steyaert, and C.N. Pace, *J. Biol. Chem.* **264**, 11621−11625 (1989).

386. N. Rosato, A. Finazzi Agro, E. Gratton, S. Stefanini, and E. Chiancone, in *Fluorescent Biomolecules: Methodologies and Applications*, D.M. Jameson and G.D. Reinhart, Eds., Plenum, New York, 1989, pp. 411−413.

387. T. Ichiye and M. Karplus, *Biochemistry* **22**, 2884−2893 (1983).

388. K.K. Turoverov, I.M. Kuznetzova, and V.M. Zaitsev, *Biophys. Chem.* **23**, 79−89 (1985).

389. L.X.-Q. Chen, R.A. Engh, A.T. Brunger, D.T. Nguyen, M. Karplus, and G.R. Fleming, *Biochemistry* **27**, 6908−6921 (1988).

390. P.H. Axelsen, C. Haydock, and F.G. Prendergast, *Biophys. J.* **54**, 249−258 (1988).

391. P.H. Axelsen, and F.G. Prendergast, *Biophys. J.* **56**, 43−66 (1989).

392. A.D. MacKerell Jr., R. Rigler, L. Nilsson, U. Hahn, and W. Saenger, *Biophys. Chem.* **26**, 247−261 (1987).

393. F. Tanaka and N. Mataga, *Biophys. J.* **51**, 487−495 (1987).

394. E. Gratton, B. Alpert, D.M. Jameson, and G. Weber, *Biophys. J.* **45**, 789−794 (1984).

395. A.A. Kasprzak, P. Chaussepied, and M.F. Morales, *Biochemistry* **28**, 9230−9238 (1989).

396. W.B. deLauder and Ph. Wahl, *Biochem. Biophys. Res. Commun.* **42**, 398−404 (1971).

397. D.G. Searcy, T. Montenany-Garestier, and C. Helene, *Biochemistry* **28**, 9058−9065 (1989).

398. O. Kuipers, M. Vincent, J.-C. Brochon, B. Verheij, G. de Haas, and J. Gallay, *SPIE Proc.* **1204**, 100−111 (1990).

399. J.R. Lakowicz, I. Gryczynski, H. Cheung, C.-K. Wang, M.L. Johnson, and N. Joshi, *Biochemistry* **27**, 9149−9160 (1989).

400. F. Schroeder, P. Butko, G. Nemecz, and T.J. Scallen, *J. Biol. Chem.* **265**, 151−157 (1990).

401. C. Pigault, A. Follenius-Wund, B. Lux, and D. Gerard, *Biochim. Biophys. Acta* **1037**, 106−114 (1990).

402. S. Campbell-Burk, D.E. Van Dyk, D. Channing, and D. Jameson, *Biophys. J.* **57**, 434a (1990).

403. C.M. Nalin, *Biophys. J.* **57**, 51a (1990).

404. N. Ohta, L. Zong, P.G. Katsoyannis, W.R. Laws, and J.B.A. Ross, *Biophys. J.* **57**, 56a (1990).

405. A.D.B. Waldman, A.R. Clarke, D.G. Wigley, K.W. Hart, W.N. Chia, D. Barstow, T. Atkinson, I. Munro, and J.J. Holbrook, *Biochim. Biophys. Acta* **913**, 66−71 (1987).

406. P. Hensley, P. Young, T. Porter, D. Porter, K. Kasyan, and J.R. Knutson, *Biophys. J.* **57**, 59a (1990).

407. J.C. McIntyre, P. Hundley, and W.D. Behnke, *Biochem. J.* **245**, 821−829 (1987).

408. C.M. Hutnik, J.P. MacManus, and A.G. Szabo, *Biochemistry* **29**, 7318−7328 (1990).

409. K.J. Willis, A.G. Szabo, M. Zuker, J.M. Ridgeway, and B. Alpert, *Biochemistry* **29**, 5270−5275 (1990).

The Use of Monoclonal Antibodies and Limited Proteolysis in Elucidation of Structure—Function Relationships in Proteins

JOHN E. WILSON *Biochemistry Department and The Neuroscience Program, Michigan State University, East Lansing, Michigan*

Methods of Biochemical Analysis, Volume 35: Protein Structure Determination, Edited by Clarence H. Suelter.
ISBN 0—471—51326—1 © 1991 John Wiley & Sons, Inc.

1. INTRODUCTION

The phrase "structure—function relationships" or its equivalent is frequently found in the literature of protein chemistry, and succinctly indicates what is a major objective for most investigators working in this area, namely, gaining an understanding of how specific structural features confer particular biological function. How is it that a certain array of amino acid residues (and in some cases, additional factors such as chelated metal ions or prosthetic groups) can harness basic thermodynamic forces to bring about a biologically relevant result that could never be accomplished on a meaningful time scale, if at all, in the absence of the protein? Aside from its intrinsic interest, an understanding of the molecular basis for such events is all the more important in a time when the techniques of protein engineering offer the potential for modifying structure, and thereby function, with a facility that would have seemed unthinkable only a few years ago. Obviously, the productive use of such methodology depends directly on a clear definition of the relationship between structure and function.

The functions of proteins vary remarkably. Even if one were to consider only catalytically active proteins (enzymes), there is an incredible array of reactions that must occur to maintain the life of even the simplest organism — and an incredible array of organisms! If such diversity were also reflected in protein structure, with each enzyme possessing a structure totally unlike that of any other enzyme, it would seem a nearly hopeless task to ever develop any coherent set of principles governing the relationship between structure and function. Fortunately, this is not the case. Some frequently recurring features become evident upon surveying known protein structures (e.g., 1—3), and certain structural motifs are often found to be associated with specific functions such as binding of DNA (4—8), nicotinamide- and flavin—adenine dinucleotides (9), or cAMP (10). Computer-based methods (e.g., 11, 11a, 11b) facilitate detection of such motifs based on characteristic amino acid sequence patterns, and recombinant DNA methods provide an effective approach for testing the predicted functional role of specific structural elements (e.g., 5, 12). It is also quite evident now that many proteins are composed of 'cassettes,' which are recombined in various ways to generate diversity in function from a relatively limited repertoire of basic structural (and functional) units, the protein kinase family (10, 13) being but one example.

In view of the increasing number of known protein structures and the growing awareness of common themes relating structure and function, it is becoming less likely that any newly determined protein structure will be truly unique in its overall structural organization. In practical terms, this means that an investigator interested in the structure of a previously uncharacterized (at least in terms of structure) protein does not start from zero. Even in the absence of any knowledge of amino acid sequence, it is not unreasonable to anticipate the presence of, for example, supersecondary structures (Fig. 1) as categorized by Chothia (2), and structural elements found to be associated with functions shared by the protein of interest and proteins of known structure.

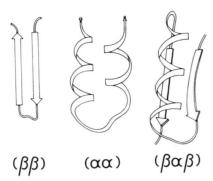

$(\beta\beta)$ $(\alpha\alpha)$ $(\beta\alpha\beta)$

Fig. 1. Supersecondary structures found in many proteins. Antiparallel β strands ($\beta\beta$) or α helices ($\alpha\alpha$); α helix packed against adjacent parallel β strands ($\beta\alpha\beta$). [Reproduced, with permission, from the *Annual Review of Biochemistry*, Vol. **53**, copyright 1984, by Annual Reviews, Inc.]

Such anticipation is greatly strengthened by demonstration of sequence similarities, and it is increasingly common to develop structures using molecular modeling techniques based on known structures of homologous proteins (e.g., 10, 14–16).

Ultimately, determination of structure (and structure–function relationships) depends on application of experimentally based methodology. The current method of choice is likely to be x-ray crystallography. Although two-dimensional NMR is finding increasing application in studies of protein structure (e.g., 17), technical considerations presently limit the general utility of this method (18). This is not to say that x-ray crystallography does not have its own limitations. In practical terms, it is by no means a given that all proteins can be crystallized in forms suitable for x-ray analysis, or that an investigator will have ready access to either the facilities or expertise required for x-ray structural analysis. Moreover, the structure determined by x-ray crystallography is effectively a static one, which contrasts markedly with the actual dynamic character of protein structure, with constant fluctuations between various conformations of similar energy (19, 20). Perhaps more importantly, it is well established that the conformation of proteins can be markedly affected by the binding of ligands, and such conformational changes can play critical roles in function. This poses another challenge in the application of x-ray crystallographic methods, for it requires that functionally important conformations be preserved in crystals suitable for x-ray analysis. Given the frequently transient nature of these structures, this can represent a major obstacle. Thus, while one can scarcely imagine that any investigator would turn down the chance to determine an x-ray structure for the protein of interest, the limitations of the method must be recognized.

The above comments should make it apparent that alternative (to x-ray crystallography or NMR) methods for gaining structural information continue to play an important role in development of structure—function relationships. Though these methods may not provide structural information with a resolution comparable to that of x-ray crystallography or NMR, they offer important advantages in terms of their general applicability. The present chapter focuses on two such methods that have proven to be extremely useful and frequently complementary, namely, limited proteolysis and the use of monoclonal antibodies. Our objective is to consider the concepts that underly the use of these methods in development of structure—function relationships, and to illustrate their fruitful application with pertinent examples from the literature. We have not attempted a comprehensive review of the many publications describing work in which monoclonal antibodies and limited proteolysis have played a major part, but have tended to focus on more recent applications. The interested reader is referred to a previous publication (21) that provides additional discussion on applications of monoclonal antibodies to structure—function studies.

1. UNDERLYING CONCEPTS

2.1. Protein Structure

A. THE STRUCTURAL HIERARCHY: PRIMARY, SECONDARY, TERTIARY, QUATERNARY

One frequently used approach (22) has been to view protein structure as a progression from an essentially linear representation (the primary structure, defined as the amino acid sequence) through localized hydrogen bonded structures such as α helices, β pleated sheets, or reverse turns (secondary structures) which in turn become associated (folded) to yield a specific three-dimensional array, referred to as the tertiary structure. In some proteins, two or more such folded subunits may associate to give the final oligomeric (quaternary) structure in which biological activity is fully expressed. Such distinctions can be useful for discussion purposes, but should not be taken to imply true independence.

Since the classic work by Anfinsen and his colleagues (23), it has been recognized that the amino acid sequence encodes the information necessary for development of higher orders of structure, and this provides a basis for attempts to predict secondary structure from amino acid sequence (e.g., 24). Though presently available methods for secondary structural prediction do not boast a remarkable accuracy (25, 26), they can certainly prove useful, particularly when several methods are used in combination (26). Further improvement in predictive accuracy can be expected as relationships between amino acid sequence and specific secondary structural features are more fully understood (27, 28).

As the number of known protein structures increased, it became evident that particular arrangements of secondary structural elements, generally referred to as *supersecondary structures*, were frequently encountered (29, 30). The physical basis for the apparent stability of these supersecondary structures and rules governing their formation have become increasingly apparent (2, 31), and initial development of computer-based prediction of supersecondary structural features has met with considerable success (11, 11a, 32).

Another level of structural organization, and one with major significance in the present context, has also been recognized. Although the overall shape of globular proteins may be described as ellipsoid, closer examination typically discloses that they are composed of smaller quasi-discrete globular regions, now generally referred to as *domains* (Fig. 2). Rose (33) has provided a concise account of the development of the concept of domains. Wetlaufer (34, 35) is credited with initial recognition of domains as fundamental units of structural organization and their likely significance in the folding process. Although a number of analytical methods have been proposed for objectively defining domains in known protein structures (33, 36–38), in practice, most investigators seem to rely on simple visual inspection to define the domains within a known protein structure.

The overall primary, secondary, and tertiary structure of a folded globular protein is thus the sum of these in the composite domains. Although the distinction between structural organization at the primary, secondary, and tertiary levels may be convenient for some purposes, it is frequently the case that consideration of protein structure in terms of domains is most advantageous.

B. THE DOMAIN AS THE FUNCTIONAL UNIT OF PROTEIN STRUCTURE

As noted above, Wetlaufer's original recognition of the domain as a basic organizational unit led him immediately to see the potential advantages if

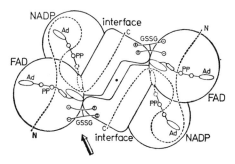

Fig. 2. Domain structure of human erythrocyte glutathione reductase. This enzyme consists of two subunits, each of which can be subdivided into three domains (32a). The cofactors, FAD and NADP, bind to discrete domains, and a third interface domain mediates interactions between the subunits; the binding site for the substrate glutathione (GSSG) is located between domains. Details of the structure of the individual domains are presented in Fig. 3. [Reprinted by permission from *Nature*, Vol. **273**, pp. 120–124. Copyright 1978, Macmillan Magazines Ltd.]

these corresponded to 'folding units' in multidomain proteins (34). Although Wetlaufer's suggestion of a 'nucleation' event within the nascent domain followed by accretion of adjacent structural features leading to the final secondary and tertiary structure does not appear to be the currently favored view of the folding process itself (39−43), there is abundant evidence to support his suggestion that domains represent independently folding units within the overall protein structure (e.g., 35, 44−49); frequently, this evidence is the demonstration that isolated fragments, derived by limited proteolysis (see below) and corresponding to the domains of the intact protein, can refold to yield structures closely resembling those of the intact native protein. Attaining the final conformation of the protein may require further, apparently rather modest, rearrangements resulting from interactions between the folded domains (46−49).

Domain organization has implications beyond its role in *acquisition* of structure. It can be viewed as a basic unit of function and evolution of proteins. There are now numerous examples of proteins in which specific functions may be associated with discrete structural domains, with glutathione reductase (32a) providing but one example (Fig. 3). Functions such as the binding of nicotinamide−adenine dinucleotides (9) or cAMP (10), or the DNA-binding "zinc finger" domain (7) have been mentioned above, and others will be considered later in this chapter. Such domains are found in many proteins, covalently linked with sequences coding for other functions, with the result that diversity in protein function results from various recombinations of these basic functional units. In other words, domains are the "cassettes" mentioned above, with duplication and recombination of the genetic information coding for these basic units permitting generation of new proteins, having new functions to be tested in the selective fire of evolution.

The original suggestion that exons might code for discrete structural or functional domains (50, 51), and hence might represent the fundamental genetic unit for recombination events leading to protein diversity, has not proven adequate as a correlation between genetic information and resulting protein structure, although some linkage between exons and "structure−function modules" may exist (52).

The above comments will, we trust, make it evident that an understanding of the domain structure and organization within a protein is critical to development of structure−function relationships. Direct determination of structure is certainly the most definitive approach, but as discussed above, this is by no means a trivial matter. How then might an investigator begin to define domain structure in a protein which, for whatever reason, direct structure determination was not a reasonable possibility? As is usually the case, several approaches might be useful. One (e.g., 10, 13, 14) rests on the well-established correlation between amino acid sequence and structure (3, 53−55), and hence function. Particularly since the development of molecular cloning techniques, acquisition of amino acid sequence information is generally quite feasible. Sequence comparisons may then disclose similarities to other proteins known

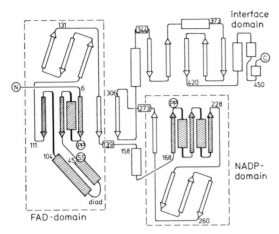

Fig. 3. Secondary structural elements and folding pattern of domains in human erythrocyte glutathione reductase. Alpha helices are represented as rectangles and β strands by arrows. The occurrence of common supersecondary structures (antiparallel β strands or α helices, βαβ units) is evident. The positions of the pyrophosphate linkages in FAD and NADP are labeled as PP; the involvement of a βαβ supersecondary structure in binding of these cofactors, as illustrated here for glutathione reductase, is frequently seen with other enzymes using FAD, NADP, or NAD as cofactors, and provides an excellent example of the association of a particular type of structure with a specific function. [Reprinted by permission from *Nature*, Vol. **273**, pp. 120–124. Copyright 1978, Macmillan Magazines Ltd.]

to contain domains associated with a particular function also expressed in the protein of interest. Few would consider it unreasonable to postulate the existence of a corresponding domain in the latter protein based on such evidence.

Another powerful method for gaining information about domain structure is differential scanning calorimetry (56–58). As noted above, domains represent independent folding units within a protein. Conversely, it is frequently the case that domains *unfold* more or less independently when subjected to increasing temperature, and the enthalpy associated with these unfolding events is the signal detected by the differential scanning calorimeter. Hence, the number of *transitions* seen in the thermogram provides an indication of the number of discrete unfolding units, presumably domains. The development of methods for deconvoluting complex thermograms into component transitions makes differential scanning calorimetry applicable to even complex multidomain proteins (see reference 57 for a very impressive example), and under favorable circumstances, one may even obtain a measure of the strength of interaction between domains (58). Perhaps the most serious factor limiting the general applicability of differential scanning calorimetery to protein structural studies is the *amount* of the protein required, typically on the order of 1–5 mg per thermogram.

Still another approach useful in defining the domain structure of proteins — an approach that requires only modest amounts of the protein and does not depend on availability of sequence information — is limited proteolysis, which will be a major focus of the present chapter. It will become evident from the discussion below that limited proteolysis is particularly useful when applied in conjunction with immunodetection methods employing monoclonal antibodies that recognize epitopes of defined position within the overall sequence.

2.2. Structural Basis for Susceptibility to Proteolytic Attack

A. ACCESSIBILITY

It is surely apparent that a protease cannot cleave a peptide bond that is not accessible. In a native protein structure, therefore, it follows that only peptide bonds located near the surface of the molecule could possibly be susceptible to proteolytic attack. Even in a small protein, however, this still implies that cleavage could occur at many sites within the overall sequence. Yet, the common experience is that cleavage is generally limited to only a few of these. What particular attributes distinguish these sites of cleavage?

Obviously, the involvement of an amino acid residue consistent with the specificity of the protease is a prerequisite for cleavage. Thus, if one uses trypsin, one cannot expect cleavage at other than lysine and arginine residues, no matter how accessible other bonds not involving these residues might be. However, it does not appear that specificity considerations are a major factor in determining the point(s) for proteolytic attack. For example, given the well-known propensity for arginine and lysines to be located at the aqueous interface, one could expect many such potential tryptic cleavage sites and yet one typically sees only one, or a few, of these actually attacked. Moreover, it has been frequently noted (e.g., 59–61) that several proteases of differing specificity cleave in close proximity, further indicating that it is something other than specificity of the protease that governs proteolysis.

Fontana et al. (62) have shown that the sites of proteolytic cleavage in thermolysin correspond to regions of high mobility, as reflected by the crystallographic temperature factors (B values), and proposed a general correlation between sites of proteolytic attack and flexibility (i.e., a relative lack of defined structure). This agrees with the observations of many investigators (see references cited by Fontana et al.) that cleavage sites in various proteins can be associated with flexible "hinge" regions, protruding surface loops, or similarly exposed segments lacking defined secondary structure. Polypeptide segments connecting structural domains typically fall in the latter category, and hence limited proteolysis has frequently been found to result in cleavage between structural domains. Indeed, it is by far the most common approach for obtaining isolated domains in amounts adequate for characterization, as illustrated by the classic study of Miller et al. (63) that led to recognition of "zinc finger" domains and their role in binding of DNA (see also 44–48, and others discussed below). Looking to the future, one might anticipate that expression of cloned cDNAs coding for particular domains may also become a fruitful approach to obtaining discrete domains for study.

Let us assume that you have isolated a previously uncharacterized protein. You have established that it is a homodimer, with subunit molecular weight 65 000. You are interested in learning something about its domain structure, and subject the native protein to limited digestion with trypsin (or subtilisin, *Staphylococcus aureus* V8 protease, or thermolysin) and observe that fragments of 12 000, 23 000, and 30 000 result. Is it reasonable to conclude that these fragments might correspond to discrete structural domains in this protein? In the opinion of this writer, absolutely! (Why else would one suggest such an experiment to address the question of interest?) There surely is abundant precedent for such an interpretation, and the size of the fragments is in the range expected for structural domains (52). Can this result be taken as unequivocal evidence that this protein consists of three domains, corresponding to the proteolytic fragments? Absolutely not! Although segments connecting domains are frequently the site of such cleavage, they are by no means the only sites that might be susceptible (62). Nor is it by any means certain that *all* segments connecting domains will be cleaved in such experiments.

Limited proteolysis experiments represent a technically simple first step, modest in demands for either experimental manipulation or protein. They can readily provide information that serves as a basis for further work that can lead to better definition of domain structure, such as isolation of proteolytically derived fragments that exhibit characteristics expected for domains, such as independent folding, or functional properties such as ligand binding. Moreover, even when cleavage sites do not correspond to segments linking domains, limited proteolysis can be extremely useful in developing structure−function correlations, as will be illustrated later in this chapter.

2.3. Antigenic Structure of Proteins

Immunological techniques hold a valued place in studies of proteins. It is not within the scope of the present chapter to review the many and diverse applications of antibodies, but rather we wish to focus particularly on their utility in developing structure−function relationships, especially in conjunction with limited proteolysis experiments. Furthermore, we will largely restrict our attention to the use of monoclonal antibodies in such studies.

The development of methodology for production of *hybridomas* capable of secreting virtually limitless amounts of homogeneous antibody (64) can surely be said to have revolutionized the application of immunological techniques to studies of protein structure (as well as many other areas of investigation). The heterogeneous antibody population in polyclonal antiserum may not be a problem, and indeed may even be advantageous, in certain applications; in others, however, this heterogeneity may complicate matters considerably. A monoclonal antibody is a homogeneous reagent, targeted to a precisely definable (at least in principle) structural feature (epitope). This element of specificity, together with other advantages inherent in monoclonal antibody technology (21), has led to the present situation where the generation and application of monoclonal antibodies are commonplace in laboratories studying protein structure and function.

B.   EPITOPES: SEGMENTAL (CONTINUOUS) VERSUS CONFORMATIONAL (DISCONTINUOUS)

It has been a common practice to distinguish between two types of epitopes that might exist on a protein (e.g., 65−68). One type, referred to variously as sequential, segmental, or continuous by various authors, implies that the epitope consists of a linear segment, probably 5−8 amino acid residues long. As noted by Berzofsky (66), this is meant to imply that the information necessary for recognition by the antibody is contained within the segment but does not imply that there is no conformational element involved in that recognition; that is, the residues within the segment presumably must be positioned in a particular spatial configuration for recognition to occur. The second type of epitope, referred to as conformational, discontinuous, or assembled topographic, is considered to include amino acid residues far apart in the primary structure (sequence) but brought into spatial proximity as a result of the folding of the protein. Thus, both types of epitopes include an element of conformation, and the distinguishing feature really is whether or not the conformational element recognized by the antibody, the epitope, is formed by amino acid residues in close proximity within the sequence. The recent article by van Regenmortel and de Marcillac (69) should be consulted for a more extensive discussion of continuous versus discontinuous epitopes.

After considering the relative dimensions of the antigen binding site on an antibody and surface features typical of a folded protein, Barlow et al. (67) suggested that all antigenic determinants 'are discontinuous to some extent.' While this may be the case, it surely is clear that amino acid residues within a limited segment represent the dominant feature in some epitopes, whereas this is just as certainly not true for other epitopes. Hence, the distinction between continuous (segmental, sequential) and discontinuous (conformational, assembled topographic) epitopes remains a useful one with important practical significance, as discussed below.

C.   EPITOPE MAPPING

The value of a monoclonal antibody is dramatically enhanced by definition of the location of its epitope within a protein's structure. It then becomes useful, for example, in identifying products of limited proteolysis. If one knows that antibody X recognizes a continuous epitope whose location has been defined within the overall sequence, a single immunoblot will provide a display of all proteolytic fragments containing that segment. Effects of antibody binding on a specific function then become interpretable in terms of involvement (directly or indirectly) of the epitopic structural region in that function. The process of defining the location of antibody binding sites (epitopes) within the protein's structure is commonly referred to as epitope mapping.

While we will not attempt to review the many experimental approaches that have been employed by various investigators, it will be useful to consider some general concepts that govern epitope mapping studies. Epitope mapping methods can be considered to fall into one of two categories, which we have called *direct* and *competitive* (21). By direct, we mean epitope mapping methods that permit assignment of the epitope to a specific segment within the overall amino acid sequence. By competitive, we mean methods that determine the location of epitopes recognized by two monoclonal antibodies *relative to each other*; operationally, this means determining whether two antibodies can bind simultaneously. Competitive mapping does not provide direct indication of the location of the epitopes within the protein's structure, though a combination of direct and competitive mapping methods may lead to this (e.g., 70).

The basic approach in most direct mapping methods is to assess the ability of the antibody to recognize (bind to) specific structural elements of the protein. Experimentally, this typically means that one measures the ability of the antibody to bind specific peptide fragments which are obtained by chemical or proteolytic cleavage of the protein, chemically synthesized based on known amino acid sequence, or more recently, obtained by recombinant DNA methodology (71–74). Immunoreactivity may be detected in a number of ways, including reactivity on immunoblots, direct immunoprecipitation, direct reactivity in ELISA (*e*nzyme-*l*inked *i*mmuno*s*orbent *a*ssay), or using the peptide segment as a competitor against the intact protein in ELISA. Although it might seem that using peptide segments would preclude detection of discontinuous epitopes by such methods, this is not always the case. There is clear evidence that even relatively short peptides may adopt conformations representative of discontinuous epitopes which are thereby detectable (75, 76). Moreover, some monoclonal antibodies that apparently recognize discontinuous epitopes based on their ability to react with native but not denatured protein, may show reactivity on immunoblots despite the fact that the protein has been subjected to the harsh conditions of SDS gel electrophoresis, presumably due to partial renaturation during the blotting or subsequent immunodetection process (21); procedures reported to enhance such renaturation have been described (77, 78).

It should be apparent that, even if reactivity of a discontinuous epitope is detected with peptide segments, its character cannot be defined with nearly the resolution that might be attained for a continuous epitope. For example, if a 30-residue peptide segment, but not shorter versions, showed reactivity with an antibody recognizing a discontinuous epitope, about the most that could be said would be that the 30-residue peptide was the shortest segment capable of adopting a recognizable conformation; in the absence of further information, the exact role of specific residues within that segment would remain undefined. (Contrast this with the reactivity of much shorter peptides with antibodies recognizing continuous epitopes, permitting definition of the epitope to segments as short as 4 residues.) Therefore, development of methods that might facilitate definition of conformational epitopes (e.g., 79, 80) by something short of x-ray crystallographic study of antibody–antigen complexes (81) are of obvious interest. At present, however, it appears likely that most of the epitopes detected by direct methods involving binding of peptide fragments are either continuous, or discontinuous epitopes involving only local elements of secondary structure (e.g., 81a).

Although epitope mapping studies employing peptide segments as representatives of continuous epitopes on the intact protein would seem to be conceptually straightforward, there are some subtleties that may affect the results. For example, in mapping epitopes recognized by monoclonal antibodies raised against an intact protein, some authors (82) have recommended using relatively long (approximately 30 residues) peptides, while others (83) have found it more advantageous to use much shorter peptides, for example, tetrapeptides. Such differences might reflect the extent to which the antibodies under consideration recognize discontinuous epitopes (which are less likely to be adequately modeled by short peptide sequences). In a study utilizing polyclonal antisera (83a), limiting conformational flexibility of a peptide by formation of an intramolecular disulfide bond has also been found useful; this would presumably be applicable to recognition by monoclonal antibodies also.

Denaturation of the antigen may occur during its adsorption to the plastic microtiter plates commonly used for ELISA, with resulting loss of reactivity with antibodies recognized conformational epitopes (e.g., 84). A modification of the standard ELISA that avoids the denaturation artifact has been suggested (85). In the latter method, *polyclonal* antibodies are adsorbed to the plastic plate and the antigen is then tethered to the plate, indirectly and without denaturation, via this immunoreaction. Since polyclonal antibodies may not be available for the antigen of interest, one might consider using alternative tethering methods, for example, limited biotinylation of the antigen and tethering it via avidin or streptavidin adsorbed to the microtiter plate.

We also note some alternative approaches to direct epitope mapping that do not involve the use of peptide fragments, and that are applicable to both discontinuous and continuous epitopes. Burnens et al. (86) proposed using the protection afforded by antibody binding against chemical modification (e.g., acetylation of lysine groups) as a means for defining epitopic regions. A somewhat similar strategy was suggested by Sheshberadaran and Payne (87), using protection against proteolysis as the basis for defining the epitope (protein "footprinting"). It should be recognized that definition of conformational epitopes by these or other (79, 80) methods presupposes knowledge of the protein's structure; in the absence of such information, it might be possible to identify discrete regions of the sequence that were involved in formation of a particular conformational epitope, but their spatial relationship within that epitope could not be appreciated. Finally, immunoelectronmicroscopy (88, 89) permits direct imaging of the disposition of bound antibody molecules relative to the molecular architecture of the protein; though more limited in both its general applicability and in the resolution with which epitopes are defined, immunoelectronmicroscopy nonetheless has proven useful in relating effects of antibody binding on function to specific definable regions of the molecule.

Competitive epitope mapping defines the relative distance between the epitopes recognized by different monoclonal antibodies by asking the question: Can two antibodies bind simultaneously? If the answer is yes (*independent binding*), then it is evident that the epitopes must be separated by a considerable distance across the molecular surface. Based on the size of the antigen binding region of IgG antibodies (90), the epitopes recognized by independently binding antibodies are separated by at least 35 Å. If simultaneous binding cannot occur (*exclusive binding*), then one interpretation would be that this is due to the close spatial proximity (<35 Å) of the epitopes. Alternatively, conformational changes induced by binding of one antibody (81, 91) might preclude binding of the second. One must be aware of this as a possible cause of exclusive binding; however, it does not appear to be a commonly encountered complication since monoclonal antibodies showing exclusive binding have frequently been found to recognize epitopes in the same or closely contiguous regions of the amino acid sequence (e.g., 70, 92–94), implying that mutually exclusive binding was indeed due to proximity rather than indirect conformational effects. The third possible outcome of competitive mapping experiments might be that one antibody hinders but does not preclude binding of the second antibody (*overlapping*); such effects are most directly interpreted as indicating sufficient proximity (≈35 Å) of the epitopes to cause steric hindrance of simultaneous binding, though again possible conformational effects should also be considered.

Experimentally, competitive mapping requires a technique capable of detecting the extent to which simultaneous binding of antibodies occurs. One approach distinguishes antigen–antibody complexes based on molecular weight. Thus, if two IgGs (molecular weight approximately 160 000) bind simultaneously to the antigen, the complex will have an apparent molecular weight of 320 000 *plus* the molecular weight of the antigen, whereas if exclusive binding occurs, the complex will be 160 000 plus the antigen molecular weight. Various methods have been found useful for fractionation of such complexes, including molecular sieve HPLC and electrophoresis on nondenaturing acrylamide gels (95). However, ELISA methods are much more commonly used (e.g., 96–99), and justifiably so considering their sensitivity and convenience.

If the position of the epitopes recognized by one or, better yet, several antibodies can be defined by direct mapping, then competitive mapping can lead to a representation of epitope distribution on the molecular surface, with defined epitopes as points of reference (e.g., 70). Even if this is not the case, competitive mapping can be very informative.

D. CORRELATION OF ANTIGENICITY WITH STRUCTURAL FEATURES

Definition of antigenic features on a protein has both practical and conceptual significance. As discussed above, defining discontinuous epitopes can be most difficult, and indeed, is impossible in the absence of structural information. On the other hand, definition of continuous epitopes by determining whether peptides, representing specific segments of the amino acid sequence, can compete for binding to antibodies reactive with the protein itself is generally much more feasible; it is, therefore, not surprising that there are many studies of this type. The fact that continuous epitopes may be the easiest to define (and we recall again that the concept of a continuous epitope on a protein surface is likely to represent an oversimplification (67)) should not be cause to forget that discontinuous epitopes exist.

Early studies, reviewed by Benjamin et al. (65) and Berzofsky (66), led to identification of certain regions of several proteins that seemed to be *immuno-dominant*, that is, to contain most of the continuous epitopes that could be defined by competition with peptides representing segments of the primary structure. In some cases, this seems to have led to the view that these regions represent the *only* antigenic regions on the protein. Such views are now discounted, and one frequently encounters statements such as "virtually the entire accessible surface of a protein may be antigenic" (66) and the surface that is "accessible" in terms of inducing antibody formation may not be limited to the "accessible surface" in the native structure since denaturation is an inevitable consequence of standard immunization procedures (21, 85, 100).

Faced with the concept of an effectively infinite number of potential epitopes on the protein surface, one might have expected epitope mappers to turn to other pursuits. That has not happened. Despite the probable existence of a multitude of discontinuous epitopes recognized by antibodies, it is clear that direct mapping methods employing peptide segments as competitors (with the protein itself) for binding to antibodies *do* identify particular regions as being particularly likely to contain continuous epitopes, or discontinuous epitopes dominated by a limited segment of the sequence which is represented by the competing peptide (67, 81a). As noted by Barlow et al. (67), such antigenic regions are likely to be distended from the molecular surface, associated with loops and/or protrusions into the surrounding aqueous space. Given the expected character of such regions, it is not surprising that they have been correlated with hydrophilicity (101, 102), surface protrusions (103–105), location of reverse turns (106), and mobility (107). Hydrophilic character (101, 102), propensity for formation of reverse turns (24, 106), and mobility (108) can be predicted (with varying degrees of success) from amino acid sequence, and this has served as the basis for prediction of antigenic regions related to such characteristics, as recently reviewed by van Regenmortel and de Marcillac (69). It is, of course, appropriate to remember that such sites represent only a subset (determined by the choice of method for their detection, i.e., competition with peptides representing continuous epitopes) of the total population of potential epitopes that might be recognized by antibodies.

Continuous epitopes are obviously much more likely to survive proteolysis, electrophoresis under denaturing conditions, and electroblotting procedures; thus antibodies recognizing continuous epitopes are of particular value in deciphering the results of limited proteolysis experiments. On the other hand, discontinuous epitopes can be expected to be much more sensitive to changes in tertiary structure, for example, during denaturation/renaturation, or in response to binding of ligands; antibodies recognizing discontinuous epitopes have found important application in studies involving such phenomena.

Before proceeding, it seems useful to make a few comments about epitopes classified on another basis, namely, their accessibility in native and denatured forms of a protein. As we have just discussed, the surface of a folded protein is likely to contain many (in principle, an infinite number of) potential epitopes, some continuous and many discontinuous. Regardless of their continuous or discontinuous character, it is apparent that these epitopes must share one property in common, namely, accessibility, an obvious prerequisite for recognition by the antibody. Thus these epitopes must be on the surface of the folded protein, and it is precisely this region that is most susceptible to change when homologous proteins from different organisms are compared (e.g., 3, 15, 55, 109). On the other hand, the internal core of homologous proteins, which is most likely to govern the overall folding pattern and to include functionally important residues, is generally much more conserved. It follows that antibodies directed against internal regions are likely to be very useful in detecting conserved features within a family of homologous proteins. Induction of antibodies against such internal epitopes obviously implies their exposure by denaturation of the protein during the immunization process; this may be done intentionally (e.g., 110), but it also occurs even if the native protein is used for immunization (e.g., 85, 100). Since recognition by these antibodies requires denaturation of the protein, it may be assumed that the corresponding epitopes are continuous in character and hence potentially definable by direct mapping techniques.

3. APPLICATIONS OF LIMITED PROTEOLYSIS AND MONOCLONAL ANTIBODY TECHNIQUES IN DEVELOPMENT OF STRUCTURE–FUNCTION RELATIONSHIPS

3.1. Brain Hexokinase

Because we think it illustrates very well how limited proteolysis and monoclonal antibody methodology can complement each other in development of concepts of protein structure and function, we begin with a discussion of work in our own laboratory on brain hexokinase. The properties of rat brain hexokinase, and mammalian hexokinases in general, have been reviewed elsewhere (111), and comments here will be limited to describing those characteristics pertinent in the present context. The interested reader may consult the review (111) for references to the numerous original publications that support this brief synopsis.

Hexokinase catalyzes the phosphorylation of glucose, using ATP as phosphoryl donor, to produce glucose 6-phosphate (G6P) and ADP. The enzyme is sensitive to inhibition by the product, G6P, and there is general agreement that this represents a major factor in regulation of its in vivo activity. In contrast to the other glycolytic enzymes, which are located in the soluble fraction of brain homogenates, much of the hexokinase is found associated with the mitochondria. This binding is reversible and sensitive to modulation by several metabolites closely associated with energy metabolism, such as G6P and P_i; reversible interactions with the mitochondria have also been suggested to play an important role in regulation of the enzyme. We will be concerned here with three functions of the enzyme: catalysis, inhibition by G6P, and reversible binding to mitochondria.

Like other mammalian hexokinases, rat brain hexokinase consists of a single polypeptide chain with molecular weight 100 000 (112). Since a protein of this size can surely be expected to consist of more than a single structural domain and since discrete segments linking structural domains are a favored target for proteases (see comments above), we initiated a study of the cleavage pattern seen during limited proteolysis with trypsin (113). Digestion of the native enzyme gave three major fragments (molecular masses of 10, 40, and 50 kDa) along with two partial cleavage intermediates (60 and 90 kDa). The origin of these fragments with respect to the overall sequence is indicated schematically as

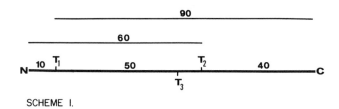

SCHEME I.

where T_1 and T_2 represent tryptic cleavage sites (T_3 is also a tryptic cleavage site, which becomes manifest under other conditions, discussed below). The original interpretation of these results was that the 10-, 40-, and 50-kDa fragments corresponded to structural domains, though more recent work makes this seem unlikely (114). This illustrates a point made earlier: Although cleavage between domains is a frequent occurrence, there is no assurance that cleavage *will* occur at that location, nor that cleavage *will not* occur at some other accessible site (which appears to be the case for T_1 and T_2). Despite the now-recognized probability that T_1 and T_2 are not located in interdomain segments, the definition of the tryptic cleavage map was fundamental to several subsequent studies that provided much information about the structure—function relationships in this enzyme.

At about the same time that the tryptic cleavage pattern was being elucidated, others in the laboratory were developing monoclonal antibodies against the enzyme (70, 115). Immunoblotting of proteolytic cleavage fragments (Figs. 4 and 5) established the location of continuous epitopes recognized by several of the antibodies (70, 113); more recently, this has been confirmed and extended using ELISA with the isolated fragments (A.D. Smith, unpublished work). Locating the epitope is invaluable since it permits identification of proteolytic fragments by simple immunoblotting methods. Hence, after affinity labeling of the enzyme with radioactive reactive analogues and subsequent proteolysis, comparison of autoradiographs and immunoblots made it evident that binding sites for *both* substrates, and hence catalytic function, are associated with the 40-kDa fragment from the C-terminal half of the molecule (116, 117); this has subsequently been confirmed by direct isolation of a catalytically active C-terminal half of the molecule (118), again by use of selective proteolysis (see below).

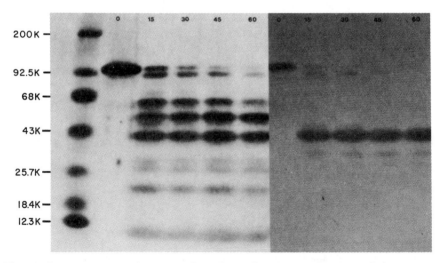

Fig. 4. Limited tryptic digestion of rat brain hexokinase: location of the epitope recognized by monoclonal antibody 5A in the C-terminal region. *Left*: The cleavage pattern seen after 0, 15, 30, 45, or 60 min of digestion with trypsin under the conditions of Polakis and Wilson (113); the SDS gel was stained for protein with Coomassie blue, and molecular weight markers are shown at the extreme left. The production of 10-, 40- and 50-kDa fragments, with 60- and 90-kDa cleavage intermediates, is evident (see text for schematic representation of cleavage pattern). *Right*: An immunoblot, prepared from an identical gel and probed with monoclonal antibody 5A. Only the 40-kDa fragment derived from the C-terminal region of the molecule (113), or larger (intact 100-kDa enzyme and the 90-kDa cleavage intermediate) species including this region, are immunoreactive. [Reprinted from (113), with permission of Academic Press.]

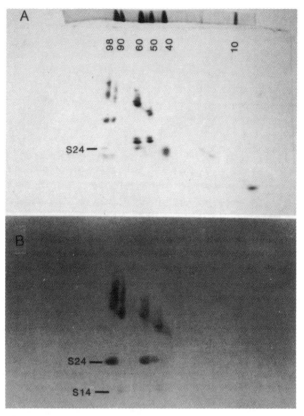

Fig. 5. Localization of the epitope recognized by monoclonal antibody 21 by two-dimensional peptide mapping of rat brain hexokinase and immunoblotting. Initial digestion with trypsin followed by SDS−gel electrophoresis gave a pattern as seen at the top of the figure. The tryptic fragments, still in the gel, were further digested with *S. aureus* V8 protease followed by electrophoresis in a second dimension; residual undigested tryptic fragments are seen along the diagonal. Smaller peptides resulting from digestion with *S. aureus* V8 protease are located beneath the diagonal; these are referred to as S peptides; for example S24 is a peptide with an estimated molecular weight of 24 kDa arising from the digestion with V8 protease. From analysis of such maps, it is possible to deduce the location of the S peptides within the overall sequence (113). *Top*: A two-dimensional gel stained with Coomassie blue. *Bottom*: An immunoblot prepared from an identical gel and probed with monoclonal antibody 21. The 50-kDa tryptic fragment, derived from the N-terminal half of the enzyme (and larger species containing this region), is immunoreactive. The location of the epitope can be further defined by the observation that the S peptides S14 and S24 are also immunoreactive; based on this information, the epitope recognized by antibody 21 must be located near the N-terminal end of the 50-kDa tryptic fragment. [Reprinted from (113), with permission of Academic Press.]

Binding of the enzyme to mitochondria depends on an intact hydrophobic N-terminal sequence (118a) that is inserted into the lipid core of the outer mitochondrial membrane (118b). Limited proteolysis with chymotrypsin results in cleavage within this N-terminal sequence with resulting loss of the ability to bind to mitochondria (118a). Independent evidence for the involvement of the N-terminal region of hexokinase in binding to mitochondria is provided by the demonstration that monoclonal antibodies mapping to the N-terminal 10-kDa tryptic fragment (produce by cleavage at T_1) prevent binding while monoclonal antibodies recognizing epitopes located elsewhere in the molecule do not (70, 115, 118a).

It is frequently observed that binding of a ligand stabilizes a protein against inactivating agents, including proteases; this is certainly the case with brain hexokinase (119–121). Conversely, denaturation of proteins by agents such as urea or guanidine increases their susceptibility to proteolysis since loss of secondary and tertiary structure results in increased accessibility of potential cleavage sites. These characteristics were exploited in work that led to isolation of discrete N- and C-terminal halves of brain hexokinase (118, 122, 123). The enzyme is partially denatured with relatively low (≈0.5 M) concentrations of guanidine and in the presence of ligands that selectively bind to either the N- or C-terminal half, addition of trypsin results in complete proteolysis of the half not stabilized by bound ligand. Identification of proteolysis products is facilitated by immunoblotting techniques using monoclonal antibodies recognizing epitopes of defined location. Selective stabilization of the C-terminal half of the molecule with substrates (or analogues) permits isolation of this half with retention of catalytic activity (118), while selective stabilization of the N-terminal half with glucose 6-phosphate or its analogues indicates that the allosteric inhibitory site is located in this region (122). The domain structure of brain hexokinase and its topological relationship to the inner and outer mitochondrial membranes, as presently understood, are represented schematically in Fig. 6.

It was noted above that more recent work (114) makes it unlikely that cleavage sites T_1 and T_2 are located in segments linking structural domains, as originally interpreted (113). When the complete deduced amino acid sequence became available (114), it was evident that there is extensive similarity between the amino acid sequences of the N- and C-terminal halves. These are also similar to the sequence of yeast hexokinase (molecular weight, 50 kDa), which clearly suggests that the 100-kDa mammalian enzyme is a product of duplication and fusion of a gene coding for an ancestral hexokinase related to the yeast enzyme. (See (118) for more extensive comments on the probable evolutionary relationship between hexokinases from mammals and other organisms.) Because similarity in sequence strongly suggests similarity in structure (3, 53–55), a model (Fig. 7) for the rat brain enzyme has been proposed (114), based on the x-ray structure of the yeast enzyme (123b). Structurally similar N- and C-terminal domains are linked by a segment that is 'buried' within the fused structure. Based on this model, cleavage sites T_1 and T_2 are not in interdomain

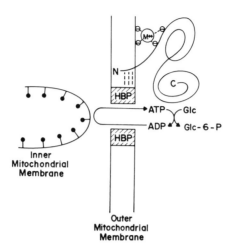

Fig. 6. Schematic representation of the domain structure of rat brain hexokinase and the topological relationship between the mitochondrially bound enzyme and mitochondrial oxidative phosphorylation. The enzyme consists of two domains. Catalytic function is associated with the C-terminal domain (116–118), while regulatory function is associated with the N-terminal region (122, 123). Binding to mitochondria requires an intact hydrophobic sequence at the extreme N-terminus (118a), which is inserted into the hydrophobic core of the outer mitochondrial membrane (118b); electrostatic interactions, mediated by divalent cations such as Mg^{2+}, are also thought to occur. Binding also involves a hexokinase binding protein (HBP), now known to be identical to the pore-forming protein of the outer mitochondrial membrane. Binding of the enzyme at the pore through which adenine nucleotides (as well as other metabolites) enter or exit the mitochondrion is thought to foster direct interactions between intramitochondrial oxidative phosphorylation and phosphorylation of glucose by the mitochondrially bound enzyme. [Reprinted with permission from (123a), coypright CRC Press, Inc., Boca Raton, FL.]

segments, but rather, in β turns located at structurally analogous surface regions of the N- and C-terminal halves. Proteolytic cleavage between N- and C-terminal domains *does* occur (at a site designated T_3, see cleavage map above) when proteolysis is done under partially denaturing conditions (118, 122, 123), presumably due to weakening of the interdomain interactions with resulting increased accessibility to this region of the molecule.

We have also examined limited tryptic cleavage patterns for hexokinases from brains of various other species, with analysis being facilitated by immunoblotting techniques employing monoclonal antibodies (124). Like the rat enzyme, the hexokinases from mouse, rabbit, and guinea pig do not show appreciable cleavage at T_3 under nondenaturing conditions. In contrast, considerable cleavage at T_3 *is* seen with the hexokinases from cat, dog, pig, cow, and sheep. These results suggest subtle differences in the interactions between the N- and C-terminal domains of hexokinases from these groups, and it is

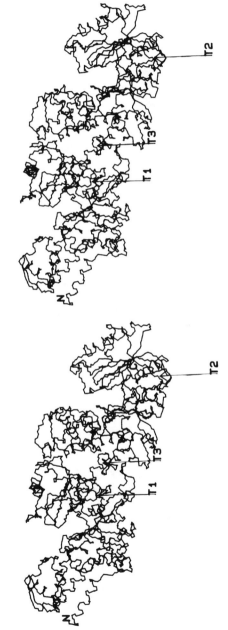

Fig. 7. Proposed structure for rat brain hexokinase. The structure was generated based on the known structure of yeast hexokinase (123b), to which the N- and C-terminal halves of the rat brain enzyme show extensive sequence similarity (114). T_1, T_2, and T_3 are tryptic cleavage sites (see text). [Reprinted from 114).]

notable that this is detected by techniques as simple as limited proteolysis and immunoblotting.

Finally, the domain structure of rat brain hexokinase has also been studied by differential scanning calorimetry (125). Consistent with the view that the N- and C-terminal halves of the molecule represent discrete domains, in both a structural (114) and functional (116–118, 122, 123) sense, two transitions are observed in the thermogram. Proteolysis and immunoblotting techniques with monoclonal antibodies recognizing defined epitopes were instrumental in identifying the domains corresponding to the two transitions, based on the following strategy. A *partial* scan, through only the first transition, resulted in irreversible denaturation of the least thermally stable domain; a rescan through the entire temperature range confirmed that the other domain, corresponding to the second transition, was indeed still present in folded form. The respective domains were identified by proteolysis of the partially denatured (by scanning only through the first transition) enzyme. The unfolded domain was completely proteolyzed while the domain retaining its structure remained intact. Based on immunoblotting experiments in which the latter fragment was selectively reactive with monoclonal antibodies directed against epitopes located in the N-terminal half of the molecule, the more stable domain (second transition) is identified as the N-terminal domain; hence, it follows that the first transition corresponds to thermal denaturation of the C-terminal domain. The stability of these domains can be influenced by binding of ligands, and, indeed, the order of denaturation can be reversed by inclusion of ligands destabilizing the N-terminal half and stabilizing the C-terminal half (125).

3.2. Multifunctional Proteins

Implicit in the concept of a *metabolic pathway* is the notion of sequential action of a series of catalytic activities. There remain some who subscribe to the view that this occurs as a random process, with metabolic intermediates diffusing about in search of useful purpose represented by the next enzyme in the pathway. On the other hand, there is clear evidence in some systems (e.g., 126) that *channeling* of metabolites occurs, though the generality of the concept of channeling is in question (e.g., 127). Presumably, if channeling is to occur, there must be some physical continuity between the enzymes involved, and thus complexes between enzymes catalyzing sequential steps in metabolic pathways are attracting considerable interest (128). Physical continuity is ensured in *multifunctional proteins* in which two or more distinct catalytic activities reside in a single polypeptide chain (or oligomers thereof). Though the kinetic and thermodynamic advantages that might accrue from such assured spatial proximity have been discussed (e.g., 129, 130), it would appear that they are not critical since catalytic activities represented in a single multifunctional enzyme frequently occur as discrete molecular species in another organisms. One cannot, of course, exclude the possibility (probability?) that the discrete enzymes are, in fact, organized as complexes under in vivo conditions (128).

Some well-studied examples of multifunctional enzymes are the fatty acid synthetase of animals (131), the CAD protein catalyzing several steps of pyrimidine biosynthesis in mammalian systems (132−135) and its counterpart in *Drosophila* (136), C_1-tetrahydrofolate synthase (137, 138), and multifunctional proteins involved in aromatic amino acid biosynthesis in yeast and fungi (130, 139, 140). In each of these cases, limited proteolysis provides fragments catalyzing discrete reactions of the metabolic sequence; this was instrumental in disclosing the multifunctional nature of these proteins and determining the relative disposition of functional units within the overall sequence. As might be expected, analysis at the DNA level can also be very helpful in the latter respect (e.g., 134, 136, 140).

The x-ray crystallographic structure of a multifunctional protein has not as yet been determined. However, the ability of proteases to cleave between regions possessing discrete catalytic functions certainly suggests that these represent distinct quasi-independent structures, analogous to structural domains. In fact, it is common to refer to the functional units of multifunctional proteins as domains, though it should be apparent that this is used in a functional sense, and that each functional domain may, in fact, consist of more than one structural domain. Such independence of the functional units in multifunctional enzymes is not unexpected since they appear to result from fusion of genes coding for homologous proteins existing as independent species (and hence, having their own intrinsic stability) in other organisms (e.g., 134, 140).

In the present context, the work of Schirch and his colleagues on C_1-tetrahydrofolate synthase has additional interest since the domain structure of this enzyme was also studied by differential scanning calorimetry (137). Two transitions are evident in the thermogram of the intact enzyme, suggesting that the enzyme is organized into two structural domains; addition of ATP shifts the temperature for the first transition to higher temperature (consistent with this transition corresponding to the ATP-binding domain) while $NADP^+$ has a similar effect on the second transition (corresponding to the $NADP^+$-binding domain). Limited proteolysis of the enzyme with chymotrypsin gives two fragments, one of which (M_r 66 000) retains 10-formyltetrahydrofolate synthetase activity while the other (M_r 36 000) retains both methylenetetrahydrofolate cyclohydrolase and methylenetetrahydrofolate dehydrogenase activities (the coexistence of such seemingly unrelated, in the chemical sense, activities in such a small protein is rather amazing!); the sum of these activities represents the overall C_1-tetrahydrofolate synthase reaction. The larger (synthetase) fragment is unstable and could not be examined by differential scanning calorimetry. The smaller fragment gives a single transition, which is shifted to higher temperature by addition of $NADP^+$; the latter result identifies this as the $NADP^+$-binding domain, corresponding to the second transition in the thermogram of the intact protein. However, the transition temperature for the isolated domain is considerably lower than that seen with the intact protein. The latter result, along with the instability of the isolated synthetase domain, suggests that interdomain interactions (presumably mediated by regions altered by proteolysis) are an important stabilizing influence in the intact protein (58). Although interdomain forces may be important for stability, they do not appear to affect catalytic function since the kinetic parameters of the isolated domains are indistinguishable from those of the intact enzyme.

3.3. Proteolytic Dissection of Functional and Structural Domains

As mentioned in the introduction to this chapter, the structural organization of a protein generally reflects function. If structure can be determined in some absolute sense, such as by x-ray crystallography, then association of function with definable structural domains may be established with some certainty. This is generally not the case, however, and hence reliance on less direct methods for defining domain structure and functions associated with (putative) domains is usually necessary.

Though it is by no means a certainty, as evident from the example of brain hexokinase discussed above, there is ample precedent for expecting that fragments obtained by limited proteolysis of the native protein may correspond to structural domains, with which specific functions such as binding of substrates or cofactors may be associated. Since structure and function of proteins are so closely correlated, it is a reasonable assumption that preservation of function in proteolytically derived fragments reflects maintenance of a structure closely resembling that in the intact protein. Domains represent quasi-independent units of structure, and hence preservation of that structure after severance of interdomain segments is not unexpected. In fact, direct evidence for preservation of native structure in isolated (by proteolysis) domains is available (e.g., 141).

In the absence of direct structural information, one cannot, with certainty, correlate fragments obtained by limited proteolysis with structural domains. Yet, especially when specific functions can be associated with such fragments, they are commonly referred to as domains. The term is being used in a functional sense (as with multifunctional proteins, discussed above), though most users of the term would probably be surprised if the domain did not turn out to be applicable in the structural sense as well.

Limited tryptic digestion of bovine adrenal tyrosine hydroxylase yields a catalytically active 34-kDa fragment derived from a central region of the intact molecule (142). The sequence in this region shows extensive similarity with corresponding regions in other mammalian hydroxylases utilizing aromatic amino acids (phenylalanine or tryptophan) as substrates (143–145). Thus, this 34-kDa tryptic fragment may represent a catalytic domain analogous to that shared by many protein kinases (10, 13).

Hanemaaijer et al. (146) used limited proteolysis to define the domain structure of the dihydrolipoyl transacetylase from the pyruvate dehydrogenase complex of *Azotobacter vinelandii*, and succeeded in isolating two discrete fragments that were associated with binding of the lipoyl prothetic group and catalytic activity, respectively. Immunoblotting techniques (using polyclonal antibodies in this case) were instrumental in deciphering proteolytic cleavage patterns.

Limited proteolysis followed by molecular sieve chromatography enabled Southerland et al. (147) to separate domains binding the heme and molybdenum prosthetic groups required for the function of liver sulfite oxidase. Trypsin, papain, and chymotrypsin all cleave in the same limited region, consistent with the cleavage site being an accessible unstructured segment typical of those linking domains. Kubo et al. (148) conducted a similar study with a still more complicated enzyme, spinach nitrate reductase, and were able to separate three fragments corresponding to domains binding molybdenum, heme, and flavin cofactors. In contrast to the results with sulfite oxidase (147), no single protease gave complete cleavage between the domains of nitrate reductase, indicating the value of exploring the action of different proteases. Based on perceptive analysis of proteolytic cleavage patterns and of residual catalytic activities associated with the various fragments, Kubo et al. were able to propose a model depicting the relative disposition and interaction of these domains in the dimeric nitrate reductase molecule (Fig. 8). It may also be noted here that extensive similarity is seen in the amino acid sequences of the heme and molybdenum binding domains of nitrate reductase, sulfite oxidase, and other proteins binding these prosthetic groups (148a, and references therein), consistent with the view that the molybdenum and heme binding domains represent functional cassettes that can be recombined in various ways to generate protein diversity (see comments above).

With both sulfite oxidase and nitrate reductase, the proteolytic fragments (domains) are readily separable by molecular sieve chromatography without the necessity for employing relatively harsh conditions such as high salt (e.g., 146) or detergents. As noted by the investigators (147, 148), this implies relatively weak interdomain interactions in these proteins, which is not always the case.

Hubbard and Klee (149) provide a very nice illustration of the use of

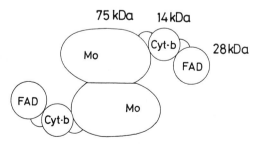

Fig. 8. Proposed domain structure of spinach nitrate reductase. This model is based on analysis of limited proteolysis experiments, which generated fragments of 14 kDa (containing the cytochrome b moiety), 28 kDa (containing bound FAD), and 75 kDa (containing molybdenum). [Reprinted, with permission, from (148).]

limited proteolysis to dissect functional (and presumably structural) domains — in this case, not involving prosthetic groups — in their recent study of calcineurin, a Ca^{2+}/calmodulin stimulated protein phosphatase found in brain. The enzyme is a heterodimer composed of two subunits, calcineurin A (61 kDa) and calcineurin B (19 kDa). Limited digestion of calcineurin with clostripain, a sulfhydryl protease from *Clostridium histolyticum*, and analysis of the functional activities associated with cleavage products led to the conclusion that four distinct functional domains exist in the calcineurin A subunit. Catalytic (phosphatase) activity is associated with a small N-terminal domain that, in the absence of calmodulin, is inhibited by interaction with a small domain at the C-terminus. Two larger domains in the central region of the chain are involved in binding of calmodulin and interactions with calcineurin B. The relative rates of cleavage at the sites linking these domains vary considerably, possibly reflecting relative accessibility; as a result, proteolysis occurs in a stepwise manner, facilitating interpretation of the results. Binding of calmodulin differentially alters the cleavage rates at the distinct sites.

Clostripain has not been widely used in studies of this type. It is instructive to note that previous limited proteolysis studies employing trypsin to digest calcineurin A had not led to isolation of fragments with retention of function. Though clostripain and trypsin both cleave at linkages involving basic amino acids, clostripain exhibits a preference for arginine. This suggests that the lack of success with trypsin might result from cleavage at functionally important (either directly or indirectly) lysine residues, and again illustrates the value of exploring the use of different proteases in studies of this type.

Lim et al. (150) deduced the amino acid sequence of yeast pyruvate carboxylase from the nucleotide sequence of the cloned gene. This enzyme binds α-keto acids (as both substrate and product), and catalyzes an ATP-dependent carboxylation using biotin as a cofactor. Noting sequence similarities between specific regions of pyruvate carboxylase and limited segments of other enzymes binding α-keto acids, ATP, and biotin, and assuming that ligands were bound to discrete domains (see comments above), Lim et al. proposed a domain structure for pyruvate carboxylase. Assuming that cleavage occurs at intra-domain segments, the limited digestion pattern seen with several different proteases is consistent with the predicted disposition of domains within the overall sequence (Fig. 9); the fragment predicted to represent the biotin binding domain does, in fact, contain the biotin cofactor.

It is interesting to note that, in the study of Lim et al., limited proteolysis was used to test a domain structure *predicted* from amino acid sequence (cf., examples above in which domain structure was deduced from limited proteolysis results with little or no sequence information); given the ever increasing number of sequences deduced from cloned cDNA or genes, such predictions may become a more frequent occurrence (e.g., 151, 151a). Indeed, analysis of functions associated with truncated proteins obtained by expression of limited segments of cloned DNA offers an alternative route to determining domain structure (152); where feasible, limited proteolysis experiments could serve as

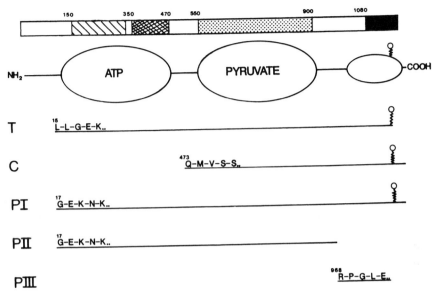

Fig. 9. Proposed domain structure of yeast pyruvate carboxylase. The domain structure was *predicted*, based on observed similarities between the amino acid sequence of various regions of pyruvate carboxylase and limited segments of other enzymes with which specific functions such as binding of ATP, α-keto acids, or biotin had been associated. Subsequent limited proteolysis experiments showed that cleavage occurred at predicted interdomain segments, which are expected to be particularly susceptible to proteolytic attack. Fragments generated by digestion with trypsin (T), chymotrypsin (C), or proteinase K (PI, PII, and PIII) are depicted schematicaly below the proposed domain structure. [Reprinted, with permission, from (150).]

a useful guide as to which truncated versions might be most likely to correspond to domains. It might also be noted that the manifestation of function (dependent on structure) in truncated proteins provides additional evidence for the independent folding of domains.

Limited proteolysis has also been extremely useful in dissecting the structural and functional organization of membrane-bound proteins. The work of Tajima and colleagues on cytochrome b_5 and cytochrome b_5 reductase (153, 154) provides an excellent example. Proteolytic digestion results in cleavage of these proteins into two functional domains, a C-terminal domain involved in binding to the microsomal membrane and an N-terminal domain with which catalytic function is associated. The observation of two distinct phases in the guanidine-induced denaturation of these proteins provides additional evidence for the two domain structure.

Brown and Black (155) compared the tryptic cleavage pattern of several cytochromes P-450, and propose a structure in which the proteins are tethered to the membrane by a short (one or two transmembrane helices) N-terminal segment, with the bulk of the protein existing as a C-terminal domain extending into the cytoplasm (an inversion of the situation with cytochrome b_5 and cytochrome b_5 reductase, discussed above). This structure differs markedly from one based on predicted secondary structure and hydrophobicity analysis of the amino acid sequence (156), not the first time for conflict between such predictions and experimental approaches (e.g., 157).

The recent studies of Waugh et al. (158) and Papp et al. (159) provide further examples of the utility of limited proteolysis in deducing the structural and functional organization of membrane proteins. In the first case, radiolabeled insulin was chemically cross-linked to the insulin receptor, which was then dissected, with each of three different proteases giving rise to the same 55-kDa radiolabeled fragment. These results, together with N-terminal sequence analysis and reactivity of the 55-kDa fragment with antibodies raised against a peptide representing a defined region of the α-subunit sequence, enabled Waugh et al. (158) to deduce the region of the receptor likely to be involved in binding of insulin. Papp et al. (159) examined the effects of limited proteolysis on various functions of the Ca^{2+} pump in the red cell membrane. Analysis of cleavage patterns (facilitated by immunoblotting with monoclonal antibodies) led to development of a model depicting the distribution of various functions within the overall sequence. Papp et al. (159) propose the existence of a 15-kDa C-terminal segment whose removal results in activation of the enzyme even in the absence of calmodulin. It is interesting to note that an analogous C-terminal *inhibitory domain* is found in calcineurin A (149), discussed above, and in myosin light chain kinase (159a), though the order (with respect to the N- to C-terminal sequence) of the inhibitory and calmodulin-binding domains may be reversed in these two proteins. Is this coincidence, or could an inhibitory domain in the C-terminal region be a common feature of calmodulin-regulated enzymes?

The above examples are success stories. Limited proteolysis does not always lead to such clear-cut definition of structural and functional regions. A case in point is the recent study of the plasma cholesterol ester transfer protein (CETP) by Hesler et al. (160). Seeking to determine whether CETP activity might be associated with some smaller domain of this 74-kDa protein, Hesler et al. quite reasonably turned to limited proteolysis experiments. Although cleavage with a variety of proteases can be demonstrated (i.e., small fragments are seen after electrophoresis under denaturing conditions), these remain tenaciously associated under nondenaturing conditions and no smaller catalytically active fragment has been isolated. However, this protein represents an atypical case. Its remarkable stability (resistant to 8 M urea!) suggests an unusually tight structure with extensive intramolecular interactions, which would not favor chances of isolating discrete units with retention of function.

3.4. Limited Proteolysis as a Probe for Conformational Change

The utility of limited proteolysis depends on accessibility of relatively few scissile bonds, and conversely, the protection of other potential cleavage sites.

Accessibility and protection are relative, and depend directly on structure. If the latter is changed, limited proteolysis may be a simple and powerful method for detecting it. It is thus not surprising that this method has frequently found a place in studies of conformational changes associated with function. If the points of cleavage have been (or can be) identified, as in many of the studies discussed above, the specific segments of the molecule affected can be defined. In that sense, limited proteolysis results may be more informative than, for example, changes in absorbance or fluorescence properties, which typically are a composite reflecting the environment at several absorbing or fluorescing residues within the molecule.

The recent study of Liu et al. (161) provides an example of the effect that ligand binding can have on proteolysis, as well as further illustrating the utility of this method in providing information about the structure of membrane proteins. The receptor for atrial natriuretic factor (ANF) has an intrinsic guanylate cyclase activity. Tryptic digestion of the 130-kDa receptor generates a 70-kDa fragment that remains associated with the membrane and retains the ability to bind ANF, but the cyclase activity is lost. In the presence of ANF, cleavage to form the 70-kDa fragment is enhanced, and the resistance of this fragment to further tryptic digestion is increased. Based on these results, Liu et al. (161) propose that the ANF receptor is composed of (at least) two domains, a 70-kDa ANF-binding domain and a 60-kDa guanylate cyclase domain, linked by a membrane spanning segment and a protease-sensitive site (in addition to trypsin, chymotrypsin and endoproteinase Glu-C also cleaved in this region). It is suggested that binding of ANF induces conformational changes that increase susceptibility to cleavage at the linking segment but tighten the structure of the 70-kDa ANF-binding domain, leading to increased resistance to proteolytic attack.

Conformational changes of a more dynamic nature can also be detected by limited proteolysis experiments. Based on altered sensitivity at defined cleavage sites in the presence of various substrates or inhibitors of transport, Gibbs et al. (162) were able to detect conformational changes associated with function of the erythrocyte glucose transport protein. In the present context, it is worth noting that the disposition of this protein in the membrane had previously been deduced based on limited tryptic digestion experiments and reactivity with antibodies generated against peptides representing defined regions of the sequence (see references cited by Gibbs et al.).

Inaba and Mohri (163) attributed altered proteolysis patterns seen with flagellar dynein to conformational changes during the catalytic cycle, and related these to the mechanism by which dynein converts the energy released by hydrolysis of ATP into the mechanical force responsible for flagellar motion.

Betton et al. (164) used proteolysis methods to detect and characterize intermediates in the denaturant-induced unfolding of horse muscle phospho-glycerate kinase. This 45-kDa enzyme consists of two domains of nearly equal size (165). In the absence of denaturants, the enzyme is resistant to proteolysis with *S. aureus* V8 protease, but in the presence of relatively low concentrations (0.7 M) of guanidine, cleavage occurs in the region linking the domains. A fragment resistant to further proteolysis and corresponding to the still partially folded C-terminal domain can be isolated. Betton et al. concluded that the latter fragment is very similar to a previously characterized (166) intermediate in the folding of the enzyme, and thus propose that folding proceeds through an intermediate in which the C-terminal domain is largely folded while the N-terminal domain lacks significant ordered structure (as reflected by its great susceptibility to proteolysis).

Two further comments might be made about the study by Betton et al. First, the inability to proteolytically cleave the segment linking the two domains in the absence of structural perturbations by low guanidine concentrations again illustrates that interdomain regions are not inevitably accessible to attack, and reflects the existence of strong interdomain interactions known to occur in phosphoglycerate kinase (165); as discussed above, similar results are seen with rat brain hexokinase (114). Second, although horse muscle phosphoglycerate kinase is resistant to proteolysis in the absence of denaturants (164), this is not the case with the pig muscle enzyme (167), which is extensively proteolyzed (cleavage *not* limited to interdomain segment!) under similar conditions. Thus limited proteolysis may detect subtle differences in the structure of even closely related homologous enzymes. The study by Jiang and Vas (167) also provides another striking example of the effects that ligand binding can have on proteolytic cleavage patterns.

3.5. Monoclonal Antibodies as Probes for Functional Sites

The above comments have focused on the utility of limited proteolysis in defining structural and functional domains and the usefulness of monoclonal antibodies in interpreting such studies. Still another example is the study of Yurchenco et al. (168) in which immunoblotting experiments, with monoclonal antibodies recognizing epitopes in each of the five domains of the α-subunit of spectrin, provided a means for confirming the alignment and distinct identity of these domains, as well as permitting the identification of smaller fragments arising from further proteolysis of these domains. Generally speaking, antibodies are used in such studies to identify proteolytic fragments by immunoblotting techniques. In the present section, we wish to turn to some examples in which the primary utility of the monoclonal antibody has been not as an aid to identification but as a probe for a specific function.

Before proceeding, a few additional comments on the underlying concepts seem appropriate. The basic idea behind most studies employing antibodies as probes for function is that binding of the antibody at or near a region of the protein critical to function is likely to disrupt that function. Most investigators would probably consider this a reasonable expectation, but it is not, of course, the only possible explanation for such effects of antibody binding. Changes in conformation of the protein may result from binding of the antibody, though the extent of such changes is likely to vary greatly with the particular protein and antibody in question (81). Thus, effects on function might be secondary, that is, the antibody binds and induces a conformational change that perturbs structure and function at a site far removed from the antibody binding site. Such effects could surely confuse matters, but fortunately they do not seem to be a common problem. The basis for the latter statement is that the percentage of monoclonal antibodies that actually affect function is relatively small (21), and when binding of an antibody does affect function, the effects are usually quite restricted in nature. This has been seen over and over with many proteins. A reasonable conclusion would be that, even if binding of the antibody does induce conformational changes, they are generally not so extensive that they grossly affect structure and dependent function. As noted above, a similar complication might attend competitive epitope mapping experiments, but here, too, this does not appear to be a common problem, with mutually exclusive antibodies frequently mapping to the same limited region of the sequence. Thus, although long-range conformational effects surely may affect binding of antibodies (169) and the possibility of such indirect effects should always be kept in mind, we believe that results obtained with numerous proteins justify a straightforward interpretation unless there is evidence to the contrary.

Monoclonal antibodies were used by Pfeiffer et al. (99) to distinguish topographicaly distinct sites involved in function of asparagine synthetase. This enzyme catalyzes the ATP-dependent synthesis of asparagine from aspartate, with either glutamine or NH_3 as nitrogen source; it also exhibits glutaminase activity (which provides the glutamine-derived nitrogen for asparagine synthesis). Three monoclonal antibodies, comprising complementation group I, inhibit asparagine synthesis from either glutamine or NH_3; in competitive mapping studies, these antibodies exhibit mutually exclusive binding, indicating that their epitopes are in close proximity. Based on competitive mapping, these epitopes are distinct (i.e., independent binding is observed) from that recognized by another antibody, the sole member of complementation group II, which inhibits glutaminase activity and asparagine synthesis with glutamine as nitrogen sources but not NH_3-linked synthesis. Pfeiffer et al. (99) conclude that the glutaminase site is topologically distinct from the site at which asparagine synthesis occurs. A third group of antibodies (complementation group III) map to a site discrete from either of the catalytic sites; these have no effect on activity.

Similar studies have been useful in defining functional regions of the photoreceptor guanyl nucleotide binding protein, transducin (170, 171); selectively perturbing functions associated with the catalytic and regulatory domains of protein kinase C (172, 172a), discrete domains of kininogens (173), or discrete subunits of phosphorylase kinase (174); discriminating between the *homologous matching* and *processive unwinding* functions of the recA protein (174a); and determining regions of IL-1β that are critical for its biological function (stimulation of thymocyte proliferation and PGE_2 release from fibroblasts) (175).

Morgan and Roth (176) have mapped epitopes recognized by several monoclonal antibodies against the human insulin receptor. Antibodies reactive with the insulin binding site in the extracellular domain block insulin binding but have little or no effect on the tyrosine kinase activity associated with the receptor. However, several monoclonal antibodies recognizing epitopes in the cytoplasmic domain do affect kinase activity. Competitive mapping studies permit division of the latter antibodies into four groups (members of each group showing mutually exclusive binding). However, there is no apparent correlation between membership in these groups and effects on tyrosine kinase function, indicating the existence of complexities apparently not present in the examples discussed above. Thus, the study of Morgan and Roth serves to illustrate that, despite the many successes, monoclonal antibodies are not infallible guides to structure−function relationships. What experimental approach *is* infallible?

3.6. Monoclonal Antibodies as Probes for Functionally Important Conformational Changes

As discussed earlier, it is likely that many, probably most, epitopes are discontinuous in nature and hence binding of the respective antibodies is intrinsically sensitive to conformation of the protein. The same might be true for an antibody recognizing a continuous epitope if the accessibility of the antigenic segment were altered by conformational change. It is thus not surprising that monoclonal antibodies have frequently been found useful in monitoring functionally relevant conformational changes. Although the conformational change reflected by antibody binding obviously involves the epitopic region, it is not necessarily the case that the event instigating the conformational change occurs directly in this same region. For example, chemical modification (169) or site-directed mutagenesis (177) of cytochrome c may affect binding of antibodies to epitopes at a considerable distance from the altered residue. Indeed, a major virtue of monoclonal antibodies is that they may be useful in assessing rather wide ranging effects on conformation. If the location of the epitope within the overall structure is known, conformational changes involving this defined region of the molecule may be detected.

Wakabayashi et al. (178) generated monoclonal antibodies against human protein C, a vitamin K-dependent enzyme involved in blood coagulation. Three of these antibodies react with the protein only in the presence of Ca^{2+} (Fig. 10) or nonphysiological analogues, Zn^{2+} or Tb^{3+}. These antibodies map to the light chain of protein C, and react only with a Ca^{2+}-induced conformation of the γ-carboxyglutamic acid-containing domain in this chain. Ca^{2+}-dependent cross-reactivity of these antibodies with other vitamin K-dependent proteins (prothrombin and factor X) suggest conservation (and hence functional importance) of a Ca^{2+}-induced conformation in the γ-carboxyglutamic acid domain of these proteins also.

The study by Dixit et al. (179) offers an interesting contrast to that of Wakabayashi et al. (178). Screening of monoclonal antibodies generated against human thrombospondin resulted in isolation of two antibodies, designated

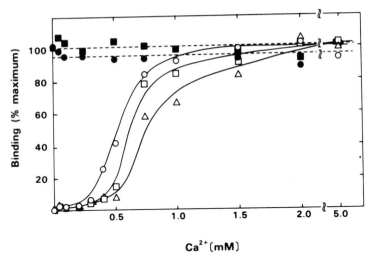

Fig. 10. Effect of Ca^{2+} on binding of various monoclonal antibodies to protein C. Binding of ^{125}I-labeled monoclonal antibodies to protein C, immobilized in the wells of a microtiter plate, was determined as a function of increasing $[Ca^{2+}]$. Binding of antibodies designated as JTC-1 (○), JTC-2 (□), and JTC-3 (△) was dependent on addition of Ca^{2+}, while this ion had no effect on binding of antibodies JTC-4 (●) or JTC-5 (■). [Reprinted, with permission, from (178).]

A6.1 and D4.6, that bind (preferentially or exclusively) to the protein only in the *absence* of Ca^{2+}, indicating that binding of Ca^{2+} induces a conformational change that adversely affects recognition of the epitopes. Based on the relationship between antibody binding and $[Ca^{2+}]$, two distinct sites can be detected. Binding of antibody D4.6 is diminished by binding of Ca^{2+} to a high affinity site (K_d less than 0.1 μM) while binding of A6.1 is decreased only at considerably higher (≈ 50 μM) Ca^{2+} levels. That these antibodies recognize discrete epitopes was confirmed by limited proteolysis and epitope mapping studies which demonstrated that the epitopes are located in different domains of the protein.

Phytochrome is a protein involved in modulating a variety of light-regulated developmental responses in plants. The protein exists in two forms. The *red-absorbing* and *far-red-absorbing* forms are interconvertible by light; that is, when either absorbs light, it is converted to the other. Cordonnier et al. (180) generated monoclonal antibodies that exhibit different affinities for the two forms; limited proteolysis and epitope mapping studies suggest that conformational change in an N-terminal domain results from the light-induced interconversion. These studies have been extended (181) with these and additional monoclonal antibodies, providing further information about the domains influenced by light-dependent conformational changes; as might be anticipated, conformational change in the domain containing the bound chromophore is also observed, in addition to the previously noted effects on the N-terminal domain.

Positive cooperativity is a kinetic effect closely linked with conformational changes (182). A discussion of the various hypotheses proposed to explain this behavior (183) is not germane here, but typically they include the postulated existence of an enzyme in (at least) two distinct conformations, differing in affinity for the substrate. Stabilization of a high-affinity form would be predicted to convert the sigmoid velocity versus [substrate] curve into classical hyperbolic Michaelis–Menten kinetic behavior. Phenylalanine hydroxylase exhibits positive cooperativity in binding of the substrate, phenylalanine. Parniak et al. (184) have described a monoclonal antibody that binds to the enzyme only after incubation with phenylalanine and converts the kinetics from sigmoid to hyperbolic. This effect might be explained by selective recognition, and hence stabilization, of the high-affinity form of the enzyme induced by the binding of substrate. The effects of the antibody are not limited to that, however, since binding of the antibody also caused some inhibition, as reflected by a decreased V_{max} value.

Folding of a protein to give the final tertiary, and in some cases quaternary, structure obviously involves conformational changes that might be detected with appropriate monoclonal antibodies. Morris (185) described a monoclonal antibody that recognizes a conformationally sensitive epitope in creatine kinase. A clever combination of chemical and proteolytic modification procedures enabled him to deduce that the epitope is formed by folding of segments located within the C-terminal region of the enzyme. Furthermore, the reactivity of this antibody with an intermediate detected during the refolding of the urea-denatured protein indicates that refolding of the C-terminal region precedes folding of the remainder of the molecule. This conclusion is further supported by proteolysis experiments which demonstrate that, in the partially refolded intermediate, the C-terminal half of the molecule is resistant to trypsin while the N-terminal half is readily digested.

Conformationally sensitive monoclonal antibodies are also useful in detecting conformational changes resulting from interdomain interactions in the β_2 subunit of tryptophan synthase (186) or from binding of ligands or interaction with the α subunit of this enzyme (187).

Riftina et al. (188) used altered reactivity with monoclonal antibodies against the α subunit of *Escherichia coli* RNA polymerase to detect conformational changes in the α subunit during assembly of the core polymerase. Although two chemically equivalent α subunits are present in the assembled polymerase, they are topologically distinct, the epitopes on one being masked while remaining accessible on the other. These and other antibody binding experiments permitted Riftina et al. (188) to make further suggestions concerning topological relationships between the various subunits comprising the core polymerase, and the route by which they are assembled.

As a final example of the use of monoclonal antibodies to detect conformational changes, we choose one quite different in character from the studies briefly described above. A common feature in these has been that the binding of the antibody itself is affected by conformational change. The study by Holowka et al. (189) illustrates quite a different strategy. Binding of IgE to its receptor has been suggested to require "bending" at a hinge region between the Fab and Fc regions of the IgE molecule. Holowka et al. tested this postulate using monoclonal antibodies (actually, Fab fragments derived from them) that recognize epitopes on these discrete regions. Labeling one of the antibodies with a fluorescent donor and the other with an acceptor provides a means to measure the distance between donor and acceptor (and thus between Fab and Fc regions) by resonance energy transfer. Decreased energy transfer is seen with the receptor-bound IgE, consistent with (but not definite proof of) the postulated bending.

Holowka et al. (189) suggest that this approach might have general utility in studying other macromolecular complexes on the cell surface. We suggest that its utility might not be restricted to membrane-bound complexes. As discussed above, competitive mapping studies can provide a reasonable representation of the distribution of epitopes on a molecular surface; if direct epitope mapping methods provide landmarks for interpretation of such maps, this distribution may be directly related to specific structural features. It seems reasonable to expect that many conformational changes might result in significant alteration of the distance between surface regions, and hence be detected by resonance energy transfer between suitably chosen monoclonal antibodies (or more likely, Fab fragments derived from them). Since fluorescence measurements can be made with both high sensitivity and rapidity, this might permit real-time measurements and hence a direct study of the kinetics of conformational changes.

3.6. Utility of Monoclonal Antibodies in Determining the Topology of Membrane-Bound Proteins

Unlike a protein in solution, a membrane-bound protein can be considered to have two surfaces, representing regions exposed on either side of the membrane. The impermeability of the membrane to macromolecular reagents (or to most small hydrophilic reagents) provides a means to distinguish the two sides, assuming that the reagent can be selectively introduced in an asymmetric manner. We have already noted the utility of proteases in detection of exposed cleavage sites in membrane-bound proteins. Reaction of monoclonal — or polyclonal (e.g., 190, 191) — antibodies with epitopes selectively accessible on one side of the membrane represents a similar, and frequently exploited, approach for determining the topology of membrane proteins. Examples include studies of rhodopsin (192), the Ca^{2+} pump of the red cell membrane (159, 193), the major outer membrane protein (MOMP) of *Chlamydia trachomatis* (194), the renal Na^+-linked glucose transporter (195), and the nicotinic acetylcholine receptor (196, 197). The results of the latter study apparently remain controversial, and the critique by McCrea et al. (157) raises some caveats meriting general consideration in such studies.

A study of the renal Na^+-linked glucose transporter (195) provides additional insight into membrane transport processes as a result of an effect discussed above, namely, the sensitivity of the binding of some monoclonal antibodies to ligand-induced conformational changes affecting their epitopes. In the presence of Na^+, the binding of eight different monoclonal antibodies is affected by glucose, suggesting that binding of glucose induces rather extensive changes in conformation of the transporter. The binding of several of these monoclonal antibodies is also affected in the presence of substrates (lactate and glutamate) transported by other Na^+-linked carriers, and one of the antibodies affects the transport of both glucose and glutamate. Mamelok et al. (198) report similar results with a monoclonal antibody that inhibits Na^+-linked transport of glucose, alanine, and glutamate. Although the molecular basis for these effects remains unclear, they may reflect *functional coupling* between different Na^+-linked transport systems (195).

Finally, we close this section with a brief discussion of another study that made novel use of monoclonal antibodies in a study of a membrane protein, the pore forming PhoE protein of *E. coli.* van der Ley et al. (199) used monoclonal antibodies recognizing epitopes on PhoE that are exposed on the extracellular surface. Cells containing surface bound antibodies become susceptible to lysis by the complement system, providing a method for selection of mutants in which the epitopes have been modified with resulting loss of antibody recognition. By analysis of mutant PhoE sequences, residues affecting the epitopes can be detected and, hence, surface-exposed regions of the amino acid sequence deduced. Three such positions (Arg-201, Gly-238, and Gly-275) were found; surface exposure of Arg-158 had previously been indicated by another experimental approach (200). Noting the regular spacing of these residues, approximately 40 residues apart, van der Ley et al. suggest that the protein exists as an antiparallel β sheet within the membrane (Fig. 11), an arrangement that would result in exposure of hydrophilic segments at the membrane surface with the observed frequency.

4. SOME CLOSING COMMENTS

One could hardly imagine a technique simpler to use than limited proteolysis. Preliminary experiments typically include some trial runs to assess reasonable protease:protein ratios, or to examine the relative virtues of different proteases for the problem of interest. A few brief incubations, perhaps some activity assays, and an SDS–gel or two, and one could be on the way to profitable use of this method. In contrast, generation and characterization of monoclonal antibodies is technically more demanding, labor-intensive, and a relatively expensive undertaking; it is likely to require at least a few months (including time required for immunization) before even preliminary assessment of success

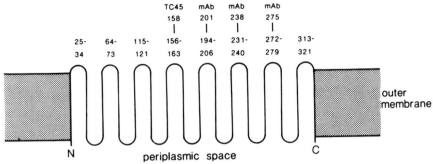

Fig. 11. Proposed model for the topology of PhoE protein in the outer membrane of *E. coli*. The numbers refer to position in the amino acid sequence of the mature PhoE protein. Residues critical for binding of various monoclonal antibodies are located at positions 201, 238, and 275; reaction of these antibodies with intact cells indicates that these residues are located on the extracellular surface of the membrane. Surface exposure of residue 158 (an arginine) was deduced from other experiments. The proposed topology was based on the observed periodicity of exposed residues (approximately 40 residue intervals), together with the hydrophilic character of these regions, consistent with a surface location. [Reprinted, with permission, from (199).]

can be made. Despite the considerable investment required, our experience has certainly been that it is worth it. These unique biological reagents have proven useful in ways we never anticipated at the time we initiated this work.

Both monoclonal antibodies and proteases (under the appropriate conditions) possess a valued quality: specificity. The ability to *selectively* modify a discrete site on a protein and then evaluate the consequences, or to selectively dissect the molecule, is a powerful approach in elucidation of structure—function relationships. The applications that have been discussed here are meant not to indicate limits, but rather to illustrate the tremendous variety of ways in which monoclonal antibodies and limited proteolysis have been fruitfully employed. It is not always the technique that imposes limits, but rather the ingenuity with which it is applied.

ACKNOWLEDGMENTS

The author's interest in applications of monoclonal antibodies and limited proteolysis methods results directly from the usefulness of these techniques in studies of brain hexokinase that have been conducted in this laboratory. It is a pleasure to acknowledge the efforts of colleagues who have been involved in these studies, especially Ken Finney, Paul Polakis, Al Smith, and Tracy White. This work has been supported by NIH Grant NS 09910.

References

1. J.S. Richardson, *Adv. Protein Chem.* **34**, 167–339 (1981).

2. C. Chothia, *Annu. Rev. Biochem.* **53**, 537–572 (1984).

3. T.E. Creighton, *Proteins: Structures and Molecular Principles*, Freeman, New York, 1983.

4. K. Struhl, *Trends Biochem. Sci.* **14**, 137–140 (1989).

5. W.H. Landschulz, P.F. Johnson, and S.L. McKnight, *Science* **243**, 1681–1688 (1989).

6. R.M. Evans and S.M. Hollenberg, *Cell* **52**, 1–3 (1988).

7. M.S. Lee, G.P. Gippert, K.V. Soman, D.A. Case, and P.E. Wright, *Science* **245**, 635–637 (1989).

8. R.G. Brennan and B.W. Matthews, *J. Biol. Chem.* **264**, 1903–1906 (1989).

9. J.J. Birktoft and L.J. Banaszak, in *Peptide and Protein Reviews*, Vol. 4, M.T.W. Hearn, Ed., Dekker, New York, 1984, pp. 1–46.

10. S.S. Taylor, *J. Biol. Chem.* **264**, 8443–8446 (1989).

11. M. Gribskov and D. Eisenberg, in *Techniques in Protein Chemistry*, T.E. Hugli, Ed., Academic, New York, 1989, pp. 108–117.

11a. R.K. Wierenga, P. Terpstra, and W.G.J. Hol, *J. Mol. Biol.* **187**, 101–107 (1986).

11b. P. Bork, *FEBS Lett.* **257**, 191–195 (1989).

12. N.D. Clarke, D.C. Lien, and P. Schimmel, *Science* **240**, 521–523 (1988).

13. S.K. Hanks, A.M. Quinn, and T. Hunter, *Science* **241**, 42–52 (1988).

14. I.T. Weber, T.A. Steitz, J. Bubis, and S.S. Taylor, *Biochemistry* **26**, 343–351 (1987).

15. M. Seville, M.G. Vincent, and K. Hahn, *Biochemistry* **27**, 8344–8349 (1988).

16. A. Wlodawer, M. Miller, M. Jaskolski, B.K. Sathyanarayana, E. Baldwin, I.T. Weber, L.M. Selk, L. Clawson, J. Schneider, and S.B.H. Kent, *Science* **245**, 616–621 (1989).

17. S.O. Smith, S. Farr-Jones, R.G. Griffin, and W.W. Bachorchin, *Science* **244**, 961–964 (1989).

18. A. Bax, *Annu. Rev. Biochem.* **58**, 223–256 (1989).

19. M. Karplus and J.A. McCammon, *Annu. Rev. Biochem.* **53**, 263–300 (1983).

20. G.A. Petsko and D. Ringe, *Annu. Rev. Biophys. Bioeng.* **13**, 331–371 (1984).

21. J.E. Wilson, in *Chemical Modification of Enzymes*, J. Eyzaguirre, Ed., Ellis Horwood, Chichester, 1987, pp. 171–181.

22. K.U. Linderström-Lang and J.A. Schellman, in *The Enzymes*, 2d ed., Vol. 1 P.D. Boyer, Ed., Academic, 1959, pp. 443–510.

23. C.B. Anfinsen, *Science* **181**, 223–230 (1973).

24. P.Y. Chou and G.D. Fasman, *Annu. Rev. Biochem.* **47**, 251–276 (1978).

25. W. Kabsch and C. Sander, *FEBS Lett.* **155**, 179–182 (1983).

26. G.E. Schulz, *Annu. Rev. Biophys. Biophys. Chem.* **17**, 1–21 (1988).

27. L.G. Presta and G.D. Rose, *Science* **240**, 1632–1641 (1988).

28. J.S. Richardson and D.C. Richardson, *Science* **240**, 1648–1652.

29. S.T. Rao and M.G Rossmann, *J. Mol. Biol.* **76**, 241–256 (1973).

30. M. Levitt and C. Chothia, *Nature* **261**, 552–558 (1976).

31. W.C.G. Hol, *Prog. Biophys. Mol. Biol.* **45**, 149–195 (1985).

32. W.R. Taylor and J.M. Thornton, *J. Mol. Biol.* **173**, 487–514 (1984).

32a. G.E. Schulz, R.H. Schirmer, W. Sachsenheimer, and E.F. Pai, *Nature* **273**, 120–124 (1978).

33. G.D. Rose, *J. Mol. Biol.* **134**, 447–470 (1979).

34. D.B. Wetlaufer, *Proc. Natl. Acad. Sci. USA* **70**, 697–701 (1973).

35. D.B. Wetlaufer, *Adv. Protein Chem.* **34**, 61–92 (1981).

36. G.M. Crippen, *J. Mol. Biol.* **126**, 315–332 (1978).

37. M.H. Zehfus and G.D. Rose, *Biochemistry* **25**, 5759–5765 (1986).

38. M.H. Zehfus, *Proteins: Structure, Function, and Genetics* **2**, 90–110 (1987).

39. R.L. Baldwin, *Trends Biochem. Sci.* **11**, 6–9 (1986).

40. T.E. Creighton, *Proc. Natl. Acad. Sci. USA* **85**, 5082–5086 (1988).

41. P.E. Wright, H.J. Dyson, and R.A. Lerner, *Biochemistry* **27**, 7167–7175 (1988).

42. J.B. Udgaonkar and R.L. Baldwin, *Nature* **335**, 694–699 (1988).

43. H. Roder, G.A. Elove, and S.W. Englander, *Nature* **335**, 700–704 (1988).

44. B. Adams, R.J. Burgess, and R.H. Pain, *Eur. J. Biochem.* **152**, 715–720 (1985).

45. J.N. Higaki and A. Light, *J. Biol. Chem.* **261**, 10606–10609 (1986).

46. S. Blond and M. Goldberg, *Proteins: Structure, Function, and Genetics* **1**, 247–255 (1986).

47. S. Blond and M. Goldberg, *Proc. Natl. Acad. Sci. USA* **84**, 1147–1151 (1987).

48. A.M. Beasty, M.R. Hurle, J.T. Manz, T. Stackhouse, J.J. Onuffer, and C.R. Matthews, *Biochemistry* **25**, 2965–2974 (1986).

49. M.R. Hurle, G.A. Michelotti, M.M. Crisanti, and C.R. Matthews, *Proteins: Structure, Function, and Genetics* **2**, 54–63 (1987).

50. W. Gilbert, *Nature* **271**, 501 (1978).

51. C.C.F. Blake, *Nature* **273**, 267 (1978).

52. T.W. Traut, *Proc. Natl. Acad. Sci. USA* **85**, 2944–2948 (1988).

53. M.G. Rossman, A. Liljas, C.I. Branden, and L.J. Banaszak, in *The Enzymes*, 3d ed., Vol. 11, P.D. Boyer, Ed., Academic, New York, pp. 61–102.

54. Keim, R.L. Heinrikson, and W.M. Fitch, *J. Mol. Biol.* **151**, 179–197 (1981).

55. C. Chothia and A.M. Lesk, *EMBO J.* **5**, 823–826 (1986).

56. J.W. Donovan, *Trends Biochem. Sci.* **9**, 340–344 (1984).

57. P.L. Privalov, *Annu. Rev. Biophys. Biophys. Chem.* **18**, 47–69 (1989).

58. J.F. Brandts, C.Q. Hu, L.-N. Lin, and M.T. Mas, *Biochemistry* **28**, 8588–8596 (1989).

59. B. Tancini, P. Dominici, M. Simmaco, M.E. Schinina, D. Barra, and C.B. Voltattorni, *Arch. Biochem. Biophys.* **260**, 569–579 (1988).

60. S. de la Viña, D. Andreu, F.J. Medrano, J.M. Nieto, and J.M. Andreu, *Biochemistry* **27**, 5352–5365 (1988).

61. T. Moreau, N. Gutman, D. Faucher, and F. Gauthier, *J. Biol. Chem.* **264**, 4298–4303 (1989).

62. A. Fontana, G. Fassina, C. Vita, D. Dalzoppo, M. Zamai, and M. Zambonin, *Biochemistry* **25**, 1847–1851 (1986).

63. J. Miller, A.D. McLachlan, and A. Klug, *EMBO J.* **4**, 1609–1614 (1985).

64. G. Kohler and C. Milstein, *Nature* **256**, 495–497 (1975).

65. D.C. Benjamin, J.A. Berzofsky, J.J. East, F.R.N. Gurd, C. Hannum, S.J. Leach, E. Margoliash, J.G. Michael, A. Miller, E.M. Prager, M. Reichlin, E.E. Sercarz, S.J. Smith-Gill, P.E. Todd, and A.C. Wilson, *Annu. Rev. Immunol.* **2**, 67–101 (1984).

66. J.A. Berzofsky, *Science* **229**, 932–940, (1985).

67. D.J. Barlow, M.S. Edwards, and J.M. Thorntoin, *Nature* **322**, 747–478 (1986).

68. H.J. Dyson, R.A. Learner, and P.E. Wright, *Annu. Rev. Biophys. Biophys. Chem.* **17**, 305–324 (1988).

69. M.H.V. van Regenmortel and G.D. de Marcillac, *Immunol. Lett.* **17**, 95–108 (1988).

70. J.E. Wilson and A.D. Smith, *J. Biol. Chem.* **260**, 12838–12843 (1985).

71. J.H. Nunberg, G. Rodgers, J.H. Gilbert, and R.M. Snead, *Proc. Natl. Acad. Sci. USA* **81**, 3675–3679 (1984).

72. V. Mehra, D. Sweetser, and R.A. Young, *Proc. Natl. Acad. Sci. USA* **83**, 7013−7017 (1986).

73. F. Lorenzo, A. Jolivet, H. Loosfelt, M. Thu Vu Hai, S. Brailly, and M. Perrot-Applanat, *Eur. J. Biochem.* **176**, 53−60 (1988).

74. J.B. Miller, S.B. Teal, and F.E. Stockdale, *J. Biol. Chem.* **264**, 13122−13130 (1989).

75. M.J. Crumpton, in *Synthetic Peptides as Antigens* (CIBA Symposium 119), Wiley, Chichester, 1986, pp. 93−106.

76. M.H.V. van Regenmortel, D. Altschuh, and A. Klug, in *Synthetic Peptides as Antigens* (CIBA Symposium 119), Wiley, Chichester, 1986, pp. 76−92.

77. S.D. Dunn, *Anal. Biochem.* **157**, 144−153 (1986).

78. H.-W. Birk and H. Koepsell, *Anal. Biochem.* **164**, 12−22 (1987).

79. H.M. Geysen, S.J. Rodda, and T.J. Mason, in *Synthetic Peptides as Antigens* (CIBA Symposium 119), Wiley, Chichester, 1986, pp. 130−149.

80. B.C. Cunningham, P. Jhurani, P. Ny, and J.A. Wells, *Science* **243**, 1330−1336 (1989).

81. D.R. Davies, S. Sheriff, and E.A. Padlan, *J. Biol. Chem.* **263**, 10541−10544 (1988).

81a. Z.Al. Moudallal, J.P. Briand, and M.H.V. van Regenmortel, *EMBO J.* **4**, 1231−1235 (1985).

82. B.A. Jameson, P.E. Rao, L.I. Kong, B.H. Hahn, G.M. Shaw, L.E. Hood, and S.B.H. Kent, *Science* **240**, 1335−1338 (1988).

83. R.S. Hodges, R.J. Heaton, J.M.R. Parker, L. Molday, and R.S. Molday, *J. Biol. Chem.* **263**, 11768−11775 (1988).

83a. J.-A. Fehrentz, A. Heitz, R. Seyer, P. Fulcrand, R. Devilliers, B. Castro, F. Heitz, and C. Carelli, *Biochemistry* **27**, 4071−4078 (1988).

84. R. Jemmerson, *Proc. Natl. Acad. Sci. USA* **84**, 9180−9184 (1987).

85. A.D. Smith and J.E. Wilson, *J. Immunol. Methods* **94**, 31−35 (1986).

86. A. Burnens, S. Demotz, G. Corradin, H. Binz, and H.R. Bosshard, *Science* **235**, 780−783 (1987).

87. H. Sheshberadaran and L.G. Payne, *Proc. Natl. Acad. Sci. USA* **85**, 1−5 (1988).

88. J. Lamy, J. Lamy, P. Billiald, P.-Y. Sizaret, G. Cavé, J. Frank, and G. Molta, *Biochemistry* **24**, 5532−5542 (1985).

89. E. Delain, M. Barray, J. Tapon-Bretaudiere, F. Pochon, P. Marynen, J.-J. Cassiman, H. Van den Berghe, and F. Van Leuven, *J. Biol. Chem.* **263**, 2981−2989 (1988).

90. S.J. Tzartos, D.E. Rand, B.L. Einarson, and J.M. Lindstrom, *J. Biol. Chem.* **256**, 8635−8645 (1981).

91. M. Ito, P.R. Pierce, R.E. Allen, and D.J. Hartshorne, *Biochemistry* **28**, 5567−5572 (1989).

92. W.J. Gullick, S. Tzartos, and J. Lindstrom, *Biochemistry* **20**, 2181−2191 (1981).

93. R. Mayne, R.D. Sanderson, H. Wiedemann, J.M. Fitch, and T.F. Linsenmayer, *J. Biol. Chem.* **258**, 5794−5797 (1984).

94. A.A. Kordossi and S.J. Tzartos, *EMBO J.* **6**, 1605−1610 (1987).

95. J.E. Wilson and A.D. Smith, *Anal. Biochem.* **143**, 179−187 (1984).

96. L. Djavadi-Ohaniance, B. Friguet, and M.E. Goldberg, *Biochemistry* **23**, 97−104 (1984).

97. D.O. Morgan and R.A. Roth, *Biochemistry* **25**, 1364−1371 (1986).

98. A.J. Dowding and Z.W. Hall, *Biochemistry* **26**, 6372−6381 (1987).

99. N.E. Pfeiffer, P.M. Mehlhaff, D.E. Wylie, and S.M. Schuster, *J. Biol. Chem.* **262**, 11565−11570 (1987).

100. H.H. Hogrefe, J.P. Griffith, M.G. Rossmann, and E. Goldberg, *J. Biol. Chem.* **262**, 13155−13160 (1987).

101. T.P. Hopp and K.R. Woods, *Proc. Natl. Acad. Sci.* **78**, 3824−3828 (1981).

102. J.M.R. Parker, D. Guo, and R.S. Hodges, *Biochemistry* **25**, 5425−5432 (1986).

103. J. Novotny, M. Handschumacher, E. Haber, R.E. Bruccoleri, W.B. Carlson, D.W. Fanning, J.A. Smith, and G.D. Rose, *Proc. Natl. Acad. Sci. USA* **83**, 226−230 (1986).

104. J.M. Thornton, M.S. Edwards, W.R. Taylor, and D.J. Barlow, *EMBO J.* **5**, 409−413 (1986).

105. G. Evin, F.-X. Galen, W.D. Carlson, M. Handschumacher, J. Novotny, J. Bouhnik, J. Menard, P. Corvol, and E. Haber, *Biochemistry* **27**, 156−164 (1988).

106. V. Krchnak, O. Mach, and A. Maly, *Anal. Biochem.* **165**, 200−207 (1987).

107. E. Westhof, D. Altschuh, D. Moras, A.C. Bloomer, A. Mondragon, A. Klug, and M.H.V. van Regenmortel, *Nature* **311**, 123−126 (1984).

108. P.A. Karplus and G.E. Schulz, *Naturwissenschaften* **72**, 212−213 (1985).

109. C.S. Craik, W.J. Rutter, and R. Fletterick, *Science* **220**, 1125−1129 (1983).

110. A.Z. Reznick, L. Rosenfelder, S. Shpund, and D. Gershon, *Proc. Natl. Acad. Sci. USA* **82**, 6114−6118 (1985).

111. J.E. Wilson, in *Regulation of Carbohydrate Metabolism*, Vol. 1, R. Beitner, Ed., CRC Press, Boca Raton, FL, 1985, pp. 45−85.

112. A.C. Chou and J.E. Wilson, *Arch. Biochem. Biophys.* **151**, 48−55 (1972).

113. P.G. Polakis and J.E. Wilson, *Arch. Biochem. Biophys.* **234**, 341−352 (1984).

114. D.A. Schwab and J.E. Wilson, *Proc. Natl. Acad. Sci. USA* **86**, 2563−2567 (1989).

115. K.G. Finney, J.L. Messer, D.L. DeWitt, and J.E. Wilson, *J. Biol. Chem.* **259**, 8232−8237 (1984).

116. M. Nemat-Gorgani and J.E. Wilson, *Arch. Biochem. Biophys.* **251**, 97−103 (1986).

117. D.M. Schirch and J.E. Wilson, *Arch. Biochem. Biophys.* **254**, 385−396 (1987).

118. T.K. White and J.E. Wilson, *Arch. Biochem. Biophys.* **274**, 375−393 (1989).

118a. P.G. Polakis and J.E. Wilson, *Arch. Biochem. Biophys.* **236**, 328−337 (1985).

118b. G. Xie and J.E. Wilson, *Arch. Biochem. Biophys.* **267**, 803−810 (1988).

119. J.E. Wilson, *Arch. Biochem. Biophys.* **185**, 88−99 (1978).

120. J.E. Wilson, *Arch. Biochem. Biophys.* **196**, 79−87 (1979).

121. M. Baijal and J.E. Wilson, *Arch. Biochem. Biophys.* **218**, 513−524 (1982).

122. T.K. White and J.E. Wilson, *Arch. Biochem. Biophys.* **259**, 402−411 (1987).

123. T.K. White and J.E. Wilson, *Arch. Biochem. Biophys.* **277**, 26−34 (1990).

123a. J.E. Wilson, in *Microcompartmentation*, D.P. Jones, Ed., CRC Press, Boca Raton, FL, 1988, pp. 171−190.

123b. R. Harrison, Ph. D. thesis, Yale University, New Haven, CT, 1985.

124. T. Ureta, A.D. Smith, and J.E. Wilson, *Arch. Biochem. Biophys.* **246**, 419−427 (1986).

125. T.K. White, J.Y. Kim, and J.E. Wilson, *Arch. Biochem. Biophys.* **276**, 510−517 (1990).

126. N.S. Cohen, C.-W. Cheung, and L. Raijman, *J. Biol. Chem.*, **262**, 203−208 (1987).

127. E. Wawra, *J. Biol. Chem.* **263**, 9908−9912 (1988).

128. P.A. Srere, *Annu. Rev. Biochem.* **56**, 89−124 (1987).

129. G.R. Welch and J.A. DeMoss, in *Microenvironments and Metabolic Compartmentation*, P.A. Srere and R.W. Estabrook, Eds., Academic, New York, 1978, pp. 323−344.

130. F.H. Gaertner, in *Microenvironments and Metabolic Compartmentation*, P.A. Srere and R.W. Estabrook, Eds., Academic, New York, 1978, pp. 345−353.

131. S. Wakil, *Biochemistry* **28**, 4523−4530 (1989).

132. J.N. Davidson, P.C. Rumsby, and J. Tamaren, *J. Biol. Chem.* **256**, 5220−5225 (1981).

133. E.A. Carrey and D.G. Hardie, *Eur. J. Biochem.* **171**, 583−588 (1988).

134. J.P. Simmer, R.E. Kelly, J.L. Scully, D.R. Grayson, A.G. Rinker, Jr., S.T. Bergh, and D.R. Evans, *Proc. Natl. Acad. Sci. USA* **86**, 4382–4386 (1989).

135. M.G. Chaparian and D.R. Evans, *FASEB J.* **2**, 2982–2989 (1989).

136. J.N. Freund and B.P. Jarsy, *J. Mol. Biol.* **193**, 1–13 (1987).

137. E. Villar, B. Schuster, D. Peterson, and V. Schirch, *J. Biol. Chem.* **260**, 2245–2252 (1985).

138. W. Strong, G. Joshi, R. Lura, N. Muthukumaraswamy, and V. Schirch, *J. Biol. Chem.* **262**, 12519–12525 (1987).

139. J.A. DeMoss, in *The Organization of Cell Metabolism*, G.R. Welch and J.S. Clegg, Eds, Plenum, New York, 1985, pp. 109–119.

140. J.R. Coggins, K. Duncan, I.A. Anton, M.R. Boocock, S. Chaudhuri, J.M. Lambert, A. Lewendon, G. Millar, D.M. Mousdale, and D.D.S. Smith, *Biochem. Soc. Trans.* **15**, 754–759 (1987).

141. M.J. Bogusky, C.M. Dobson, and R.A.G. Smith, *Biochemistry* **28**, 6728–6735 (1989).

142. C. Abate, J.A. Smith, and T.H. Joh, *Biochem. Biophys. Res. Commun.* **151**, 1446–1453 (1988).

143. F.D. Ledley, A.G. DiLella, S. Kwok, and S.L.C. Woo, *Biochemistry* **24**, 3390–3394 (1985).

144. H. Dahl and J.F.B. Mercer, *J. Biol. Chem.* **216**, 4148–4153 (1986).

145. H.C. Grenett, F.D. Ledley, C.C. Reed, and S.L.C. Woo, *Proc. Natl. Acad. Sci. USA* **84**, 5530–5534 (1987).

146. R. Hanemaaijer, A. de Kok, J. Jolles, and C. Veeger, *Eur. J. Biochem.* **169**, 245–252 (1987).

147. W.M. Southerland, D.R. Winge, and K.V. Rajagopalan, *J. Biol. Chem.* **253**, 8747–8752 (1978).

148. Y. Kubo, N. Ogura, and H. Nakagawa, *J. Biol. Chem.* **263**, 19684–19689 (1988).

148a. P.J. Neame and M.J. Barber, *J. Biol. Chem.* **264**, 20894–20901 (1989).

149. M.J. Hubbard and C.B. Klee, *Biochemistry* **28**, 1868–1874 (1989).

150. F. Lim, C.P. Morris, F. Occhiodoro, and J.C. Wallace, *J. Biol. Chem.* **263**, 11493–11497 (1988).

151. S. Jia and J.L. Wang, *J. Biol. Chem.* **263**, 6009–6011 (1988).

151a. M.E. Baker, *FEBS Lett.* **244**, 31–33 (1989).

152. M. Jasin, L. Regan, and P. Schimmel, *Nature* **306**, 441–447 (1983).

153. S. Tajima, K. Enomoto, and R. Sato, *J. Biochem. (Tokyo)* **84**, 1573–1586 (1978).

154. S. Tajima, K. Mihara, and R. Sato, *Arch. Biochem. Biophys.* **198**, 137–144 (1979).

155. C.A. Brown and S.D. Black, *J. Biol. Chem.* **264**, 4442–4444 (1989).

156. G.E. Tarr, S.D. Black, V.S. Fujita, and M.J. Coon, *Proc. Natl. Acad. Sci. USA* **80**, 6552–6556 (1983).

157. P.D. McCrea, D.M. Engelman, and J.-L. Popot, *Trends Biochem. Sci.* **13**, 289–290 (1988).

158. S.M. Waugh, E.E. DiBella, and P.F. Pilch, *Biochemistry* **28**, 3448–3455 (1989).

159. B. Papp, B. Sarkadi, A. Enyedi, A.J. Caride, J.T. Penniston, and G. Gardos, *J. Biol. Chem.* **264**, 4577–4582 (1989).

159a. A.M. Edelman, K. Takio, D.K. Blumenthal, R.S. Hansen, K.A. Walsh, K. Titani, and E.G. Krebs, *J. Biol. Chem.* **260**, 11275–11285 (1985).

160. C.B. Hesler, M.L. Brown, D.S. Feuer, Y.L. Marcel, R.W. Milne, and A.R. Tall, *J. Biol. Chem.* **264**, 11317–11325 (1989).

161. B. Liu, S. Meloche, N. McNicoll, C. Lord, and A. DeLéan, *Biochemistry* **28**, 5599–5605 (1989).

162. A.F. Gibbs, D. Chapman, and S.A. Baldwin, *Biochem. J.* **256**, 421–427 (1988).

163. K. Inaba and H. Mohri, *J. Biol. Chem.* **264**, 8384–8388 (1989).

164. J.-M. Betton, M. Desmadril, and J.M. Yon, *Biochemistry* **28**, 5421–5428 (1989).

165. R.D. Banks, C.C.F. Blake, P.R. Evans, R. Haser, D.W. Rice, G.H. Hardy, M. Merrett, and A.W. Phillips, *Nature* **279**, 773–777 (1979).

166. A. Mitraki, J.M. Betton, M. Desmadril, and J.M. Yon, *Eur. J. Biochem.* **163**, 29–34 (1987).

167. S.X. Jiang and M. Vas, *FEBS Lett.* **231**, 151–154 (1988).

168. P.D. Yurchenco, D.W. Speicher, J.S. Morrow, W.J. Knowles, and V.T. Marchesi, *J. Biol. Chem.* **257**, 9102–9107 (1982).

169. H.M. Cooper, R. Jemmerson, D.F. Hunt, P.R. Griffin, J.R. Yates III, J. Shabanowitz, N.-Z. Zhu, and Y. Paterson, *J. Biol. Chem.* **262**, 11591–11597 (1987).

170. D. Deretic and H.E. Hamm, *J. Biol. Chem.* **262**, 10839–10847 (1987).

171. S.E. Navon and B.K.-K. Fung, *J. Biol. Chem.* **263**, 489–496 (1988).

172. D. Mochly-Rosen and D.E. Koshland, Jr., *J. Biol. Chem.* **262**, 2291–2297 (1987).

172a. K.L. Leach, E.A. Powers, J.C. McGuire, L. Dong, S. Kiley, and S. Jaken, *J. Biol. Chem.* **263**, 13223–13230 (1988).

173. H. Ishiguro, S. Higashiyama, I. Ohkubo, and M. Sasaki, *Biochemistry* **26**, 7021–7029 (1987).

174. Z. Hessova, R. Thieleczek, M. Varsányi, F.W. Falkenberg, and L.M.G. Heilmeryer, Jr., *J. Biol. Chem.* **260**, 10111–10117 (1985).

174a. O. Makino, S. Ikawa, Y. Shibata, H. Maeda, T. Ando, and T. Shibata, *J. Biol. Chem.* **262**, 12237–12246 (1987).

175. A. Massone, C. Baldari, S. Censini, M. Bartalini, D. Nucci, D. Borascli, and J.L. Telford, *J. Immunol.* **140**, 3812–3816 (1988).

176. D.O. Morgan and R.A. Roth, *Biochemistry* **25**, 1364–1371 (1986).

177. J.F. Collawn, C.J.A. Wallace, A.E.I. Proudfoot, and Y. Paterson, *J. Biol. Chem.* **263**, 8625–8634 (1988).

178. K. Wakabayashi, Y. Sakata, and N. Aoki, *J. Biol. Chem.* **261**, 11097–11105 (1986).

179. V.M. Dixit, N.J. Galvin, K.M. O'Rourk, and W.A. Frazier, *J. Biol. Chem.* **261**, 1962–1966 (1986).

180. M.-M. Cordonnier, H. Greppin, and L.H. Pratt, *Biochemistry* **24**, 3246–3253 (1985).

181. L.H. Pratt, M.-M. Cordonnier, and J.C. Lagarias, *Arch. Biochem. Biophys.* **267**, 723–735 (1988).

182. J. Monod, J.-P. Changeux, and F. Jacob, *J. Mol. Biol.* **6**, 306–329 (1963).

183. I.H. Segel, *Enzyme Kinetics*, Wiley, New York, 1975, pp. 346–464.

184. M.A. Parniak, I.G. Jennings, and R.G.H. Cotton, *Biochem. J.* **257**, 383–388 (1989).

185. G.E. Morris, *Biochem. J.* **257**, 461–469 (1989).

186. B. Friguet, L. Djavadi-Ohaniance, and M.E. Goldberg, *Eur. J. Biochem.* **160**, 593–597 (1986).

187. L. Dyavadi-Ohaniance, B. Friguet, and M.E. Goldberg, *Biochemistry* **25**, 2502–2508 (1986).

188. F. Riftina, E. DeFalco, and J.S. Krakow, *Biochemistry* **28**, 3299–3305 (1989).

189. D. Holowka, D.H. Conrad, and B. Baird, *Biochemistry* **24**, 6260–6267 (1985).

190. A. Davies, K. Meeran, M.T. Cairns, and S.A. Baldwin, *J. Biol. Chem.* **262**, 9347–9352 (1987).

191. H.C. Haspel, M.G. Rosenfeld, and O.M. Rosen, *J. Biol. Chem.* **263**, 398–403 (1988).

192. R.S. Molday and D. MacKenzie, *Biochemistry* **22**, 653–660 (1983).

193. A.K. Grover, *J. Biol. Chem.* **263**, 19510–19512 (1988).

194. W. Baehr, Y.-X. Zhang, T. Joseph, H. Su, F.E. Nano, K.D.E. Everett, and H.D. Caldwell, *Proc. Natl. Acad. Sci. USA* **85**, 1–5 (1988).

195. H. Koepsell, K. Korn, A. Raszeja-Specht, S. Bernotat-Danielowski, and D. Ollig, *J. Biol. Chem.* **263**, 18419–18429 (1988).

196. M. Ratnam, P.B. Sargent, V. Sarin, J.L. Fox, D.L. Nguyen, J. Rivier, M. Criado, and J. Lindstrom, *Biochemistry* **25**, 2621−2632 (1986).

197. M. Ratnam, D.L. Nguyen, J. Rivier, P.B. Sargent, and J. Lindstrom, *Biochemistry* **25**, 2633−2643 (1986).

198. R.D. Mamelok, D. Liu, and S. Tse, *Kidney Int.* **25**, 309 (1984).

199. P. van der Ley, M. Struyvé, and J. Tommassen, *J. Biol. Chem.* **261**, 12222−12225 (1986).

200. J. Korteland, N. Overbeeke, P. DeGraaff, P. Overduin, and B. Lugtenberg, *Eur. J. Biochem.* **152**, 691−697 (1985).

AUTHOR INDEX

SUBJECT INDEX

Acetylcholine receptor, topology of, 241
Acetyl tryptophanamide, fluorescence of, 147, 149
Acrylamide, fluorescence quenching by, 150, 169–176
Adrenocorticotropin (ACTH), fluorescence of, 150, 163
Alcohol dehydrogenase:
  fluorescence of, 168, 171, 174, 183–185
  fluorescence lifetime of, 168
  fluorescence quenching of, 171–172, 183–185
  phosphorescence of, 190–192
Alpha helix:
  right handed, 35, 37
  screw sense, 35, 37
Amino acid composition, polarity parameter and, 108
Anilinonaphthalene sulfonate (ANS), 93
Anisotropy decay associated spectra, 184
Antibodies:
  monoclonal, 207
    advantages of, 215, 243
    probes for conformational changes, 238–241
    probes for specific function, 236
    use in determining protein topology, 241
Antigen–antibody complexes, fluorescence of, 189
Antigenicity, correlation with structural features of protein, 220
Antigenic regions, predictions of, 220
Asparagine synthetase, functional sites of, 237
Associative memory methods, predicting protein structure, 71

Atrial natriuretic factor (ANF) receptor, proteolysis of, 235
Azurin:
  fluorescence of, 147, 150, 158, 160, 168, 170, 174
  fluorescence anisotropy of, 160
  fluorescence lifetime of, 150, 158–160, 168
  fluorescence quenching of, 170

Beta-alpha-beta structures, right handed, 36
Beta barrels:
  hydrogen bonding patterns, 39
  right handed tilt, 37
  schematic depiction of, 39
Beta sheet, right handed, 35, 37
Bombesin, fluorescence of, 150, 163, 165
Brain hexokinase, see Hexokinase
Brookhaven protein data bank, 2
Brownian dynamics, see Molecular dynamics

Ca$_2$ pump, topology of, 241
Calcineurin:
  domain structure of, 232
  proteolysis of, 232
CD spectrum, predicting structure and, 64
Chirality of protein structure:
  geometric, 4
  topologic, 4
Cholesterol ester transfer protein, proteolysis of, 234
Combinatoric methods, predicting protein structure by, 63–66

277

Cumulative Author Index, Volumes 1–35 and Supplemental Volume

Cumulative Subject Index, Volumes 1–35 and Supplemental Volume